Thomas Jefferson Lee

**Tables and formulae useful in surveying, geodesy and practical astronomy**

Thomas Jefferson Lee

**Tables and formulae useful in surveying, geodesy and practical astronomy**

ISBN/EAN: 9783337275600

Printed in Europe, USA, Canada, Australia, Japan

Cover: Foto ©berggeist007 / pixelio.de

More available books at **www.hansebooks.com**

# Professional Papers

OF

# The Corps of Engineers

OF

## THE UNITED STATES ARMY.

PUBLISHED

BY AUTHORITY OF THE SECRETARY OF WAR.

HEADQUARTERS CORPS OF ENGINEERS.

1873.

# TABLES AND FORMULÆ

USEFUL IN

# SURVEYING, GEODESY,

AND

# PRACTICAL ASTRONOMY,

INCLUDING

## ELEMENTS FOR THE PROJECTION OF MAPS,

AND

## INSTRUCTIONS FOR FIELD MAGNETIC OBSERVATIONS.

*Third Edition, Revised and Enlarged.*

WASHINGTON:
GOVERNMENT PRINTING OFFICE.
1873.

OFFICE OF THE CHIEF OF ENGINEERS,
*Washington, February* 24, 1873.

GENERAL: In preparing this third edition of a volume compiled in 1849 for the use of the Corps of Topographical Engineers, when an officer in that corps, I have made such additions and corrections as experience has suggested and the requirements of the service seemed to demand.

Intended more especially for field-use by officers engaged in surveys or explorations, and aspiring to no further merit than that of accuracy, utility, and convenience, it is submitted with the hope that it may continue to be favorably received by those who may have occasion to refer to it.

The revision was undertaken at the suggestion of others, and not without reluctance. I beg indulgence for its defects.

Very respectfully,

THOS. J. LEE.

Brigadier-General A. A. HUMPHREYS,
*Chief of Engineers, United States Army.*

In the additions and corrections introduced into this volume, besides the authorities named in the text, I have availed myself, in the article on the Gauging of Rivers, of notes by MAJOR ABBOT, Corps of Engineers, on the practical gauging of rivers, printed in the proceedings of the Essayons Club at Willet's Point.

The article on Trigonometrical Leveling is taken principally from Appendix No. 7 of the Coast-Survey report for 1868, by ASSISTANT R. D. CUTTS, United States Coast Survey.

I have also made use of Appendices Nos. 9, 10, and 11 of Coast-Survey report of 1866, by ASSISTANT C. A. SCHOTT, United States Coast Survey, in the revision of the portions relating to the use of the transit-instrument, of the zenith-telescope, and the determination of astronomical azimuths.

The article on Longitude by Lunar Culminations was prepared in 1858 for the Topographical Bureau by PROFESSOR BARTLETT, United States Military Academy.

I am indebted to CAPTAIN ERNST and LIEUTENANT MERCUR, Corps of Engineers, and to ASSISTANTS WOODWARD and WRIGHT, Survey of the Lakes, for suggestions and corrections ; to LIEUTENANT MERCUR for the article on Probable Errors, &c.; and to CAPTAIN RAYMOND, Corps of Engineers, for the valuable contribution on Magnetic Field Observations, which forms the Appendix ; and I have to regret that LIEUTENANT-COLONEL WILLIAMSON was prevented by absence from the country from preparing a suitable set of hypsometrical tables, which he had consented to do.

T. J. L.

# CONTENTS.

## PART I.—MISCELLANEOUS.

## PART II.—GEODESY.

## PART III.—ASTRONOMY.

## APPENDIX.

## ERRATA

*To the 3d edition of Tables and Formulæ, Professional Papers No. 12.*

Page  10. For Sec. XXXIV, read Sec. XXXVII.

Page  14. *Limestone*, for 197.5, read 179.5, and for 11.355, read 12.462.

Page  47. *In Chord-Deflection*, for is "therefore double", read "may be assumed as double" the tangent deflection.

Page  85. Hassler's expansion of iron bar, for 0.000240687260, read 0.000250687260.

Page  96. In value $u''$ Cos. Z, strike out $Cos^2 L'$ from the numerator.

Page  99. For log. $\delta$ M=2.9060529, read log. $\delta$ M=2.9060531.

$\delta$ Z=2.7619618,  log. $\delta$ Z=2.7619620.

Page 134. In 4th column, 5th line of table, for 449.4, read 499.4.

Page 135. In 7th column, 4th line of table, for 331.0, read 231.0.

Page 142. For 0.000000667, read 0.0000000667.

Page 159. Omit minus sign at foot of last column.

Page 168. Top line of last column for +0.0958, read +0.0956.

Page 168. 10th line of table, for +0.9967, read +0.0967.

Page 170. At head of table, read $D_{iv}$=60384.3 log. $\frac{h}{H} \times$ &c.

Page 229. In last line and last figure, for .06, read .09.

Page 250. In lines 13, 15 and 18. increase the values by one tenth of a second.

Page 268. In value of $\tau$ in example, for 2m. 30.3s, read 2m. 20.3s.

Page 288. For E, read $\epsilon$.

Page 288. For p, p' etc., read p', p'' etc.

Page 289. For $\kappa\epsilon > v$, read $\kappa\epsilon < v$.

Page 290. For 1 and 10 in first example, read 1 and 11.

# ERRATA.

Page 10.   For "See XXXIV" read "See XXXVII."

Page 85.   Hassler's expansion of iron bar, for "0.000240687260" read "0.000250687260."

Page 288.   For E read $\epsilon$.

Page 288.   For $p$, $p'$, &c., read $p'$, $p''$, &c.

# TABLES AND FORMULÆ.

## PART I.

## MISCELLANEOUS.

# TRIGONOMETRY.

## I.—*Equivalent Expressions.*

$$\sin^2 x + \cos^2 x = 1$$

$$\sin x = \cos x \tan x$$

$$= \frac{\cos x}{\cot x}$$

$$= \sqrt{1 - \cos^2 x}$$

$$= \sqrt{\frac{1}{1 + \cot^2 x}}$$

$$= 2 \sin \tfrac{1}{2} x \cos \tfrac{1}{2} x$$

$$= \frac{\tan x}{\sqrt{1 + \tan^2 x}}$$

$$= \frac{1}{\operatorname{cosec} x}$$

$$\cos x = \frac{\sin x}{\tan x}$$

$$= \sin x \cot x$$

$$= \sqrt{1 - \sin^2 x}$$

$$= 1 - 2 \sin^2 \tfrac{1}{2} x$$

$$= \cos^2 \tfrac{1}{2} x - \sin^2 \tfrac{1}{2} x$$

$$= \frac{1}{\sec x}$$

$$\tan x = \frac{\sin x}{\cos x}$$

$$= \frac{1}{\cot x}$$

$$= \frac{\sin x}{\sqrt{1 - \sin^2 x}}$$

$$= \frac{1 - \cos 2x}{\sin 2x}$$

$$= \frac{\sin 2x}{1 + \cos 2x}$$

### I.—*Equivalent Expressions*—Continued.

$$\cot x = \frac{1}{\tan x}$$

$$\sec x = \frac{1}{\cos x}$$

$$\operatorname{cosec} x = \frac{1}{\sin x}$$

$$\operatorname{versin} x = 1 - \cos x$$

$$= 2 \sin^2 \tfrac{1}{2} x$$

$$\text{co-versin } x = 1 - \sin x$$

$$\operatorname{chord} x = 2 \sin \tfrac{1}{2} x$$

$$\sin (A \pm B) = \sin A \cos B \pm \sin B \cos A$$

$$\cos (A \pm B) = \cos A \cos B \mp \sin A \sin B$$

$$\sin 2A = 2 \sin A \cos A$$

$$\cos 2A = 2 \cos^2 A - 1$$

$$= 1 - 2 \sin^2 A$$

$$= \cos^2 A - \sin^2 A$$

$$2 \cos^2 \tfrac{1}{2} A = 1 + \cos A$$

$$2 \sin^2 \tfrac{1}{2} A = 1 - \cos A$$

$$\tan (A \pm B) = \frac{\tan A \pm \tan B}{1 \mp \tan A \tan B}$$

$$\tan \tfrac{1}{2} A = \sqrt{\frac{1 - \cos A}{1 + \cos A}}$$

$$= \frac{1 - \cos A}{\sin A}$$

$$\sin A \pm \sin B = 2 \sin \tfrac{1}{2} (A \pm B) \cos \tfrac{1}{2} (A \mp B)$$

$$\cos A + \cos B = 2 \cos \tfrac{1}{2} (A + B) \cos \tfrac{1}{2} (A - B)$$

$$\cos A - \cos B = 2 \sin \tfrac{1}{2} (A + B) \sin \tfrac{1}{2} (B - A)$$

$$\sin^2 A - \sin^2 B = \sin (A + B) \sin (A - B)$$

$$\cos^2 A - \sin^2 B = \cos (A + B) \cos (A - B)$$

I.—*Equivalent Expressions*—Continued.

$$\tan A \pm \tan B = \frac{\sin (A \pm B)}{\cos A \cos B}$$

$$\cot A \pm \cot B = \frac{\sin (A \pm B)}{\sin A \sin B}$$

$$\frac{\sin A + \sin B}{\sin A - \sin B} = \frac{\tan \frac{1}{2} (A + B)}{\tan \frac{1}{2} (A - B)}$$

$$\frac{1 \pm \sin A}{1 \mp \sin A} = \tan^2 (45° \pm \tfrac{1}{2} A)$$

$$\frac{1 \pm \sin A}{\cos A} = \tan (45° \pm \tfrac{1}{2} A)$$

II.—*Solution of Plane Triangles.*

In the following formulæ, A, B, C, represent the angles, and *a*, *b*, *c*, the sides opposite, respectively.

1. Any plane triangle:

$$a^2 = b^2 + c^2 - 2\ b\ c \cos A$$

$$\frac{\sin A}{a} = \frac{\sin B}{b} = \frac{\sin C}{c}$$

$$\frac{\tan \frac{1}{2} (A + B)}{\tan \frac{1}{2} (A - B)} = \frac{\cot \frac{1}{2} C}{\tan \frac{1}{2} (A - B)} = \frac{a + b}{a - b}$$

$$\sin \tfrac{1}{2} A = \left\{ \frac{(s - b)\ (s - c)}{b\ c} \right\}^{\frac{1}{2}}$$

$$\cos \tfrac{1}{2} A = \left\{ \frac{s\ (s - a)}{b\ c} \right\}^{\frac{1}{2}}$$

$$s = \frac{a + b + c}{2}$$

II.—*Solution of Plane Triangles*—Continued.

2. Right-angled triangles:

Making A = 90° in the preceding, they become.

$$a^2 = b^2 + c^2$$
$$b = a \sin B = a \cos C,$$
$$c = a \sin C = a \cos B$$

$$\tan B = \frac{b}{c}$$

$$\tan C = \frac{c}{o}$$

III.—*Solution of Spherical Triangles.*

*a, b, c*, represent the arcs, and Λ, B, C, the angles opposite.

1. Oblique spherical triangles:

$$\frac{\sin A}{\sin a} = \frac{\sin B}{\sin b} = \frac{\sin C}{\sin c}$$

$$\begin{cases} \cos a = \dfrac{\cos b \sin (c + \varphi)}{\sin \varphi} \\ \cot \varphi = \tan b \cos A \end{cases}$$

$$\begin{cases} \cos A = \dfrac{\cos B \sin (C - \varphi)}{\sin \varphi} \\ \cot \varphi = \tan B \cos a \end{cases}$$

$$\begin{cases} \cot a \tan b = \dfrac{\sin (C + \varphi)}{\sin \varphi} \\ \cot \varphi = \dfrac{\cot A}{\cos b} \end{cases}$$

*Napier's Analogies.*

$$\tan \tfrac{1}{2} (a + b) = \tan \tfrac{1}{2} c \frac{\cos \tfrac{1}{2}(A - B)}{\cos \tfrac{1}{2}(A + B)}$$

$$\tan \tfrac{1}{2} (a - b) = \tan \tfrac{1}{2} c \frac{\sin \tfrac{1}{2}(A - B)}{\sin \tfrac{1}{2}(A + B)}$$

$$\tan \tfrac{1}{2} (A + B) = \cot \tfrac{1}{2} C \frac{\cos \tfrac{1}{2}(a - b)}{\cos \tfrac{1}{2}(a + b)}$$

$$\tan \tfrac{1}{2} (A - B) = \cot \tfrac{1}{2} C \frac{\sin \tfrac{1}{2}(a - b)}{\sin \tfrac{1}{2}(a + b)}$$

III.—*Solution of Spherical Triangles*—Continued.

$$\sin^2 \tfrac{1}{2}\, a = \frac{\sin S \sin (A - S)}{\sin B \sin C}$$

$$\cos^2 \tfrac{1}{2}\, a = \frac{\sin (B - S) \sin (C - S)}{\sin B \sin C}$$

$$\tan^2 \tfrac{1}{2}\, a = \frac{\sin S \sin (A - S)}{\sin (B - S) \sin (C - S)}$$

$$\sin^2 \tfrac{1}{2}\, A = \frac{\sin (s - b) \sin (s - c)}{\sin b \sin c}$$

$$\cos^2 \tfrac{1}{2}\, A = \frac{\sin s \sin (s - a)}{\sin b \sin c}$$

$$\tan^2 \tfrac{1}{2}\, A = \frac{\sin (s - b) \sin (s - c)}{\sin s \sin (s - a)}$$

In which S and *s* represent the half-sum of the three angles diminished by 90° and the half-sum of the three sides, respectively.

2. Right-angled spherical triangles, *a* being the hypothenuse:

$\cos a = \cos b \cos c$      $\cot B = \cot b \sin c$

$\cos a = \cot B \cot C$      $\cot C = \cot c \sin b$

$\cos B = \sin C \cos b$      $\tan b = \tan B \sin c$

$\cos C = \sin B \cos c$      $\tan c = \tan C \sin b$

$\tan b = \tan a \cos C$      $\sin b = \sin a \sin B$

$\tan c = \tan a \cos B$      $\sin c = \sin a \sin C$

IV.—*Multiple Arcs.*

$$\sin 2x = 2 \sin x \cos x$$

$$\sin 3x = 2 \sin x \, . \, \cos 2x + \sin x$$

$$\cos 2x = 2 \cos x \, . \, \cos x - 1$$

$$\cos 3x = 2 \cos x \, . \, \cos 2x - \cos x$$

$$\tan 2x = \frac{2 \tan x}{1 - \tan^2 x}$$

$$\tan 3x = \frac{\tan x + \tan 2x}{1 - \tan x \, . \, \tan 2x}$$

## V.—*Trigonometrical Series.*

$$\sin A = A - \frac{A^3}{2.3} + \frac{A^5}{2.3.4.5} - \frac{A^7}{2.3.....7} + \text{etc.}$$

$$\cos A = 1 - \frac{A^2}{2} + \frac{A^4}{2.3.4} - \frac{A^6}{2.....6} + \text{etc.}$$

$$\tan A = A + \frac{A^3}{3} + \frac{2\,A^5}{3.5} + \frac{17\,A^7}{3^2.5.7} + \text{etc.}$$

$$\text{arc } A = \sin A + \frac{\sin^3 A}{2.3} + \frac{3\,\sin^5 A}{2.4.5} + \frac{3.5\,\sin^7 A}{2.4.6.7} + \text{etc.}$$

$$= \tan A - \tfrac{1}{3}\tan^3 A + \tfrac{1}{5}\tan^5 A - \tfrac{1}{7}\tan^7 A. + \text{etc.}$$

$$\log \sin A = \log A + \log \left(1 - \frac{x^2}{6} + \frac{x^4}{120} - \text{etc.}\right)$$

$$= \log A - M\left(\frac{x^2}{6} + \frac{x^4}{180} + \frac{x^6}{2835}\right)$$

$$M = \text{logarithmic modulus}$$

$$= 0.43429\ 45........$$

$$\log M = 9.63778\ 43113.....$$

### *Differentials of Trigonometrical Lines.*

$$d \sin x = + d\,x \cos x$$

$$d \cos x = - d\,x \sin x$$

$$d \tan x = + \frac{d\,x}{\cos^2 x}$$

$$d \cot x = - \frac{d\,x}{\sin^2 x}$$

$$d \sin^2 x = + 2\,d\,x \sin x \cos x$$

$$d \cot^2 x = - 2\,d\,x \sin x \cos x$$

$$d \tan^2 x = + \frac{2\,d\,x \tan x}{\cos^2 x}$$

$$d \cot^2 x = - \frac{2\,d\,x \cot x}{\sin^2 x}$$

## VI.—*Ratio of the Circumference of a Circle to its Diameter.*

$$\pi = 3.14159\ 26535\ 898\ldots\ldots$$
$$\log \pi = 0.49714\ 98726\ 941\ldots\ldots$$

The radius being unity, the number of degrees in an arc equal

to radius $= r^\circ = \dfrac{180^\circ}{\pi} = \dfrac{1}{\text{arc } 1^\circ} = 57^\circ.29578 = 57^\circ\ 17'\ 44''.8.$

The number of minutes $= r' = \dfrac{10800'}{\pi} = \dfrac{1}{\text{arc } 1'}$, or

$\dfrac{1}{\sin 1'} = 3437'.74677.$

The number of seconds $= r'' = \dfrac{648000''}{\pi} = \dfrac{1}{\sin 1''}$

$= 206264''.80625.$

$$\log r^\circ = 1.75812\ 26324\ 09172$$
$$\text{comp } \log r^\circ = 8.24187\ 73675\ 90828$$

$$\log r' = 3.53627\ 38827\ 92816$$
$$\text{comp } \log r' = 6.46372\ 61172\ 07184 = \log \sin 1'$$

$$\log r'' = 5.31442\ 51331\ 76459$$
$$\text{comp } \log r'' = 4.68557\ 48668\ 23541 = \log \sin 1''$$

Let $a$ be the length of an arc of a circle whose radius is 1, and $a''$ the number of seconds in that arc, as

$$r'' = \frac{1}{\sin 1''} \text{ and } R : r'' :: a : a'' \text{ or } a'' = r'' a ; \ a = a'' \sin 1''$$

In an equation, therefore, any arc $a$ of a circle whose radius is 1 is expressed in seconds by changing $a$ into $a'' \sin 1''$.

### *Signs of Trigonometrical Lines.*

| Quadrants. | Sin. | Cos. | Tan. | Cot. | Sec. | Cosec. |
|---|---|---|---|---|---|---|
| 1, 5, 9, | + | + | + | + | + | + |
| 2, 6, 10, | + | − | − | − | − | + |
| 3, 7, 11, | − | − | + | + | − | − |
| 4, 8, 12, &c. | − | + | − | − | + | − |

### VII.—*Weights and Measures of the United States.*

The standards of length and weight of this country and Great Britain are theoretically identical. The United States gallon and bushel represent old English measures.

The standard of linear dimensions, adopted by the Treasury Department in the construction of standards for distribution to the custom-houses and States, is a brass scale of 82 inches in length, made in London by Troughton, which formed part of the instruments collected in 1815 by Mr. Hassler for the Survey of the Coast, and was supposed identical with the Schuckburg scale, one of the old English standards. The standard temperature is 62° Fahrenheit, and the yard-measure is between the 27th and 63d inches of its scale. This length has not been legalized by act of Congress. (See XXXIV.)

### *Linear Measure.*

The *unit of linear measure* is the *yard*. The yard is divided into 3 feet, and the foot subdivided into 12 inches. The multiples of the yard are the *pole* or *perch*, the *furlong*, and the *mile;* but the pole and furlong are now scarcely ever used, itinerary distances being reckoned in miles and yards.

The following are the relations:

| Inches. | Feet. | Yards. | Poles. | Furlongs. | Miles. |
|---|---|---|---|---|---|
| 1 | 0. 083 | 0. 028 | 0. 00505 | 0. 00012626 | 0. 0000157828 |
| 12 | 1. | 0. 333 | 0. 06060 | 0. 00151515 | 0. 00018939 |
| 36 | 3. | 1. | 0. 1818 | 0. 004545 | 0. 00056818 |
| 198 | 16. 5 | 5. 5 | 1. | 0. 025 | 0. 003125 |
| 7920 | 660. | 220. | 40. | 1. | 0. 125 |
| 63360 | 5280. | 1760. | 320. | 8. | 1. |

$$\log 5280 = 3.7226339$$

$$\log 1760 = 3.2455127$$

VII.—*Weights and Measures of the United States*—Continued.

## *Square Measure.*

In *square measure* the yard is subdivided, as in general measure, into *feet* and *inches;* 144 square inches being equal to a square foot. For land-measure the multiples of the yard are the *pole*, the *rood*, and the *acre*. Very large surfaces, as of whole countries, are expressed in square miles.

The following are the relations of square measure :

| Sq. feet. | Sq. yards. | Poles. | Roods. | Acres. | Sq. miles. |
|---|---|---|---|---|---|
| 1. | 0.1111 | 0.00367309 | 0.00009l827 | 0.000022957 | |
| 9. | 1. | 0.0330579 | 0.000826448 | 0.000206612 | |
| 272.25 | 30.25 | 1. | 0.025 | 0.00625 | |
| 10890. | 1210. | 40. | 1. | 0.25 | |
| 43560. | 4840. | 160. | 4. | 1. | |
| 27878400. | 3097600. | 102400. | 2560. | 640. | 1. |

## *Measure of Capacity.*

The *units of capacity measure* are the *gallon* for *liquid* and the *bushel* for *dry* measure. The gallon is a vessel containing 58372.2 grains (8.3389 pounds avoirdupois) of the standard pound of distilled water, at the temperature of maximum density of water, the vessel being weighed in air in which the barometer is 30 inches at 62° Fahrenheit. The bushel is a measure containing 543391.89 standard grains (77.6274 pounds avoirdupois) of distilled water, at the temperature of maximum density of water, and barometer 30 inches at 62° Fahrenheit.

The gallon is thus the wine-gallon, (of 231 cubic inches,) nearly; and the bushel, the Winchester bushel, nearly.

The temperature of maximum density of water was determined by Mr. Hassler to be 39°.83 Fahrenheit.

|  DRY MEASURES. | | | LIQUIDS. | |
|---|---|---|---|---|
| Pint | = $\frac{1}{64}$ bushel. | Gill | | = $\frac{1}{32}$ gall. |
| Quart | = 2 pints | = $\frac{1}{32}$ bushel. | Pint = 4 . gills | = $\frac{1}{8}$ gall. |
| Peck | = 8 quarts | = $\frac{1}{4}$ bushel. | Quart = 2 pints | = $\frac{1}{4}$ gall. |
| Bushel | = 4 pecks | = 1 bushel. | Gallon = 4 quarts | = 1 gall. |
| | | | Barrel = $31\frac{1}{2}$ gallons | = $31\frac{1}{2}$ galls. |
| | | | Hhd. = 2 barrels | = 63 galls. |

The only *legalized unit* of weight or measure is a *troy-pound*,

### VII.—*Weights and Measures of the United States*—Continued.

(act of May 19, 1828,) copied by Captain Kater, in 1827, from the imperial troy-pound of England, for the use of the Mint of the United States, and there deposited. This pound is a standard at 30 inches of the barometer and 62° of the Fahrenheit thermometer.

The standard *avoirdupois-pound*, as determined by Mr. Hassler, is the weight of 27.7015 cubic inches of distilled water. It is greater than the troy-pound in the proportion of 7000 to 5760; that is, the avoirdupois-pound is equivalent in weight to 7000 grains troy.

#### *Weights.*

| AVOIRDUPOIS. | | | TROY. | | |
|---|---|---|---|---|---|
| Dram | | = $\frac{1}{256}$ lb. | Grain | | = $\frac{1}{5760}$ lb. |
| Ounce | = 16 drs. | = $\frac{1}{16}$ lb. | Pennyweight | = 20 grs. | = $\frac{1}{240}$ lb. |
| Pound | = 16 ozs. | = 1 lb. | Ounce | = 24 dwt. | = $\frac{1}{12}$ lb. |
| Quarter | = 25 lbs. | = 28 lbs. | Pound | = 12 ozs. | = 1 lb. |
| Hundred-wt. | = 4 qrs. | = 112 lbs. | | | |
| Ton | = 20 cwt. | = 2240 lbs. | | | |
| Short ton | | = 2000 lbs. | | | |

### VIII.—*Miscellaneous.*

*Length.*—Gunter's chain = 66 feet = 4 poles = 100 links of 7.92 inches.

  1 fathom = 6 feet; 1 cable-length = 120 fathoms.

  1 hand = 4 inches; 1 palm = 3 inches; 1 span = 9 inches.

*Solid.*—1 cubic foot = 1728 cubic inches.

  1 cubic yard = 27 cubic feet = 46656 cubic inches.

  1 reduced foot (board-measure) = 1 square foot × 1 inch thick = 144 cubic inches.

  1 perch of masonry = 1 perch (16½ feet) long × 1 foot high × 1½ foot thick = 24.75 cubic feet; 25 cubic feet has generally been adopted for convenience.

  1 cord fire-wood = 8 feet long × 4 feet high × 4 feet deep = 128 cubic feet.

  1 chaldron coal = 36 bushels = 57.25 cubic feet.

*Paper.*—24 sheets = 1 quire.

  20 quires = 1 ream = 480 sheets.

## IX.—*Weights and Volumes of various Substances.*

### METALS.

| Substances. | Cubic foot. | Cubic inch. |
|---|---|---|
| | *Pounds.* | *Pounds.* |
| Brass. { Copper....67 } { Zinc.....33 } | 488. 75 | .2829 |
| Brass, gun-metal | 543. 75 | .3147 |
| Copper, cast | 547. 25 | .3179 |
| plates | 543. 625 | .3167 |
| Iron, cast | 450. 437 | .2607 |
| gun-metal | 466. 5 | .27 |
| wrought bars | 486. 75 | .2816 |
| Lead, cast | 709. 5 | .4106 |
| rolled | 711. 75 | .4119 |
| Mercury, 60° | 848. 7487 | .491174 |
| Steel, plates | 487. 75 | .2823 |
| soft | 489. 562 | .2833 |
| Tin | 455. 687 | .2637 |
| Zinc, cast | 428. 812 | .2482 |
| rolled | 449. 437 | .2601 |

### WOODS.

| Substances. | Cubic foot. | Cubic feet in a ton. |
|---|---|---|
| | *Pounds.* | |
| Ash | 52. 812 | 42. 414 |
| Cedar | 35. 062 | 63. 886 |
| Chestnut | 38. 125 | 58. 754 |
| Hickory, pig-nut | 49. 5 | 45. 252 |
| shell-bark | 43. 125 | 51. 942 |
| Lignum-vitæ | 83. 312 | 26. 886 |
| Mahogany, Honduras | { 35. { 66. 437 | 64. 33. 714 |
| Oak, Canadian | 54. 5 | 41. 101 |
| English | 58. 25 | 38. 455 |
| live, seasoned | 66. 75 | 33. 558 |
| white, dry | 53. 75 | 41. 674 |
| upland | 42. 937 | 52. 169 |

## IX.—*Weights and Volumes of various Substances*—Continued.

### Woods—Continued.

| Substances. | Cubic foot. | Cubic feet in a ton. |
|---|---|---|
| | *Pounds.* | |
| Pine, yellow .......................... .... | 33. 812 | 66. 248 |
| Spruce ......... ......................... .... | 31. 25 | 71. 68 |
| Walnut, black, dry ...................... .... | 31. 25 | 71. 68 |
| Willow, dry ...... ..................... .... | 30. 375 | 73. 744 |

### MISCELLANEOUS.

| Substances. | Cubic foot. | Cubic feet in a ton. |
|---|---|---|
| | *Pounds.* | |
| Air ..................................... .... | . 075291 | ........... |
| Brick, fire ......... .... .................... | 137. 562 | 16. 284 |
|     mean ......... : ...................... | 102. | 21. 961 |
| Coal, anthracite ........................ | 89. 75 | 24. 958 |
| | 102. 5 | 21. 854 |
|     bituminous, mean...................... | 80. | 28. |
|     cannel ............................... .... | 94. 875 | 23. 609 |
|     Cumberland ...................... .... | 84. 687 | 26. 451 |
| Coke.................................... | 62. 5 | 35. 84 |
| Cotton, bale, mean..... ..................... | 14. 5 | 154. 48 |
| | 20. | 114. |
|     pressed ........ ............... | 25. | 89. 6 |
| Earth, clay ......... .... ................. | 120. 625 | 18. 569 |
|     common soil.......................... | 137. 125 | 16. 335 |
|         gravel .................... | 109. 312 | 20. 49 |
|     dry sand .....................·...... | 120. | 18. 667 |
|     loose ......:...................... | 93. 75 | 23. 893 |
| Granite, Quincy ........................ | 165. 75 | 13. 514 |
|     Susquehanna...................... | 169. | 13. 254 |
| Limestone .................... .............:...... | 197. 25 | 11. 355 |
| Marble, mean........................... | 167. 875 | 13. 343 |
| Mortar, dry, mean...................... | 97. 98 | 22. 862 |
| Water, fresh ............................ | 62. 5 | 35. 84 |
|     salt ............................... | 64. 125 | 34. 931 |
| Steam ..................... .................. | . 036747 | ........... |

## X.—*The Army-Ration.*

TABLE SHOWING THE WEIGHT AND BULK OF 1000 RATIONS.

| One thousand rations of— | Net weight. | Gross weight. | Bulk. | 100 rations consist of— |
|---|---|---|---|---|
| | *Pounds.* | *Pounds.* | *Barrels.* | |
| Pork ... ... ... | 750. | 1218.75 | 3.75 | 75 lbs. or } |
| Bacon ... ... | 750. | 903.19 | 4.90 | 75 lbs. } |
| Flour ... ... | 1125. | 1234.06 | 5.74 | 112.5 lbs. or ⎫ |
| Pilot-bread ... | 750. | 921.69 | 9.03 | 75 lbs. or ⎬ |
| Do ... ... | 1000. | 1228.91 | 12.05 | 100 lbs. in the field. ⎭ |
| Beans ... ... | 155. | 177.32 | 0.71 | 8 quarts, or } |
| Rice ... ... | 100. | 114.50 | 0.46 | 10 lbs. } |
| Coffee, green .. | 100. | 122. | 0.65 | 10 lbs. |
| roasted. | 80. | 108. | 0.83 | 8 lbs. |
| Sugar ... ... | 150. | 161. | 0.6 | 15 lbs. |
| Vinegar ... ... | 92.5 | 107.50 | 0.33 | 4 quarts. |
| Candles ... ... | 15. | 17.50 | 0.09 | 1½ lbs. |
| Soap ... ... ... | 40. | 46.89 | 0.19 | 4 lbs. |
| Salt ... ... ... | 33.75 | 38.63 | 0.16 | 2 quarts. |

### *Forage.*

14 lbs. hay or fodder ⎫ ⎧ per horse ⎧ hay, when pressed, 11 lbs. to cub. ft.
12 quarts oats, or ⎬ ⎨ ⎨ 32 lbs. to bushel, 25.71 to cub. ft.
8 quarts corn ⎭ ⎩ per day ⎩ 56 lbs. to bushel, 45.02 to cub. ft.

Three beeves or 15 sheep consume the forage of 2 horses.

### *Weights of Grain per Bushel.*

| | | | |
|---|---|---|---|
| Wheat ... ... | 60 lbs. ‖ | Oats ... ... | 32 lbs. |
| Corn and rye ... | 56 lbs. ‖ | Barley ... ... | 48 lbs. |

A box 16 × 16.8 × 8. inches contains 1 bushel ⎫
12 × 11.2 × 8. " ½ bushel ⎬ dry measure.
8 × 8.4 × 8. " 1 peck ⎭
6 × 6 × 6.4 " 1 gallon ⎫ liquid measure.
4 × 4 × 3.6 " 1 quart ⎭

## XI.—*Metric System.*

By an act of Congress, approved July 28, 1866, the *metric system* of weights and measures is made optional in the United States; and the act provides that the tables in a schedule annexed shall be recognized "as establishing, in terms of the weights and measures now in use in the United States, the equivalents of the weights and measures expressed therein in terms of the metric system; and said tables may be lawfully used for computing, determining, and expressing, in customary weights and measures, the weights and measures of the metric system."

*Schedule annexed to act of July 28, 1866.*

### MEASURES OF LENGTH.

| Metric denominations. | Values in metres. | Equivalents in denominations in use. |
|---|---|---|
| Myriametre............... | 10000. | 6.2137 miles. |
| Kilometre ............... | 1000. | 0.62137 mile, or 3280 feet and 10 inches. |
| Hectometre............... | 100. | 328 feet and 1 inch. |
| Decametre............... | 10. | 393.7 inches. |
| Metre.................... | 1. | 39.37 inches. |
| Decimetre............... | 0.1 | 3.937 inches. |
| Centimetre ............... | 0.01 | 0.3937 inch. |
| Millimetre ............... | 0.001 | 0.0394 inch. |

### MEASURES OF SURFACE.

| Metric denominations. | Values in square metres. | Equivalents in denominations in use. |
|---|---|---|
| Hectare ..................... | 10000 | 2.471 acres. |
| Are ........................ | 100 | 119.6 square yards. |
| Centare .................... | 1 | 1550 square inches. |

### MEASURES OF CAPACITY.

| Names. | No. of litres. | Cubic measure. | Dry measure. | Liquid or wine measure. |
|---|---|---|---|---|
| Kilolitre or stere. | 1000. | 1 cubic metre ...... | 1.308 cubic yards . | 264.17 gallons. |
| Hectolitre ....... | 100. | 0.1 cubic metre .... | 2 bus. and 3.35 pks. | 26.417 gallons. |
| Decalitre ........ | 10. | 10 cubic decimetres | 9.08 quarts........ | 2.6417 gallons. |
| Litre ........... | 1. | 1 cubic decimetre .. | 0.908 quart........ | 1.0567 quarts. |
| Decilitre......... | 0.1 | 0.1 cubic decimetre | 6.1022 cubic inches | 0.845 gill. |
| Centilitre ........ | 0.01 | 10 cubic centimetres | 0.6102 cubic inch.. | 0.338 fluid-ounce. |
| Millilitre......... | 0.001 | 1 cubic centimetre . | 0.061 cubic inch... | 0.27 fluid-drachm. |

## XI.—*Metric System*—Continued.

### WEIGHTS.

| Metric denominations and values. | | | Equivalents in denominations in use. |
|---|---|---|---|
| Names. | Number of grammes. | Weight of what quantity of water at maximum density. | Avoirdupois weight. |
| Millier or tonneau.. | 1000000. | 1 cubic metre ............... | 2204.6 pounds. |
| Quintal ............ | 100000. | 1 hectolitre ................. | 220 46 pounds. |
| Myriagramme ..... | 10000. | 10 litres.................... | 22.046 pounds. |
| Kilogramme, or kilo | 1000. | 1 litre...................... | 2.2046 pounds. |
| Hectogramme ....\ | 100. | 1 decilitre ................. | 3.5274 ounces. |
| Decagramme ...... | 10. | 10 cubic centimetres ........ | 0.3527 ounce. |
| Gramme.......... | 1. | 1 cubic centimetre .......... | 15.432 grains. |
| Decigramme...... | 0.1 | 0.1 cubic centimetre ........ | 1.5432 grains. |
| Centigramme ..... | 0.01 | 10 cubic millimetres........ | 0.1543 grain. |
| Milligramme...... | 0.001 | 1 cubic millimetre.......... | 0.0154 grain. |

### ADDITIONAL METRICAL EQUIVALENTS.

| | | | |
|---|---|---|---|
| 1 surveyor's chain in metres.. | = 20.11662 | .......... | log = 1.3035550 |
| 1 metre in surveyor's chain .. | = 0.04971 | .......... | log = 8.6964450 |

| 1 square foot in square metres | = 0.09290 | .......... | log = 8.9680221 |
| 1 acre in hectares ........... | = 0.40467 | .......... | log = 9.6071100 |
| 1 square mile in hectares..... | = 258.994 | .......... | log = 2.4132900 |

| 1 square metre in square feet. | = 10.76410 | .......... | log = 1.0319779 |
| 1 hectare in acres ........... | = 2.47109 | .......... | log = 0.3928900 |
| 1 hectare in square miles..... | = 0.00386 | .......... | log = 7.5867100 |

| 1 cubic foot in steres ........ | = 0.02831 | .......... | log = 8.4520332 |
| 1 cord in steres............. | = 3.62445 | .......... | log = 0.5592432 |

| 1 stere in cubic feet ......... | = 35.31561 | .......... | log = 1.5479668 |
| 1 stere in cords............. | = 0.27590. | .......... | log = 9.4407568 |

| 1 grain in grammes ......... | = 0.064798 | .......... | log = 8.8115680 |

## XII.—Foreign Measures of Length.

TABLE OF RELATIONS BETWEEN THE LINEAR MEASURES OF SEVERAL COUNTRIES, WITH CORRESPONDING LOGARITHMS.

| Metre. | France. | England and Russia. | Prussia and Denmark. | Bavaria. | Saxony. | Baden and Switzerland. | Austria. | Spain and Mexico. |
|---|---|---|---|---|---|---|---|---|
| 1 | *Paris feet.* 3.078444 / 0.4883313 | *Feet.* 3.280899 / 0.5159929 | *Feet.* 3.186199 / 0.5037390 | *Feet.* 3.426310 / 0.5348266 | *Feet.* 3.531197 / 0.5479220 | *Feet.* 3.333333 / 0.5228987 | *Vienna feet.* 3.163446 / 0.5001605 | *Feet.* 3.537877 / 0.5487427 |
| 0.3248594 / 9.5116687 | 1 | 1.065765 / 0.0276616 | 1.035003 / 0.0149417 | 1.113000 / 0.0464954 | 1.147072 / 0.0595907 | 1.082798 / 0.0345475 | 1.027612 / 0.0118292 | 1.149242 / 0.0604114 |
| 0.3047945 / 9.4840071 | 0.938293 / 9.9723384 | 1 | 0.971136 / 9.9872801 | 1.044320 / 0.0188337 | 1.076290 / 0.0319291 | 1.015982 / 0.0068859 | 0.964201 / 9.9841676 | 1.078325 / 0.0327498 |
| 0.3138535 / 9.4967270 | 0.966181 / 9.9850583 | 1.029722 / 0.0127199 | 1 | 1.075359 / 0.0315536 | 1.108279 / 0.0446490 | 1.046178 / 0.0196058 | 0.992859 / 9.9968875 | 1.110375 / 0.0454697 |
| 0.2918592 / 9.4651734 | 0.898472 / 9.9535047 | 0.957561 / 9.9811663 | 0.929922 / 9.9684464 | 1 | 1.036012 / 0.0139954 | 0.972864 / 9.9880521 | 0.923281 / 9.9653339 | 1.032562 / 0.0139161 |
| 0.2831901 / 9.4520780 | 0.871785 / 9.9404093 | 0.929118 / 9.9680709 | 0.902300 / 9.9553510 | 0.970297 / 9.9869046 | 1 | 0.943967 / 9.9749567 | 0.895856 / 9.9523385 | 1.001892 / 0.0008207 |
| 0.3000000 / 9.4771213 | 0.923533 / 9.9654525 | 0.984270 / 9.9931141 | 0.955860 / 9.9803942 | 1.027893 / 0.0119479 | 1.050359 / 0.0250433 | 1 | 0.949034 / 9.9772817 | 1.061361 / 0.0258630 |
| 0.3161109 / 9.4998395 | 0.973130 / 9.9881708 | 1.037128 / 0.0158324 | 1.007193 / 0.0031125 | 1.083094 / 0.0346661 | 1.116250 / 0.0477615 | 1.053703 / 0.0227183 | 1 | 1.118361 / 0.0485822 |
| 0.2826553 / 9.4512573 | 0.870139 / 9.9393896 | 0.927364 / 9.9672502 | 0.900597 / 9.9543393 | 0.968465 / 9.9860839 | 0.998112 / 9.9991793 | 0.942184 / 9.9741360 | 0.894165 / 9.9514178 | 1 |

XIII.—*Table of Relations between Itinerary Measures of Several Countries, with the Corresponding Logarithms.*

| France. | England. | Prussia and Denmark. | Austria. | Russia. | Spain and Mexico. | Germany. | England and France. |
|---|---|---|---|---|---|---|---|
| Myriamètre = 10000 M. | Statute mile = 5280'. | Mile = 24000'. | Mile = 24000'. | Verst = 3500'. | Jud. league = 15000'. | Geo. mile, 15 = 1 deg. | Naut. league, 20 = 1 deg. |
| — | 6.213824 / 0.7933590 | 1.327583 / 0.1236617 | 1.318103 / 0.1199492 | 9.373997 / 0.9719248 | 2.358584 / 0.3726514 | 1.347680 / 0.1295809 | 1.796907 / 0.2545256 |
| 0.1609315 / 9.2066410 | — | 0.213650 / 9.329708 | 0.212124 / 9.326593 | 1.508571 / 0.1785059 | 0.379570 / 9.579294 | 0.216884 / 9.336279 | 0.289179 / 9.461666 |
| 0.7532485 / 9.8769383 | 4.680554 / 0.6702977 | — | 0.992859 / 9.996873 | 7.060950 / 0.8486631 | 1.776600 / 0.2495897 | 1.015138 / 0.0065451 | 1.353518 / 0.1314039 |
| 0.7586663 / 9.8800508 | 4.714219 / 0.6734097 | 1.007193 / 0.0031125 | — | 7.111736 / 0.8519756 | 1.789379 / 0.2527922 | 1.022440 / 0.0096376 | 1.363253 / 0.1345764 |
| 0.1066781 / 9.0280758 | 0.662879 / 9.821431 | 0.141624 / 9.151189 | 0.140613 / 9.148244 | — | 2.516092 / 0.4007266 | 0.143768 / 9.157620 | 0.191691 / 9.282608 |
| 0.4439831 / 9.6473486 | 2.634556 / 0.4207076 | 0.562873 / 9.750410 | 0.558853 / 9.7472978 | 0.397442 / 9.5992734 | — | 0.571394 / 9.7569355 | 0.761868 / 9.8818742 |
| 0.7420158 / 9.8704131 | 4.610755 / 0.6637721 | 0.985088 / 9.9934749 | 0.978053 / 9.9903624 | 6.955654 / 0.8423380 | 1.750107 / 0.2436645 | — | 1.333333 / 0.1249387 |
| 0.5565118 / 9.7454744 | 3.458667 / 0.5388334 | 0.738816 / 9.8685361 | 0.733540 / 9.8654236 | 5.216740 / 0.7173922 | 1.312586 / 0.1181258 | 0.750000 / 9.8750613 | — |

1 English or French geographical mile = $\frac{1}{60}$ of a degree of longitude at the equator = 2028.7 English yards.

| | Eng. stat. miles. | | | Eng. stat. miles. |
|---|---|---|---|---|
| Modern Roman mile.... | = 0.925 | | Portugal league ........ .. | = 3.841 |
| Tuscan mile .... ........ | = 1.027 | | Flanders league.......... | = 3.900 |
| Old Scottish mile ........ | = 1.127 | | Spanish common league . | = 4.214 |
| Irish mile.............. | = 1.273 | | Hungarian mile.......... | = 5.178 |
| French posting league .. | = 2.422 | | Swedish mile .......... | = 6.648 |

*Table for converting Metres into Toises and French and English Feet and Inches.*

| Metres. | Toises. | French. | | | English. | |
|---|---|---|---|---|---|---|
| | | Feet. | Inches. | Lines. | Feet. | Inches. |
| 1 | 0.51307 | 3 | 0 | 11,296 | 3 | 3.3708 |
| 2 | 1.02615 | 6 | 1 | 10.592 | 6 | 6.7416 |
| 3 | 1.53922 | 9 | 2 | 9.888 | 9 | 10.1124 |
| 4 | 2.05230 | 12 | 3 | 9.184 | 13 | 1.4832 |
| 5 | 2.56537 | 15 | 4 | 8.480 | 16 | 4.8539 |
| 6 | 3.07844 | 18 | 5 | 7.776 | 19 | 8.2247 |
| 7 | 3.59152 | 21 | 6 | 7.072 | 22 | 11.5955 |
| 8 | 4.10459 | 24 | 7 | 6.368 | 26 | 2.9663 |
| 9 | 4.61767 | 27 | 8 | 5.664 | 29 | 6.3371 |
| 10 | 5.13074 | 30 | 9 | 4.960 | 32 | 9.7079 |
| 20 | 10.26148 | 61 | 6 | 9.920 | 65 | 7.4158 |
| 30 | 15.39222 | 92 | 4 | 2.880 | 98 | 5.1237 |
| 40 | 20.52296 | 123 | 1 | 7.840 | 131 | 2.8316 |
| 50 | 25.65370 | 153 | 11 | 0.800 | 164 | 0.5395 |
| 60 | 30.78444 | 184 | 8 | 5.760 | 196 | 10.2474 |
| 70 | 35.91519 | 215 | 5 | 10.720 | 229 | 7.9553 |
| 80 | 41.04593 | 246 | 3 | 3.680 | 262 | 5.6632 |
| 90 | 46.17667 | 277 | 0 | 8.640 | 295 | 3.3711 |
| 100 | 51.30741 | 307 | 10 | 1 600 | 328 | 1.0790 |
| 200 | 102.61481 | 615 | 8 | 3.200 | 656 | 2.1580 |
| 300 | 153.92222 | 923 | 6 | 4.800 | 984 | 3.2370 |
| 400 | 205.22963 | 1231 | 4 | 6.400 | 1312 | 4.3160 |
| 500 | 256.53704 | 1539 | 2 | 8.000 | 1640 | 5.3950 |
| 600 | 307.84444 | 1847 | 0 | 9.600 | 1968 | 6.4740 |
| 700 | 359.15185 | 2154 | 10 | 11.200 | 2296 | 7.5530 |
| 800 | 410.45926 | 2462 | 9 | 0.800 | 2624 | 8.6320 |
| 900 | 461.76667 | 2770 | 7 | 2.400 | 2952 | 9.7110 |
| 1000 | 513.07407 | 3078 | 5 | 4.000 | 3280 | 10.7900 |
| 2000 | 1026.14815 | 6156 | 10 | 8.000 | 6561 | 9.5800 |
| 3000 | 1539.22222 | 9235 | 4 | 0.000 | 9842 | 8.3700 |
| 4000 | 2052.29630 | 12313 | 9 | 4.000 | 13123 | 7.1600 |
| 5000 | 2565.37037 | 15392 | 2 | 8.000 | 16404 | 5.9500 |
| 6000 | 3078.44444 | 18470 | 8 | 0.000 | 19685 | 4.7400 |
| 7000 | 3591.51852 | 21549 | 1 | 4.000 | 22966 | 3.5300 |
| 8000 | 4104.59259 | 24627 | 6 | 8.000 | 26247 | 2.3200 |
| 9000 | 4617.66667 | 27706 | 0 | 0.000 | 29528 | 1.1100 |
| 10000 | 5130.74074 | 30784 | 5 | 4.000 | 32808 | 11.9000 |

*Table for converting English Feet into French Toises, Metres, and Feet.*

| English feet. | Toises. | Metres. | French. | | |
|---|---|---|---|---|---|
| | | | Feet. | Inches. | Lines. |
| 1 | 0.15638 | 0.30479 | 0 | 11 | 3.114 |
| 2 | 0.31276 | 0.60959 | 1 | 10 | 6.228 |
| 3 | 0.46915 | 0.91438 | 2 | 9 | 9.343 |
| 4 | 0.62553 | 1.21918 | 3 | 9 | 0.457 |
| 5 | 0.78191 | 1.52397 | 4 | 8 | 3.571 |
| 6 | 0.93829 | 1.82877 | 5 | 7 | 6.685 |
| 7 | 1.09468 | 2.13356 | 6 | 6 | 9.799 |
| 8 | 1.25106 | 2.43836 | 7 | 6 | 0.913 |
| 9 | 1.40744 | 2.74315 | 8 | 5 | 4.028 |
| 10 | 1.56382 | 3.04794 | 9 | 4 | 7.142 |
| 20 | 3.12764 | 6.09589 | 18 | 9 | 2.284 |
| 30 | 4.69146 | 9.14383 | 28 | 1 | 9.425 |
| 40 | 6.25529 | 12.19178 | 37 | 6 | 4.567 |
| 50 | 7.81911 | 15.23972 | 46 | 10 | 11.709 |
| 60 | 9.38293 | 18.28767 | 56 | 3 | 6.851 |
| 70 | 10.94675 | 21.33561 | 65 | 8 | 1.993 |
| 80 | 12.51057 | 24.38536 | 75 | 0 | 9.134 |
| 90 | 14.07439 | 27.43150 | 84 | 5 | 4.276 |
| 100 | 15.63822 | 30.47945 | 93 | 9 | 11.418 |
| 200 | 31.27643 | 60.95850 | 187 | 7 | 10.836 |
| 300 | 46.91465 | 91.43835 | 281 | 5 | 10.254 |
| 400 | 62.55286 | 121.91780 | 375 | 3 | 9.672 |
| 500 | 78.19108 | 152.39725 | 469 | 1 | 9.090 |
| 600 | 93.82929 | 182.87670 | 562 | 11 | 8.508 |
| 700 | 109.46751 | 213.35615 | 656 | 9 | 7.926 |
| 800 | 125.10572 | 243.83559 | 750 | 7 | 7.344 |
| 900 | 140.74394 | 274.31504 | 844 | 5 | 6.762 |
| 1000 | 156.38215 | 304.79449 | 938 | 3 | 6.180 |
| 2000 | 312.76431 | 609.58899 | 1876 | 7 | 0.360 |
| 3000 | 469.14646 | 914.38348 | 2814 | 10 | 6.539 |
| 4000 | 625.52861 | 1219.17797 | 3753 | 2 | 0.719 |
| 5000 | 781.91076 | 1523.97246 | 4691 | 5 | 6.899 |
| 6000 | 938.29292 | 1828.76696 | 5629 | 9 | 1.079 |
| 7000 | 1094.67507 | 2133.56145 | 6568 | 0 | 7.259 |
| 8000 | 1251.05722 | 2438.35594 | 7506 | 4 | 1.438 |
| 9000 | 1407.43937 | 2743.15044 | 8444 | 7 | 7.618 |
| 10000 | 1563.82153 | 3047.94493 | 9382 | 11 | 1.798 |

### XIV.—*Analytical Expressions for different Lines, Surfaces, and Solids.*

#### 1.—LINES.

Ratio of diagonal to side of square $= \sqrt{2} = 1.414 = \frac{10}{7}$, nearly.

Log $\sqrt{2} = 0.1505149978$

Side of inscribed square : R :: $\sqrt{2}$ : 1

Side of inscribed equilateral triangle : R :: $\sqrt{3}$ : 1

Side of inscribed regular hexagon $=$ R

Side of inscribed regular decagon $= \frac{1}{2}$ R $(- 1 + \sqrt{5}) = 0.618$ R

#### *Circle.*

Ratio of circumference to diameter $= 3.1415926 = \frac{355}{113}$, nearly.

Length of an arc $= \dfrac{a \pi r}{180}$, $r$ being the radius of the circle and $a$ the number of degrees in the arc; or nearly $= \dfrac{8 \, c' - c}{3}$,

$c$ being the chord of the arc, and $c'$ (the chord of half the arc)
$= \sqrt{\frac{1}{4} c^2 + \text{ver sin}^2}$

#### *Ellipse.*

Circumference $= \frac{100}{200} \pi \sqrt{\frac{1}{2} (a^2 + b^2)}$ nearly; $a$ and $b$ being the axes.

*Lengths of Circular Arcs, taking the Base of Segments as Unity.*

| Ver sin. | Length. | Ver sin. | Length. | Ver sin. | Length. | Ver sin. | Length. | Ver sin. | Length. |
|---|---|---|---|---|---|---|---|---|---|
| .01 | 1.000 | .11 | 1.032 | .21 | 1.114 | .31 | 1.239 | .41 | 1.401 |
| .02 | 1.000 | .12 | 1.038 | .22 | 1.124 | .32 | 1.254 | .42 | 1.418 |
| .03 | 1.000 | .13 | 1.044 | .23 | 1.135 | .33 | 1.269 | .43 | 1.437 |
| .04 | 1.000 | .14 | 1.051 | .24 | 1.147 | .34 | 1.284 | .44 | 1.455 |
| .05 | 1.000 | .15 | 1.059 | .25 | 1.159 | .35 | 1.300 | .45 | 1.474 |
| .06 | 1.006 | .16 | 1.067 | .26 | 1.171 | .36 | 1.316 | .46 | 1.493 |
| .07 | 1.014 | .17 | 1.075 | .27 | 1.184 | .37 | 1.332 | .47 | 1.512 |
| .08 | 1.018 | .18 | 1.084 | .28 | 1.197 | .38 | 1.349 | .48 | 1.531 |
| .09 | 1.020 | .19 | 1.093 | .29 | 1.212 | .39 | 1.366 | .49 | 1.551 |
| .10 | 1.026 | .20 | 1.103 | .30 | 1.225 | .40 | 1.383 | .50 | 1.571 |

### XIV.—*Analytical Expressions, &c.*—Continued.

#### 2.—SURFACES.

1. Triangle in terms of—

its base and its altitude . . . . . . . . . . . . . . $= \dfrac{b \, A}{2}$

two sides and the included angle . . . . . . . . $= \dfrac{a \, b \sin C}{2}$

its three sides . . . . . . . . $= [s \, (s - a) \, (s - b) \, (s - c)]^{\frac{1}{2}}$

where

$\quad\quad A =$ the altitude;

$\quad\quad a, b, c =$ the three sides;

$\quad\quad\quad C =$ the angle included between $a$ and $b$; and

$\quad\quad s = \dfrac{a + b + c}{2}$

2. Parallelogram in terms of—

its base and its altitude . . . . . . . . . . . . . . $= b \, A$

two sides and the included angle . . . . . . . . . $= a \, b \sin C$

two sides and their corresponding diagonal

$$= 2 \, [s \, (s - a) \, (s - b) \, (s - c)]^{\frac{1}{2}}$$

where

$\quad C =$ the angle included between two adjacent sides $a, b$;

$\quad c =$ the diagonal opposite; and

$\quad s = \dfrac{a + b + c}{2}$

3. Trapezium in terms of—

its two parallel bases and its altitude . . . . . . $= \dfrac{B + b}{2} \, A$

its two parallel bases, one of its oblique sides, and the angle between one of these bases and this side . . . $= \dfrac{B + b}{2} \, l \sin C$

where

$\quad\quad A =$ the distance between the two parallel bases $B, b$;

$\quad\quad l =$ the length of one of the oblique sides; and

$\quad\quad C =$ the angle between one of these bases and this side.

4. Any quadrilateral $=$ half the product of its two diagonals multiplied by the sine of the included angle.

### XIV.—*Analytical Expressions, &c.*—Continued.

5. Regular polygon . . . . . . . . . . . . $= \dfrac{n \left(\dfrac{a}{2}\right)^2}{\tan \dfrac{180°}{n}}$

where

$n =$ the number of sides; and

$a =$ the length of one of them.

6. Circle . . . . . . . . . . . . . . . . . . . $= \pi R^2$

7. Ellipse . . . . . . . . . . . . . . . . . $= \pi a b$

     $a$ and $b$ being the semi-axes.

8. Right cylinder, exclusive of its bases . . . . $= 2 \pi R A$

9. Sphere . . . . . . . . . . . . . . . . . . . $= 4 \pi R^2$

10. Zone . . . . . $= 4 \pi R^2 \sin \frac{1}{2} (L' - L) \cos \frac{1}{2} (L' + L)$

11. Right cone . . . . . . . . . . . . . . . . $= \pi R L$

12. Frustum of cone with parallel bases . . . $= \pi l (R + r)$

where

R and $r =$ the radii of the bases of these solids; and

L and $l =$ the lengths of their generating elements.

13. Spherical quadrilateral, formed by two parallels of latitude and two meridians

$$= \frac{\pi}{90°} (M' - M) R^2 \sin \tfrac{1}{2} (L' - L) \cos \tfrac{1}{2} (L' + L)$$

where

R $=$ the radius of the sphere;

L, L' $=$ the latitudes of the bases of the zone, $+$ when north, $-$ when south; and

M', M $=$ the longitudes of the extreme meridians of the quadrilateral, (M' $-$ M) being expressed in degrees and decimals.

In the place of R, the normal N, of the mean latitude $\left(\dfrac{L' + L}{2}\right)$, can be used.

## XIV.—*Analytical Expressions, &c.*—Continued.

### 3.—SOLIDS.

14. Prism . . . . . . . . . . . . . . . . . . . . = B A

where

      B = the area of the base; and

      A = the altitude.

15. Rectangular parallelopepidon . . . . . . $= p \times q \times r$

    Cube . . . . . . . . . . . . . . . $= p^3$

      where $p, q, r,$ = the lengths of the three contiguous edges.

16. Pyramid . . . . . . . . . . . . . . . . $= \dfrac{B\ A}{3}$

      The area, B, being found from No. 5.

17. Right cylinder . . . . . . . . . . . . . $= \pi\ R^2\ A$

18. Right cone . . . . . . . . . . . . . . $= \tfrac{1}{3} \pi\ R^2\ A$

19. Sphere . . . . . . . . . . . . . . . . $= \tfrac{4}{3} \pi\ R^3$

20. Prismoid, or solid figure, similar to that which is formed in excavations or embankments of roads, terminated by parallel cross-sections.

Solid content = area of each end, added to four times the middle area, and the sum multiplied by the length divided by 6, or

$$= \left\{ (b + r\,h')\,h' + (b + r\,h)\,h + 4\left(b + r \times \frac{h+h'}{2}\right)\frac{h+h'}{2} \right\} \frac{l}{6}$$

where

      $b$ = the breadth at the bottom of the cutting;

      $h$ = the perpendicular depth of cutting at higher end;

      $h'$ = the perpendicular depth of cutting at lower end;

      $l$ = the length of the solid; and

      $r$ = the ratio of the perpendicular height of the slope to its horizontal base.

1. Arithmetical:

$$a = z - (n - 1)d \qquad z = a + (n - 1)d$$

$$d = \frac{z - a}{n - 1} \qquad n = \frac{z - a}{d} + 1 \qquad s = \frac{a + z}{2} n$$

2. Geometrical:

$$a = \frac{z}{r^{n-1}} \qquad z = a r^{n-1} \qquad r = \left(\frac{z}{a}\right)^{\frac{1}{n-1}}$$

$$n = \log. \frac{r z}{a} \div \log. r \qquad s = \frac{r z - a}{r - 1} \qquad = \frac{r^n - 1}{r - 1} a$$

$$a = \frac{r - 1}{r^n - 1} s \qquad a = z r - (r - 1) s \qquad r = \frac{s - a}{s - z}$$

where

$a =$ the least term;     $r =$ the common ratio;
$z =$ the greatest term;     $n =$ the number of terms;
$d =$ the common diff.;     $s =$ the sum of the terms.

## XVI.—*Force of Gravity.*

The velocity acquired at the end of one second by a body falling in vacuo, at the level of the sea, in the latitude of London $= 32.1915$ feet.

The force of gravity at the latitude of $45° = 32.17$ feet per second being represented by $g$, for any other latitude, $l$,

$$g' = g (1 - 0.002588 \cos 2 l)$$

If $g$ represents the force of gravity at the height $h$, and $r$ the radius of the earth, the force of gravity at the level of the sea

$$= g' = g \left(1 + \frac{5}{4}\frac{h}{r}\right)$$

Length, in inches, of a pendulum vibrating seconds at the level of the sea:

| | | | |
|---|---|---|---|
| Equator . . . . | $= 39.0152$ | London, lat. 51° 31' - - | $= 39.1393$ |
| New York, lat. 40° 43' | $= 39.1017$ | Spitzbergen, lat. 75° 50' | $= 39.2147$ |

## XVII.—*Land-Surveying with Compass and Chain.*

### *To calculate the Area or Content of Land.*

If the sum of each adjacent pair of distances perpendicular to a meridian (*departures*) assumed without the survey be multiplied by the northing or southing between them in succession round the figure in the same order, the difference between the sum of the *north* products and the sum of the *south* products will be double the area of the tract.

The *meridian distance* of a course is the distance of the middle point of that course from an assumed meridian.

Hence, the double meridian distance of the first course is equal to its departure.

And the double meridian distance of any course is equal to the double meridian distance of the preceding course, plus its departure, plus the departure of the course itself, having regard to the algebraic sign of each.

Then, to find the area—

1. Multiply the double meridian distance of each course by its northing or southing.
2. Place all the *plus* products in one column, and all the *minus* products in another.
3. Add up each column separately, and take their difference. This difference will be *double* the area of the land.

In *balancing* the work, the error for each particular course is found by the proportion—

> As the sum of the courses is to the error of latitude, (or departure,) so is each particular course to its correction.

When a bearing is due east or west, the error of latitude is nothing, and the course must be subtracted from the sum of the courses before balancing the columns of latitude. And so with the departures.

EXAMPLE.—It is required to find the content of a piece of land, of which the following are the field-notes:

| Sta. | Course. | Dist. | Sta. | Course. | Dist. |
|---|---|---|---|---|---|
| 1 | North 46½° west. | 20. chains. | 4 | South 56° east .. | 27.60 chains. |
| 2 | North 51¾° east. | 13.80 chains. | 5 | South 33½° west. | 18.80 chains. |
| 3 | East............ | 21.25 chains. | 6 | North 74½° west. | 30.95 chains. |

## XVII.—*Land-Surveying, &c.*—Continued.

*Calculation.*

| Stations | Courses | Dist., chains | Diff. lat. N. + | Diff. lat. S. − | Departure E. + | Departure W. − | Balanced Latitude | Balanced Departure | D.M.D. + | Area. + | Area. − |
|---|---|---|---|---|---|---|---|---|---|---|---|
| 1 | N. 46¼° W | 20.00 | 13.77 | | | 14.51 | + 13.88 | − 14.56 | 14.56 | 202.0928 | |
| 2 | N. 51¼° E | 13.80 | 8.54 | | 10.84 | | + 8.61 | + 10.81 | 10.81 | 93.0741 | |
| 3 | East | 21.25 | | | 21.25 | | | + 21.20 | 42.82 | | |
| 4 | S. 56° E | 27.60 | | 15.44 | 22.88 | | − 15.29 | + 22.82 | 86.84 | | 1327.7836 |
| 5 | S. 33¼° W | 18.80 | | 15.72 | | 10.31 | − 15.63 | − 10.36 | 99.30 | | 1552.0590 |
| 6 | N. 74¼° W | 30.95 | 8.27 | | | 29.83 | + 8.43 | − 29.91 | 59.03 | 497.6229 | |
| Sums | | 132.40 | 30.58 | 31.16 | 54.97 | 54.65 | | | | 792.7898 | 2879.8426 |
| | | | | 30.58 | 54.65 | | | | | | 792.7898 |
| | | | | 0.58 | 0.32, error in westling | | | | | | 2)2087.0528 |
| | | | | | | | | | | | 1043.5264 |

Error in northing........ 0.58

Answer 104 A.   1 R.   16 P.

100000 square links of Gunter's chain = 1 acre.

1 square chain = 66 feet square = 1/10 acre.

## XVIII.—Table showing Differences of Latitude and Departures.

| Minutes | Distance | 0° Lat. | 0° Dep. | 1° Lat. | 1° Dep. | 2° Lat. | 2° Dep. | Distance | Minutes |
|---|---|---|---|---|---|---|---|---|---|
| 0 | 1 | 1.00000 | 0.00000 | 0.99984 | 0.01745 | 0.99939 | 0.03490 | 1 | |
| | 2 | 2.00000 | 0.00000 | 1.99969 | 0.03490 | 1.99878 | 0.06980 | 2 | |
| | 3 | 3.00000 | 0.00000 | 2.99954 | 0.05235 | 2.99817 | 0.10470 | 3 | |
| | 4 | 4.00000 | 0.00000 | 3.99939 | 0.06980 | 3.99756 | 0.13960 | 4 | |
| | 5 | 5.00000 | 0.00000 | 4.99923 | 0.08726 | 4.99695 | 0.17450 | 5 | 60 |
| | 6 | 6.00000 | 0.00000 | 5.99908 | 0.10471 | 5.99634 | 0.20940 | 6 | |
| | 7 | 7.00000 | 0.00000 | 6.99893 | 0.12216 | 6.99573 | 0.24430 | 7 | |
| | 8 | 8.00000 | 0.00000 | 7.99878 | 0.13961 | 7.99512 | 0.27920 | 8 | |
| | 9 | 9.00000 | 0.00000 | 8.99862 | 0.15707 | 8.99451 | 0.31410 | 9 | |
| 15 | 1 | 0.99999 | 0.00436 | 0.99976 | 0.02181 | 0.99922 | 0.03925 | 1 | |
| | 2 | 1.99998 | 0.00872 | 1.99952 | 0.04363 | 1.99845 | 0.07851 | 2 | |
| | 3 | 2.99997 | 0.01308 | 2.99928 | 0.06544 | 2.99768 | 0.11777 | 3 | |
| | 4 | 3.99996 | 0.01745 | 3.99904 | 0.08725 | 3.99691 | 0.15703 | 4 | |
| | 5 | 4.99995 | 0.02181 | 4.99881 | 0.10907 | 4.99614 | 0.19629 | 5 | 45 |
| | 6 | 5.99994 | 0.02617 | 5.99857 | 0.13089 | 5.99537 | 0.23555 | 6 | |
| | 7 | 6.99993 | 0.03054 | 6.99833 | 0.15270 | 6.99460 | 0.27481 | 7 | |
| | 8 | 7.99992 | 0.03490 | 7.99809 | 0.17452 | 7.99383 | 0.31407 | 8 | |
| | 9 | 8.99991 | 0.03926 | 8.99785 | 0.19633 | 8.99306 | 0.35333 | 9 | |
| 30 | 1 | 0.99996 | 0.00872 | 0.99965 | 0.02617 | 0.99904 | 0.04361 | 1 | |
| | 2 | 1.99992 | 0.01745 | 1.99931 | 0.05235 | 1.99809 | 0.08723 | 2 | |
| | 3 | 2.99988 | 0.02617 | 2.99897 | 0.07853 | 2.99714 | 0.13085 | 3 | |
| | 4 | 3.99984 | 0.03490 | 3.99862 | 0.10470 | 3.99619 | 0.17447 | 4 | |
| | 5 | 4.99981 | 0.04363 | 4.99828 | 0.13088 | 4.99524 | 0.21809 | 5 | 30 |
| | 6 | 5.99977 | 0.05235 | 5.99794 | 0.15706 | 5.99428 | 0.26171 | 6 | |
| | 7 | 6.99973 | 0.06108 | 6.99760 | 0.18323 | 6.99333 | 0.30533 | 7 | |
| | 8 | 7.99969 | 0.06981 | 7.99725 | 0.20941 | 7.99238 | 0.34895 | 8 | |
| | 9 | 8.99965 | 0.07853 | 8.99691 | 0.23559 | 8.99143 | 0.39257 | 9 | |
| 45 | 1 | 0.99991 | 0.01308 | 0.99953 | 0.03053 | 0.99884 | 0.04797 | 1 | |
| | 2 | 1.99982 | 0.02617 | 1.99906 | 0.06107 | 1.99769 | 0.09595 | 2 | |
| | 3 | 2.99974 | 0.03926 | 2.99860 | 0.09161 | 2.99654 | 0.14393 | 3 | |
| | 4 | 3.99965 | 0.05235 | 3.99813 | 0.12215 | 3.99539 | 0.19191 | 4 | |
| | 5 | 4.99957 | 0.06544 | 4.99766 | 0.15269 | 4.99424 | 0.23989 | 5 | 15 |
| | 6 | 5.99948 | 0.07853 | 5.99720 | 0.18323 | 5.99309 | 0.28786 | 6 | |
| | 7 | 6.99940 | 0.09162 | 6.99673 | 0.21376 | 6.99193 | 0.33584 | 7 | |
| | 8 | 7.99931 | 0.10471 | 7.99626 | 0.24430 | 7.99078 | 0.38382 | 8 | |
| | 9 | 8.99922 | 0.11780 | 8.99580 | 0.27484 | 8.98963 | 0.43180 | 9 | |
| Minutes | Distance | Dep. | Lat. | Dep. | Lat. | Dep. | Lat. | Distance | Minutes |
| | | 89° | | 88° | | 87° | | | |

*Differences of Latitude and Departures*—Continued.

| Minutes. | Distance. | 3° | | 4° | | 5° | | Distance. | Minutes. |
|---|---|---|---|---|---|---|---|---|---|
| | | Lat. | Dep. | Lat. | Dep. | Lat. | Dep. | | |
| | 1 | 0.99863 | 0.05233 | 0.99756 | 0.06975 | 0.99619 | 0.08715 | 1 | |
| | 2 | 1.99726 | 0.10467 | 1.99512 | 0.13951 | 1.99238 | 0.17431 | 2 | |
| | 3 | 2.99589 | 0.15700 | 2.99269 | 0.20926 | 2.98858 | 0.26146 | 3 | |
| | 4 | 3.99452 | 0.20934 | 3.99025 | 0.27902 | 3.98477 | 0.34862 | 4 | |
| 0 | 5 | 4.99315 | 0.26168 | 4.98782 | 0.34878 | 4.98097 | 0.43577 | 5 | 60 |
| | 6 | 5.99178 | 0.31401 | 5.98538 | 0.41853 | 5.97716 | 0.52293 | 6 | |
| | 7 | 6.99041 | 0.36635 | 6.98294 | 0.48829 | 6.97336 | 0.61008 | 7 | |
| | 8 | 7.98904 | 0.41868 | 7.98051 | 0.55805 | 7.96955 | 0.69724 | 8 | |
| | 9 | 8.98767 | 0.47102 | 8.97807 | 0.62780 | 8.96575 | 0.78440 | 9 | |
| | 1 | 0.99839 | 0.05669 | 0.99725 | 0.07410 | 0.09580 | 0.09150 | 1 | |
| | 2 | 1.99678 | 0.11338 | 1.99450 | 0.14821 | 1.99160 | 0.18300 | 2 | |
| | 3 | 2.99517 | 0.17007 | 2.99175 | 0.22232 | 2.98741 | 0.27450 | 3 | |
| | 4 | 3.99356 | 0.22677 | 3.98900 | 0.29643 | 3.98321 | 0.36600 | 4 | |
| 15 | 5 | 4.99195 | 0.28346 | 4.98625 | 0.37054 | 4.97902 | 0.45750 | 5 | 45 |
| | 6 | 5.99035 | 0.34015 | 5.98350 | 0.44465 | 5.97482 | 0.54900 | 6 | |
| | 7 | 6.98874 | 0.39684 | 6.98075 | 0.51875 | 6.97063 | 0.64051 | 7 | |
| | 8 | 7.98713 | 0.45354 | 7.97800 | 0.59286 | 7.96643 | 0.73201 | 8 | |
| | 9 | 8.98552 | 0.51023 | 8.97525 | 0.66697 | 8.96224 | 0.82351 | 9 | |
| | 1 | 0.99813 | 0.06104 | 0.99691 | 0.07845 | 0.99539 | 0.09584 | 1 | |
| | 2 | 1.99626 | 0.12209 | 1.99383 | 0.15691 | 1.99079 | 0.19169 | 2 | |
| | 3 | 2.99440 | 0.18314 | 2.99075 | 0.23537 | 2.98618 | 0.28753 | 3 | |
| | 4 | 3.99253 | 0.24419 | 3.98766 | 0.31383 | 3.98158 | 0.38338 | 4 | |
| 30 | 5 | 4.99067 | 0.30524 | 4.98458 | 0.39229 | 4.97698 | 0.47922 | 5 | 30 |
| | 6 | 5.98880 | 0.36629 | 5.98150 | 0.47075 | 5.97237 | 0.57507 | 6 | |
| | 7 | 6.98694 | 0.42733 | 6.97842 | 0.54921 | 6.96777 | 0.67092 | 7 | |
| | 8 | 7.98507 | 0.48838 | 7.97533 | 0.62767 | 7.96316 | 0.76676 | 8 | |
| | 9 | 8.98321 | 0.54943 | 8.97225 | 0.70613 | 8.95856 | 0.86261 | 9 | |
| | 1 | 9.99785 | 0.06540 | 0.99656 | 0.08280 | 0.99496 | 0.10018 | 1 | |
| | 2 | 1.99571 | 0.13080 | 1.99313 | 0.16561 | 1.98993 | 0.20037 | 2 | |
| | 3 | 2.99357 | 0.19620 | 2.98969 | 0.24842 | 2.98490 | 0.30056 | 3 | |
| | 4 | 3.99143 | 0.26161 | 5.98626 | 0.33123 | 3.97987 | 0.40075 | 4 | |
| 45 | 5 | 4.98929 | 0.32701 | 4.98282 | 0.41404 | 4.97484 | 0.50094 | 5 | 15 |
| | 6 | 5.98715 | 0.39241 | 5.97939 | 0.49684 | 5.96981 | 0.60112 | 6 | |
| | 7 | 6.98501 | 0.45782 | 6.97595 | 0.57965 | 6.96477 | 0.70131 | 7 | |
| | 8 | 7.98287 | 0.52322 | 7.97252 | 0.66246 | 7.95974 | 0.80150 | 8 | |
| | 9 | 8.98073 | 0.58862 | 8.96908 | 0.74527 | 8.95471 | 0.90169 | 9 | |
| Minutes. | Distance. | Dep. | Lat. | Dep. | Lat. | Dep. | Lat. | Distance. | Minutes. |
| | | 86° | | 85° | | 84° | | | |

## *Differences of Latitude and Departures*—Continued.

| Minutes | Distance | 6° Lat. | 6° Dep. | 7° Lat. | 7° Dep. | 8° Lat. | 8° Dep. | Distance | Minutes |
|---|---|---|---|---|---|---|---|---|---|
| 0 | 1 | 0.99452 | 0.10452 | 0.99254 | 0.12186 | 0.99026 | 0.13917 | 1 | 60 |
|   | 2 | 1.98904 | 0.20905 | 1.98509 | 0.24373 | 1.98053 | 0.27834 | 2 |  |
|   | 3 | 2.98356 | 0.31358 | 2.97763 | 0.36560 | 2.97080 | 0.41751 | 3 |  |
|   | 4 | 3.97808 | 0.41811 | 3.97018 | 0.48747 | 3.96107 | 0.55669 | 4 |  |
|   | 5 | 4.97261 | 0.52264 | 4.96273 | 0.60934 | 4.95134 | 0.69586 | 5 |  |
|   | 6 | 5.96713 | 0.62717 | 5.95519 | 0.73121 | 5.94160 | 0.83503 | 6 |  |
|   | 7 | 6.96165 | 0.73169 | 6.94782 | 0.85308 | 6.93187 | 0.97421 | 7 |  |
|   | 8 | 7.95617 | 0.83622 | 7.94038 | 0.97495 | 7.92214 | 0.11338 | 8 |  |
|   | 9 | 8.95069 | 0.94075 | 8.93291 | 0.09682 | 8.91241 | 0.25255 | 9 |  |
| 15 | 1 | 0.99405 | 0.10886 | 0.99200 | 0.12619 | 0.98965 | 0.14349 | 1 | 45 |
|   | 2 | 1.98811 | 0.21773 | 1.98400 | 0.25239 | 1.97930 | 0.28698 | 2 |  |
|   | 3 | 2.98216 | 0.32660 | 2.97601 | 0.37859 | 2.96895 | 0.43047 | 3 |  |
|   | 4 | 3.97622 | 0.43546 | 3.96801 | 0.50479 | 3.95860 | 0.57397 | 4 |  |
|   | 5 | 4.97028 | 0.54433 | 4.96002 | 0.63099 | 4.94825 | 0.71746 | 5 |  |
|   | 6 | 5.96433 | 0.65320 | 5.95202 | 0.75719 | 5.93790 | 0.86095 | 6 |  |
|   | 7 | 6.95839 | 0.76206 | 6.94403 | 0.88339 | 6.92755 | 1.00444 | 7 |  |
|   | 8 | 7.95245 | 0.87093 | 7.93603 | 1.00959 | 7.91721 | 1.14794 | 8 |  |
|   | 9 | 8.94650 | 0.97980 | 8.92804 | 1.13579 | 8.90686 | 1.29143 | 9 |  |
| 30 | 1 | 0.99357 | 0.11320 | 0.99144 | 0.13052 | 0.98901 | 0.14780 | 1 | 30 |
|   | 2 | 1.98714 | 0.22640 | 1.98288 | 0.26105 | 1.97803 | 0.29561 | 2 |  |
|   | 3 | 2.98071 | 0.33960 | 2.97433 | 0.39157 | 2.96704 | 0.44342 | 3 |  |
|   | 4 | 3.97428 | 0.45281 | 3.96577 | 0.52210 | 3.95606 | 0.59123 | 4 |  |
|   | 5 | 4.96786 | 0.56601 | 4.95722 | 0.65263 | 4.94508 | 0.73904 | 5 |  |
|   | 6 | 5.96143 | 0.67921 | 5.94866 | 0.78315 | 5.93409 | 0.88685 | 6 |  |
|   | 7 | 6.95500 | 0.79242 | 6.94011 | 0.91368 | 6.92311 | 1.03466 | 7 |  |
|   | 8 | 7.94857 | 0.90562 | 7.93155 | 1.04420 | 7.91212 | 1.18247 | 8 |  |
|   | 9 | 8.94214 | 1.01882 | 8.92300 | 1.17473 | 8.90114 | 1.33028 | 9 |  |
| 45 | 1 | 0.99306 | 0.11753 | 0.99086 | 0.13485 | 0.98836 | 0.15212 | 1 | 15 |
|   | 2 | 1.98613 | 0.23507 | 1.98173 | 0.26970 | 1.97672 | 0.30424 | 2 |  |
|   | 3 | 2.97920 | 0.35261 | 2.97259 | 0.40455 | 2.96508 | 0.45637 | 3 |  |
|   | 4 | 3.97227 | 0.47014 | 3.96346 | 0.53940 | 3.95344 | 0.60849 | 4 |  |
|   | 5 | 4.96534 | 0.58768 | 4.95432 | 0.67425 | 4.94180 | 0.76061 | 5 |  |
|   | 6 | 5.95841 | 0.70522 | 5.94519 | 0.80910 | 5.93016 | 0.91274 | 6 |  |
|   | 7 | 6.95147 | 0.82276 | 6.93606 | 0.94395 | 6.91853 | 1.06486 | 7 |  |
|   | 8 | 7.94454 | 0.94029 | 7.92692 | 1.07880 | 7.90689 | 1.21698 | 8 |  |
|   | 9 | 8.93761 | 1.05783 | 8.91779 | 1.21365 | 8.89525 | 1.36911 | 9 |  |
| Minutes | Distance | Dep. | Lat. | Dep. | Lat. | Dep. | Lat. | Distance | Minutes |
|   |   | 83° | | 82° | | 81° | | | |

*Differences of Latitude and Departures*—Continued.

| Minutes. | Distance. | 9° | | 10° | | 11° | | Distance. | Minutes. |
|---|---|---|---|---|---|---|---|---|---|
| | | Lat. | Dep. | Lat. | Dep. | Lat. | Dep. | | |
| 0 | 1 | 0.98768 | 0.15643 | 0.98480 | 0.17364 | 0.98162 | 0.19081 | 1 | 60 |
| | 2 | 1.97537 | 0.31286 | 1.96961 | 0.34729 | 1.96325 | 0.38162 | 2 | |
| | 3 | 2.96306 | 0.46930 | 2.95442 | 0.52094 | 2.94488 | 0.57243 | 3 | |
| | 4 | 3.95075 | 0.62573 | 3.93923 | 0.69459 | 3.92650 | 0.76324 | 4 | |
| | 5 | 4.93844 | 0.78217 | 4.92403 | 0.86824 | 4.90813 | 0.95405 | 5 | |
| | 6 | 5.92612 | 0.93860 | 5.90884 | 1.04188 | 5.88976 | 1.14486 | 6 | |
| | 7 | 6.91381 | 1.09504 | 6.89365 | 1.21553 | 6.87139 | 1.33566 | 7 | |
| | 8 | 7.90150 | 1.25147 | 7.87846 | 1.38018 | 7.85301 | 1.52648 | 8 | |
| | 9 | 8.88919 | 1.40791 | 8.86327 | 1.56283 | 8.83464 | 1.71729 | 9 | |
| 15 | 1 | 0.98699 | 0.16074 | 0.98404 | 0.17794 | 0.98078 | 0.19509 | 1 | 45 |
| | 2 | 1.97399 | 0.32148 | 1.96808 | 0.35588 | 1.96157 | 0.39018 | 2 | |
| | 3 | 2.96098 | 0.48222 | 2.95212 | 0.53383 | 2.94235 | 0.58527 | 3 | |
| | 4 | 3.94798 | 0.64297 | 3.93616 | 0.71177 | 3.92314 | 0.78036 | 4 | |
| | 5 | 4.93498 | 0.80371 | 4.92020 | 0.88971 | 4.90392 | 0.97545 | 5 | |
| | 6 | 5.92197 | 0.96445 | 5.90424 | 1.06766 | 5.88471 | 1.17054 | 6 | |
| | 7 | 6.90897 | 1.12519 | 6.88828 | 1.24560 | 6.86549 | 1.36563 | 7 | |
| | 8 | 7.89597 | 1.28594 | 7.87232 | 1.42354 | 7.84628 | 1.56072 | 8 | |
| | 9 | 8.88296 | 1.44668 | 8.85636 | 1.60149 | 8.82706 | 1.75581 | 9 | |
| 30 | 1 | 0.98628 | 0.16504 | 0.98325 | 0.18223 | 0.97992 | 0.19936 | 1 | 30 |
| | 2 | 1.97257 | 0.33009 | 1.96650 | 0.36447 | 1.95984 | 0.39873 | 2 | |
| | 3 | 2.95885 | 0.49514 | 2.94976 | 0.54670 | 2.93977 | 0.59810 | 3 | |
| | 4 | 3.94514 | 0.66019 | 3.93301 | 0.72894 | 3.91969 | 0.79747 | 4 | |
| | 5 | 4.93142 | 0.82523 | 4.91627 | 0.91117 | 4.89962 | 0.99683 | 5 | |
| | 6 | 5.91771 | 0.99028 | 5.89952 | 1.09341 | 5.87954 | 1.19620 | 6 | |
| | 7 | 6.90399 | 1.15533 | 6.88278 | 1.27564 | 6.85947 | 1.39557 | 7 | |
| | 8 | 7.89028 | 1.32038 | 7.86603 | 1.45788 | 7.83939 | 1.59494 | 8 | |
| | 9 | 8.87657 | 1.48542 | 8.84929 | 1.64011 | 8.81932 | 1.79431 | 9 | |
| 45 | 1 | 0.98555 | 0.16935 | 0.98245 | 0.18652 | 0.97904 | 0.20364 | 1 | 15 |
| | 2 | 1.97111 | 0.33870 | 1.96490 | 0.37304 | 1.95809 | 0.40728 | 2 | |
| | 3 | 2.95666 | 0.50805 | 2.94735 | 0.55957 | 2.93713 | 0.61092 | 3 | |
| | 4 | 3.94222 | 0.67740 | 3.92980 | 0.74609 | 3.91618 | 0.81456 | 4 | |
| | 5 | 4.92778 | 0.84675 | 4.91225 | 0.93262 | 4.89522 | 1.01820 | 5 | |
| | 6 | 5.91333 | 1.01610 | 5.89470 | 1.11914 | 5.87427 | 1.22185 | 6 | |
| | 7 | 6.89889 | 1.18545 | 6.87715 | 1.30566 | 6.85331 | 1.42549 | 7 | |
| | 8 | 7.88444 | 1.35480 | 7.85960 | 1.49219 | 7.83236 | 1.62913 | 8 | |
| | 9 | 8.87000 | 1.52415 | 8.84205 | 1.67871 | 8.81140 | 1.83277 | 9 | |
| Minutes. | Distance. | Dep. | Lat. | Dep. | Lat. | Dep. | Lat. | Distance. | Minutes. |
| | | 80° | | 79° | | 78° | | | |

## Differences of Latitude and Departures—Continued.

| Minutes | Distance | 12° Lat. | 12° Dep. | 13° Lat. | 13° Dep. | 14° Lat. | 14° Dep. | Distance | Minutes |
|---|---|---|---|---|---|---|---|---|---|
|   | 1 | 0.97814 | 0.20791 | 0.97437 | 0.22495 | 0.97029 | 0.24192 | 1 |   |
|   | 2 | 1.95629 | 0.41582 | 1.94874 | 0.44990 | 1.94059 | 0.48384 | 2 |   |
|   | 3 | 2.93444 | 0.62373 | 2.92311 | 0.67485 | 2.91088 | 0.72576 | 3 |   |
|   | 4 | 3.91259 | 0.83164 | 3.89748 | 0.89980 | 3.88118 | 0.96768 | 4 |   |
| 0 | 5 | 4.89073 | 1.03955 | 4.87185 | 1.12475 | 4.85147 | 1.20961 | 5 | 60 |
|   | 6 | 5.86888 | 1.24747 | 5.84622 | 1.34970 | 5.82177 | 1.45153 | 6 |   |
|   | 7 | 6.84703 | 1.45538 | 6.82059 | 1.57465 | 6.79206 | 1.69345 | 7 |   |
|   | 8 | 7.82518 | 1.66329 | 7.79496 | 1.79960 | 7.76236 | 1.93537 | 8 |   |
|   | 9 | 8.80332 | 1.87120 | 8.76933 | 2.02455 | 8.73266 | 2.17729 | 9 |   |
|   | 1 | 0.97723 | 0.21217 | 0.97337 | 0.22920 | 0.96923 | 0.24615 | 1 |   |
|   | 2 | 1.95446 | 0.42435 | 1.94675 | 0.45840 | 1.93846 | 0.49230 | 2 |   |
|   | 3 | 2.93169 | 0.63653 | 2.92013 | 0.68760 | 2.90769 | 0.73845 | 3 |   |
|   | 4 | 3.90892 | 0.84871 | 3.89351 | 0.91680 | 3.87692 | 0.98461 | 4 |   |
| 15 | 5 | 4.88615 | 1.06088 | 4.86689 | 1.14600 | 4.84615 | 1.23076 | 5 | 45 |
|   | 6 | 5.86338 | 1.27306 | 5.84027 | 1.37520 | 5.81538 | 1.47691 | 6 |   |
|   | 7 | 6.84061 | 1.48524 | 6.81365 | 1.60440 | 6.78461 | 1.72307 | 7 |   |
|   | 8 | 7.81784 | 1.69742 | 7.78703 | 1.83360 | 7.75384 | 1.96922 | 8 |   |
|   | 9 | 8.79507 | 1.90959 | 8.76041 | 2.06280 | 8.72307 | 2.21537 | 9 |   |
|   | 1 | 0.97629 | 0.21644 | 0.97237 | 0.23344 | 0.96814 | 0.25038 | 1 |   |
|   | 2 | 1.95259 | 0.43288 | 1.94474 | 0.46689 | 1.93629 | 0.50076 | 2 |   |
|   | 3 | 2.92888 | 0.64932 | 2.91711 | 0.70033 | 2.90444 | 0.75114 | 3 |   |
|   | 4 | 3.90518 | 0.86576 | 3.88948 | 0.93378 | 3.87259 | 1.00152 | 4 |   |
| 30 | 5 | 4.88148 | 1.08220 | 4.86185 | 1.16722 | 4.84073 | 1.25190 | 5 | 30 |
|   | 6 | 5.85777 | 1.29864 | 5.83422 | 1.40067 | 5.80888 | 1.50228 | 6 |   |
|   | 7 | 6.83407 | 1.51508 | 6.80659 | 1.63411 | 6.77703 | 1.75266 | 7 |   |
|   | 8 | 7.81036 | 1.73152 | 7.77896 | 1.86756 | 7.74518 | 2.00304 | 8 |   |
|   | 9 | 8.78666 | 1.94796 | 8.75133 | 2.10100 | 8.71332 | 2.25342 | 9 |   |
|   | 1 | 0.97534 | 0.22069 | 0.97134 | 0.23768 | 0.96704 | 0.25460 | 1 |   |
|   | 2 | 1.95068 | 0.44139 | 1.94268 | 0.47537 | 1.93409 | 0.50920 | 2 |   |
|   | 3 | 2.92602 | 0.66209 | 2.91402 | 0.71305 | 2.90113 | 0.76380 | 3 |   |
|   | 4 | 3.90136 | 0.88278 | 3.88536 | 0.95074 | 3.86818 | 1.01840 | 4 |   |
| 45 | 5 | 4.87671 | 1.10348 | 4.85671 | 1.18843 | 4.83523 | 1.27301 | 5 | 15 |
|   | 6 | 5.85205 | 1.32418 | 5.82805 | 1.42611 | 5.80227 | 1.52761 | 6 |   |
|   | 7 | 6.82739 | 1.54488 | 6.79939 | 1.66380 | 6.76932 | 1.78221 | 7 |   |
|   | 8 | 7.80273 | 1.76557 | 7.77073 | 1.90148 | 7.73636 | 2.03681 | 8 |   |
|   | 9 | 8.77808 | 1.98627 | 8.74207 | 2.13917 | 8.70341 | 2.29141 | 9 |   |
| Minutes | Distance | Dep. | Lat. | Dep. | Lat. | Dep. | Lat. | Distance | Minutes |
|   |   | 77° |   | 76° |   | 75° |   |   |   |

*Differences of Latitude and Departures*—Continued.

| Minutes. | Distance. | 15° Lat. | 15° Dep. | 16° Lat. | 16° Dep. | 17° Lat. | 17° Dep. | Distance. | Minutes. |
|---|---|---|---|---|---|---|---|---|---|
| | 1 | 0.96592 | 0.25881 | 0.96126 | 0.27563 | 0.95630 | 0.29237 | 1 | |
| | 2 | 1.93185 | 0.51763 | 1.92252 | 0.55127 | 1.91260 | 0.58474 | 2 | |
| | 3 | 2.89777 | 0.77645 | 2.88378 | 0.82691 | 2.86891 | 0.87711 | 3 | |
| | 4 | 3.86370 | 1.03527 | 3.84504 | 1.10254 | 3.82521 | 1.16948 | 4 | |
| 0 | 5 | 4.82962 | 1.29409 | 4.80630 | 1.37818 | 4.78152 | 1.46185 | 5 | 60 |
| | 6 | 5.79555 | 1.55291 | 5.76757 | 1.65382 | 5.73782 | 1.75423 | 6 | |
| | 7 | 6.76148 | 1.81173 | 6.72883 | 1.92946 | 6.69413 | 2.04660 | 7 | |
| | 8 | 7.72740 | 2.07055 | 7.69009 | 2.20509 | 7.65043 | 2.33897 | 8 | |
| | 9 | 8.69333 | 2.32937 | 8.65135 | 2.48073 | 8.60674 | 2.63134 | 9 | |
| | 1 | 0.96478 | 0.26303 | 0.96005 | 0.27982 | 0.95502 | 0.29654 | 1 | |
| | 2 | 1.92957 | 0.52606 | 1.92010 | 0.55965 | 1.91004 | 0.59308 | 2 | |
| | 3 | 2.89436 | 0.78909 | 2.88015 | 0.83948 | 2.86506 | 0.88962 | 3 | |
| | 4 | 3.85914 | 1.05212 | 3.84020 | 1.11931 | 3.82008 | 1.18616 | 4 | |
| 15 | 5 | 4.82393 | 1.31515 | 4.80025 | 1.39914 | 4.77510 | 1.48270 | 5 | 45 |
| | 6 | 5.78872 | 1.57818 | 5.76030 | 1.67897 | 5.73012 | 1.77924 | 6 | |
| | 7 | 6.75351 | 1.84121 | 6.72035 | 1.95880 | 6.68514 | 2.07579 | 7 | |
| | 8 | 7.71829 | 2.10424 | 7.68040 | 2.23863 | 7.64016 | 2.37233 | 8 | |
| | 9 | 8.68308 | 2.36728 | 8.64045 | 2.51846 | 8.59518 | 2.66887 | 9 | |
| | 1 | 0.96363 | 0.26723 | 0.95882 | 0.28401 | 0.95371 | 0.30070 | 1 | |
| | 2 | 1.92726 | 0.53447 | 1.91764 | 0.56803 | 1.90743 | 0.60141 | 2 | |
| | 3 | 2.89089 | 0.80171 | 2.87646 | 0.85204 | 2.86115 | 0.90211 | 3 | |
| | 4 | 3.85452 | 1.06895 | 3.83528 | 1.13606 | 3.81486 | 1.20282 | 4 | |
| 30 | 5 | 4.81815 | 1.33619 | 4.79410 | 1.42007 | 4.76858 | 1.50352 | 5 | 30 |
| | 6 | 5.78178 | 1.60343 | 5.75292 | 1.70409 | 5.72230 | 1.80423 | 6 | |
| | 7 | 6.74541 | 1.87066 | 6.71174 | 1.98810 | 6.67601 | 2.10494 | 7 | |
| | 8 | 7.70904 | 2.13790 | 7.67056 | 2.27212 | 7.62973 | 2.40564 | 8 | |
| | 9 | 8.67267 | 2.40514 | 8.62938 | 2.55613 | 8.58345 | 2.70635 | 9 | |
| | 1 | 0.96245 | 0.27144 | 0.95757 | 0.28819 | 0.95239 | 0.30486 | 1 | |
| | 2 | 1.92491 | 0.54288 | 1.91514 | 0.57639 | 1.90479 | 0.60972 | 2 | |
| | 3 | 2.88736 | 0.81432 | 2.87271 | 0.86458 | 2.85718 | 0.91459 | 3 | |
| | 4 | 3.84982 | 1.08576 | 3.83028 | 1.15278 | 3.80958 | 1.21945 | 4 | |
| 45 | 5 | 4.81227 | 1.35720 | 4.78785 | 1.44098 | 4.76197 | 1.52432 | 5 | 15 |
| | 6 | 5.77473 | 1.62864 | 5.74542 | 1.72917 | 5.71437 | 1.82918 | 6 | |
| | 7 | 6.73718 | 1.90008 | 6.70299 | 2.01737 | 6.66677 | 2.13405 | 7 | |
| | 8 | 7.69964 | 2.17152 | 7.66057 | 2.30557 | 7.61916 | 2.43891 | 8 | |
| | 9 | 8.66209 | 2.44296 | 8.61814 | 2.59376 | 8.57156 | 2.74377 | 9 | |
| Minutes. | Distance. | Dep. | Lat. | Dep. | Lat. | Dep. | Lat. | Distance. | Minutes. |
| | | 74° | | 73° | | 72° | | | |

## *Differences of Latitude and Departures*—Continued.

| Minutes | Distance | 18° Lat. | 18° Dep. | 19° Lat. | 19° Dep. | 20° Lat. | 20° Dep. | Distance | Minutes |
|---|---|---|---|---|---|---|---|---|---|
| | 1 | 0.95105 | 0.30901 | 0.94551 | 0.32556 | 0.93969 | 0.34202 | 1 | |
| | 2 | 1.90211 | 0.61803 | 1.89103 | 0.65113 | 1.87938 | 0.68404 | 2 | |
| | 3 | 2.85316 | 0.92705 | 2.83655 | 0.97670 | 2.81907 | 1.02606 | 3 | |
| | 4 | 3.80422 | 1.23606 | 3.78207 | 1.30227 | 3.75877 | 1.36808 | 4 | |
| 0 | 5 | 4.75528 | 1.54508 | 4.72759 | 1.62784 | 4.69846 | 1.71010 | 5 | 60 |
| | 6 | 5.70633 | 1.85410 | 5.67311 | 1.95340 | 5.63815 | 2.05212 | 6 | |
| | 7 | 6.65739 | 2.16311 | 6.61863 | 2.27897 | 6.57784 | 2.39414 | 7 | |
| | 8 | 7.60845 | 2.47213 | 7.56414 | 2.60454 | 7.51754 | 2.73616 | 8 | |
| | 9 | 8.55950 | 2.78115 | 8.50966 | 2.93011 | 8.45723 | 3.07818 | 9 | |
| | 1 | 0.94969 | 0.31316 | 0.94408 | 0.32969 | 0.93819 | 0.34611 | 1 | |
| | 2 | 1.89939 | 0.62632 | 1.88817 | 0.65938 | 1.87638 | 0.69223 | 2 | |
| | 3 | 2.84909 | 0.93949 | 2.83226 | 0.98907 | 2.81457 | 1.03835 | 3 | |
| | 4 | 3.79879 | 1.25265 | 3.77635 | 1.31876 | 3.75276 | 1.38446 | 4 | |
| 15 | 5 | 4.74849 | 1.56581 | 4.72044 | 1.64845 | 4.69095 | 1.73058 | 5 | 45 |
| | 6 | 5.69819 | 1.87898 | 5.66453 | 1.97814 | 5.62914 | 2.07670 | 6 | |
| | 7 | 6.64789 | 2.19214 | 6.60862 | 2.30783 | 6.56733 | 2.44281 | 7 | |
| | 8 | 7.59759 | 2.50531 | 7.55271 | 2.63752 | 7.50553 | 2.76893 | 8 | |
| | 9 | 8.54729 | 2.81847 | 8.49680 | 2.96721 | 8.44372 | 3.11505 | 9 | |
| | 1 | 0.94832 | 0.31730 | 0.94264 | 0.33380 | 0.93667 | 0.35020 | 1 | |
| | 2 | 1.89664 | 0.63460 | 1.88528 | 0.66761 | 1.87334 | 0.70041 | 2 | |
| | 3 | 2.84497 | 0.95191 | 2.82792 | 1.00142 | 2.81001 | 1.05062 | 3 | |
| | 4 | 3.79329 | 1.26921 | 3.77056 | 1.33522 | 3.74668 | 1.40082 | 4 | |
| 30 | 5 | 4.74161 | 1.58652 | 4.71320 | 1.66903 | 4.68336 | 1.75103 | 5 | 30 |
| | 6 | 5.68994 | 1.90382 | 5.65584 | 2.00284 | 5.62003 | 2.10124 | 6 | |
| | 7 | 6.63826 | 2.22113 | 6.59849 | 2.33664 | 6.55670 | 2.45145 | 7 | |
| | 8 | 7.58658 | 2.53843 | 7.54113 | 2.67045 | 7.49337 | 2.80165 | 8 | |
| | 9 | 8.53491 | 2.85574 | 8.48377 | 3.00426 | 8.43004 | 3.15186 | 9 | |
| | 1 | 0.94693 | 0.32143 | 0.94117 | 0.33791 | 0.93513 | 0.35429 | 1 | |
| | 2 | 1.89386 | 0.64287 | 1.88235 | 0.67583 | 1.87027 | 0.70858 | 2 | |
| | 3 | 2.84079 | 0.96431 | 2.82352 | 1.01375 | 2.80540 | 1.06287 | 3 | |
| | 4 | 3.78772 | 1.28575 | 3.76470 | 1.35166 | 3.74054 | 1.41716 | 4 | |
| 45 | 5 | 4.73465 | 1.60719 | 4.70588 | 1.68958 | 4.67567 | 1.77145 | 5 | 15 |
| | 6 | 5.68158 | 1.92863 | 5.64705 | 2.02750 | 5.61081 | 2.12574 | 6 | |
| | 7 | 6.62851 | 2.25007 | 6.58823 | 2.36541 | 6.54594 | 2.48003 | 7 | |
| | 8 | 7.57544 | 2.57151 | 7.52940 | 2.70333 | 7.48108 | 2.83432 | 8 | |
| | 9 | 8.52237 | 2.89295 | 8.47058 | 3.04125 | 8.41621 | 3.18861 | 9 | |
| Minutes | Distance | Dep. | Lat. | Dep. | Lat. | Dep. | Lat. | Distance | Minutes |
| | | 71° | | 70° | | 69° | | | |

*Differences of Latitude and Departures*—Continued.

| Minutes | Distance | 21° Lat. | 21° Dep. | 22° Lat. | 22° Dep. | 23° Lat. | 23° Dep. | Distance | Minutes |
|---|---|---|---|---|---|---|---|---|---|
| 0 | 1 | 0.93358 | 0.35836 | 0.92718 | 0.37460 | 0.92050 | 0.39073 | 1 | 60 |
| | 2 | 1.86716 | 0.71673 | 1.85436 | 0.74921 | 1.84100 | 0.78146 | 2 | |
| | 3 | 2.80074 | 1.07510 | 2.78155 | 1.12381 | 2.76151 | 1.17219 | 3 | |
| | 4 | 3.73432 | 1.43347 | 3.70873 | 1.49842 | 3.68201 | 1.56292 | 4 | |
| | 5 | 4.66790 | 1.79183 | 4.63591 | 1.87303 | 4.60252 | 1.95365 | 5 | |
| | 6 | 5.60148 | 2.15020 | 5.56310 | 2.24763 | 5.52302 | 2.34438 | 6 | |
| | 7 | 6.53506 | 2.50857 | 6.49028 | 2.62224 | 6.44353 | 2.73511 | 7 | |
| | 8 | 7.46864 | 2.86694 | 7.41747 | 2.99685 | 7.36403 | 3.12584 | 8 | |
| | 9 | 8.40222 | 3.22531 | 8.34465 | 3.37145 | 8.28454 | 3.51657 | 9 | |
| 15 | 1 | 0.93200 | 0.36243 | 0.92554 | 0.37864 | 0.91879 | 0.39474 | 1 | 45 |
| | 2 | 1.86401 | 0.72487 | 1.85108 | 0.75729 | 1.83758 | 0.78948 | 2 | |
| | 3 | 2.79602 | 1.08731 | 2.77662 | 1.13594 | 2.75637 | 1.18423 | 3 | |
| | 4 | 3.72803 | 1.44975 | 3.70216 | 1.51459 | 3.67516 | 1.57897 | 4 | |
| | 5 | 4.66004 | 1.81219 | 4.62770 | 1.89324 | 4.59395 | 1.97372 | 5 | |
| | 6 | 5.59204 | 2.17462 | 5.55324 | 2.27189 | 5.51274 | 2.36846 | 6 | |
| | 7 | 6.52405 | 2.53706 | 6.47878 | 2.65054 | 6.43153 | 2.76320 | 7 | |
| | 8 | 7.45606 | 2.89950 | 7.40432 | 3.02918 | 7.35032 | 3.15795 | 8 | |
| | 9 | 8.38807 | 3.26194 | 8.32986 | 3.40783 | 8.26912 | 3.55269 | 9 | |
| 30 | 1 | 0.93041 | 0.36650 | 0.92388 | 0.38268 | 0.91706 | 0.39874 | 1 | 30 |
| | 2 | 1.86083 | 0.73300 | 1.84776 | 0.76536 | 1.83412 | 0.79749 | 2 | |
| | 3 | 2.79125 | 1.09950 | 2.77164 | 1.14805 | 2.75118 | 1.19624 | 3 | |
| | 4 | 3.72167 | 1.46600 | 3.69552 | 1.53073 | 3.66824 | 1.59499 | 4 | |
| | 5 | 4.65208 | 1.83250 | 4.61940 | 1.91341 | 4.58530 | 1.99374 | 5 | |
| | 6 | 5.58250 | 2.19900 | 5.54328 | 2.29610 | 5.50236 | 2.39249 | 6 | |
| | 7 | 6.51292 | 2.56550 | 6.46716 | 2.67878 | 6.41942 | 2.79124 | 7 | |
| | 8 | 7.44334 | 2.93200 | 7.39104 | 3.06146 | 7.33648 | 3.18999 | 8 | |
| | 9 | 8.37375 | 3.29851 | 8.31492 | 3.44415 | 8.25354 | 3.58874 | 9 | |
| 45 | 1 | 0.92881 | 0.37055 | 0.92220 | 0.38671 | 0.91531 | 0.40274 | 1 | 15 |
| | 2 | 1.85762 | 0.74111 | 1.84440 | 0.77342 | 1.83062 | 0.80549 | 2 | |
| | 3 | 2.78643 | 1.11167 | 2.76660 | 1.16013 | 2.74593 | 1.20824 | 3 | |
| | 4 | 3.71524 | 1.48222 | 3.68880 | 1.54684 | 3.66124 | 1.61098 | 4 | |
| | 5 | 4.64405 | 1.85278 | 4.61100 | 1.93355 | 4.57655 | 2.01373 | 5 | |
| | 6 | 5.57286 | 2.22334 | 5.53320 | 2.32026 | 5.49186 | 2.41648 | 6 | |
| | 7 | 6.50167 | 2.59390 | 6.45540 | 2.70697 | 6.40718 | 2.81922 | 7 | |
| | 8 | 7.43048 | 2.96445 | 7.37760 | 3.09368 | 7.32249 | 3.22197 | 8 | |
| | 9 | 8.35929 | 3.33501 | 8.29980 | 3.48039 | 8.23780 | 3.62472 | 9 | |
| Minutes | Distance | Dep. | Lat. | Dep. | Lat. | Dep. | Lat. | Distance | Minutes |
| | | 68° | | 67° | | 66° | | | |

## Differences of Latitude and Departures—Continued.

| Minutes | Distance | 24° Lat. | 24° Dep. | 25 Lat. | 25 Dep. | 26° Lat. | 26° Dep. | Distance | Minutes |
|---|---|---|---|---|---|---|---|---|---|
|   | 1 | 0.91354 | 0.40673 | 0.90630 | 0.42261 | 0.89879 | 0.43837 | 1 |   |
|   | 2 | 1.82709 | 0.81347 | 1.81261 | 0.84523 | 1.79758 | 0.87674 | 2 |   |
|   | 3 | 2.74063 | 1.22020 | 2.71892 | 1.26785 | 2.69638 | 1.31511 | 3 |   |
|   | 4 | 3.65418 | 1.62694 | 3.62523 | 1.69047 | 3.59517 | 1.75348 | 4 |   |
| 0 | 5 | 4.56772 | 2.03368 | 4.53153 | 2.11309 | 4.49397 | 2.19185 | 5 | 60 |
|   | 6 | 5.48127 | 2.44041 | 5.43784 | 2.53570 | 5.39276 | 2.63022 | 6 |   |
|   | 7 | 6.39481 | 2.84715 | 6.34415 | 2.95832 | 6.29155 | 3.06859 | 7 |   |
|   | 8 | 7.30836 | 3.25389 | 7.25046 | 3.38094 | 7.19035 | 3.50696 | 8 |   |
|   | 9 | 8.22190 | 3.66062 | 8.15677 | 3.80356 | 8.08914 | 3.94533 | 9 |   |
|   | 1 | 0.91176 | 0.41071 | 0.90445 | 0.42656 | 0.89687 | 0.44228 | 1 |   |
|   | 2 | 1.82352 | 0.82143 | 1.80891 | 0.85313 | 1.79374 | 0.88457 | 2 |   |
|   | 3 | 2.73528 | 1.23215 | 2.71336 | 1.27970 | 2.69061 | 1.32686 | 3 |   |
|   | 4 | 3.64704 | 1.64287 | 3.61782 | 1.70627 | 3.58749 | 1.76915 | 4 |   |
| 15 | 5 | 4.55881 | 2.05359 | 4.52227 | 2.13284 | 4.48436 | 2.21144 | 5 | 45 |
|   | 6 | 5.47057 | 2.46431 | 5.42673 | 2.55941 | 5.38123 | 2.65373 | 6 |   |
|   | 7 | 6.38233 | 2.87503 | 6.33118 | 2.98598 | 6.27810 | 3.09602 | 7 |   |
|   | 8 | 7.29409 | 3.28575 | 7.23564 | 3.41254 | 7.17498 | 3.53830 | 8 |   |
|   | 9 | 8.20585 | 3.69647 | 8.14009 | 3.83911 | 8.07185 | 3.98059 | 9 |   |
|   | 1 | 0.90996 | 0.41469 | 0.90258 | 0.43051 | 0.89493 | 0.44619 | 1 |   |
|   | 2 | 1.81992 | 0.82938 | 1.80517 | 0.86102 | 1.78986 | 0.89239 | 2 |   |
|   | 3 | 2.72988 | 1.24407 | 2.70775 | 1.29153 | 2.68480 | 1.33859 | 3 |   |
|   | 4 | 3.63984 | 1.65877 | 3.61034 | 1.72204 | 3.57973 | 1.78479 | 4 |   |
| 30 | 5 | 4.54980 | 2.07346 | 4.51292 | 2.15255 | 4.47467 | 2.23098 | 5 | 30 |
|   | 6 | 5.45976 | 2.48815 | 5.41551 | 2.58306 | 5.36960 | 2.67718 | 6 |   |
|   | 7 | 6.36972 | 2.90285 | 6.31809 | 3.01357 | 6.26454 | 3.12338 | 7 |   |
|   | 8 | 7.27969 | 3.31754 | 7.22068 | 3.44408 | 7.15947 | 3.56958 | 8 |   |
|   | 9 | 8.18965 | 3.73223 | 8.12326 | 3.87459 | 8.05440 | 4.01578 | 9 |   |
|   | 1 | 0.90814 | 0.41866 | 0.90069 | 0.43444 | 0.89297 | 0.45009 | 1 |   |
|   | 2 | 1.81628 | 0.83732 | 1.80139 | 0.86889 | 1.78595 | 0.90019 | 2 |   |
|   | 3 | 2.72442 | 1.25598 | 2.70209 | 1.30333 | 2.67893 | 1.35029 | 3 |   |
|   | 4 | 3.63257 | 1.67464 | 3.60279 | 1.73778 | 3.57191 | 1.80039 | 4 |   |
| 45 | 5 | 4.54071 | 2.09330 | 4.50349 | 2.17222 | 4.46489 | 2.25049 | 5 | 15 |
|   | 6 | 5.44885 | 2.51196 | 5.40418 | 2.60667 | 5.35787 | 2.70059 | 6 |   |
|   | 7 | 6.35700 | 2.93062 | 6.30488 | 3.04111 | 6.25085 | 3.15068 | 7 |   |
|   | 8 | 7.26514 | 3.34928 | 7.20558 | 3.47556 | 7.14383 | 3.60078 | 8 |   |
|   | 9 | 8.17328 | 3.76794 | 8.10628 | 3.91000 | 8.03681 | 4.05088 | 9 |   |
| Minutes | Distance | Dep. | Lat. | Dep. | Lat. | Dep. | Lat. | Distance | Minutes |
|   |   | 65° | | 64° | | 63° | | | |

*Differences of Latitude and Departures*—Continued.

| Minutes. | Distance. | 27° Lat. | 27° Dep. | 28° Lat. | 28° Dep. | 29° Lat. | 29° Dep. | Distance. | Minutes. |
|---|---|---|---|---|---|---|---|---|---|
| 0 | 1 | 0.89100 | 0.45399 | 0.88294 | 0.46947 | 0.87462 | 0.48481 | 1 | 60 |
|  | 2 | 1.78201 | 0.90798 | 1.76589 | 0.93894 | 1.74924 | 0.96962 | 2 |  |
|  | 3 | 2.67301 | 1.36197 | 2.64884 | 1.40841 | 2.62386 | 1.45443 | 3 |  |
|  | 4 | 3.56402 | 1.81596 | 3.53179 | 1.87788 | 3.49848 | 1.93924 | 4 |  |
|  | 5 | 4.45503 | 2.26995 | 4.41473 | 2.34735 | 4.37310 | 2.42405 | 5 |  |
|  | 6 | 5.34603 | 2.72394 | 5.29768 | 2.81682 | 5.24772 | 2.90886 | 6 |  |
|  | 7 | 6.23704 | 3.17793 | 6.18063 | 3.28630 | 6.12234 | 3.39367 | 7 |  |
|  | 8 | 7.12805 | 3.63193 | 7.06358 | 3.75577 | 6.99696 | 3.87848 | 8 |  |
|  | 9 | 8.01905 | 4.08591 | 7.94652 | 4.22524 | 7.87156 | 4.36329 | 9 |  |
| 15 | 1 | 0.88901 | 0.45787 | 0.88089 | 0.47332 | 0.87249 | 0.48862 | 1 | 45 |
|  | 2 | 1.77803 | 0.91574 | 1.76178 | 0.94664 | 1.74499 | 0.97724 | 2 |  |
|  | 3 | 2.66705 | 1.37362 | 2.64267 | 1.41996 | 2.61748 | 1.46566 | 3 |  |
|  | 4 | 3.55606 | 1.83149 | 3.52356 | 1.89328 | 3.48998 | 1.95448 | 4 |  |
|  | 5 | 4.44508 | 2.28937 | 4.40445 | 2.36660 | 4.36248 | 2.44310 | 5 |  |
|  | 6 | 5.33410 | 2.74724 | 5.28534 | 2.83992 | 5.23497 | 2.93172 | 6 |  |
|  | 7 | 6.22311 | 3.20511 | 6.16623 | 3.31324 | 6.10747 | 3.42034 | 7 |  |
|  | 8 | 7.11213 | 3.66299 | 7.04712 | 3.78656 | 6.97996 | 3.90896 | 8 |  |
|  | 9 | 8.00115 | 4.12086 | 7.92801 | 4.25988 | 7.85246 | 4.39759 | 9 |  |
| 30 | 1 | 0.88701 | 0.46174 | 0.87881 | 0.47715 | 0.87035 | 0.49242 | 1 | 30 |
|  | 2 | 1.77402 | 0.92349 | 1.75763 | 0.95431 | 1.74071 | 0.98484 | 2 |  |
|  | 3 | 2.66103 | 1.38524 | 2.63645 | 1.43147 | 2.61106 | 1.47727 | 3 |  |
|  | 4 | 3.54804 | 1.84699 | 3.51526 | 1.90863 | 3.48142 | 1.96969 | 4 |  |
|  | 5 | 4.43505 | 2.30874 | 4.39408 | 2.38579 | 4.35177 | 2.46211 | 5 |  |
|  | 6 | 5.32206 | 2.77049 | 5.27290 | 2.86295 | 5.22213 | 2.95454 | 6 |  |
|  | 7 | 6.20907 | 3.23224 | 6.15171 | 3.34011 | 6.09248 | 3.44696 | 7 |  |
|  | 8 | 7.09608 | 3.69398 | 7.03053 | 3.81727 | 6.96284 | 3.93938 | 8 |  |
|  | 9 | 7.98309 | 4.15573 | 7.90935 | 4.29442 | 7.83320 | 4.43181 | 9 |  |
| 45 | 1 | 0.88498 | 0.46561 | 0.87672 | 0.48098 | 0.86819 | 0.49621 | 1 | 15 |
|  | 2 | 1.76997 | 0.93122 | 1.75345 | 0.96197 | 1.73639 | 0.99243 | 2 |  |
|  | 3 | 2.65496 | 1.39684 | 2.63018 | 1.44296 | 2.60459 | 1.48864 | 3 |  |
|  | 4 | 3.53995 | 1.86245 | 3.50690 | 1.92395 | 3.47279 | 1.98486 | 4 |  |
|  | 5 | 4.42493 | 2.32807 | 4.38363 | 2.40494 | 4.34099 | 2.48108 | 5 |  |
|  | 6 | 5.30992 | 2.79368 | 5.26036 | 2.88593 | 5.20919 | 2.97729 | 6 |  |
|  | 7 | 6.19491 | 3.25930 | 6.13708 | 3.36692 | 6.07739 | 3.47351 | 7 |  |
|  | 8 | 7.07990 | 3.72491 | 7.01381 | 3.84791 | 6.94559 | 3.96973 | 8 |  |
|  | 9 | 7.96488 | 4.19053 | 7.89054 | 4.32889 | 7.81378 | 4.46594 | 9 |  |
| Minutes. | Distance. | Dep. | Lat. | Dep. | Lat. | Dep. | Lat. | Distance. | Minutes. |
|  |  | 62° |  | 61° |  | 60° |  |  |  |

### Differences of Latitude and Departures—Continued.

| Minutes | Distance | 30° Lat. | Dep. | 31 Lat. | Dep. | 32 Lat. | Dep. | Distance | Minutes |
|---|---|---|---|---|---|---|---|---|---|
| | 1 | 0.86602 | 0.50000 | 0.85716 | 0.51503 | 0.84804 | 0.52991 | 1 | |
| | 2 | 1.73205 | 1.00000 | 1.71433 | 1.03007 | 1.69609 | 1.05983 | 2 | |
| | 3 | 2.59807 | 1.50000 | 2.57150 | 1.54511 | 2.54414 | 1.58975 | 3 | |
| | 4 | 3.46410 | 2.00000 | 3.42866 | 2.06015 | 3.39219 | 2.11967 | 4 | |
| 0 | 5 | 4.33012 | 2.50000 | 4.28583 | 2.57519 | 4.24024 | 2.64959 | 5 | 60 |
| | 6 | 5.19615 | 3.00000 | 5.14300 | 3.09022 | 5.08828 | 3.17951 | 6 | |
| | 7 | 6.06217 | 3.50000 | 6.00017 | 3.60526 | 5.93633 | 3.70943 | 7 | |
| | 8 | 6.92820 | 4.00000 | 6.85733 | 4.12030 | 6.78438 | 4.23935 | 8 | |
| | 9 | 7.79422 | 4.50000 | 7.71450 | 4.63534 | 7.63243 | 4.76927 | 9 | |
| | 1 | 0.86383 | 0.50377 | 0.85491 | 0.51877 | 0.84572 | 0.53361 | 1 | |
| | 2 | 1.72767 | 1.00754 | 1.70982 | 1.03754 | 1.69145 | 1.06722 | 2 | |
| | 3 | 2.59150 | 1.51132 | 2.56473 | 1.55631 | 2.53718 | 1.60084 | 3 | |
| | 4 | 3.45534 | 2.01509 | 3.41964 | 2.07509 | 3.38291 | 2.13445 | 4 | |
| 15 | 5 | 4.31917 | 2.51887 | 4.27456 | 2.59386 | 4.22863 | 2.66807 | 5 | 45 |
| | 6 | 5.18301 | 3.02264 | 5.12947 | 3.11263 | 5.07436 | 3.20168 | 6 | |
| | 7 | 6.04684 | 3.52641 | 5.98438 | 3.63141 | 5.92009 | 3.73530 | 7 | |
| | 8 | 6.91068 | 4.03019 | 6.83920 | 4.15018 | 6.76582 | 4.26891 | 8 | |
| | 9 | 7.77451 | 4.53396 | 7.69420 | 4.66895 | 7.61155 | 4.80253 | 9 | |
| | 1 | 0.86162 | 0.50753 | 0.85264 | 0.52249 | 0.84339 | 0.53730 | 1 | |
| | 2 | 1.72325 | 1.01507 | 1.70528 | 1.04499 | 1.68678 | 1.07460 | 2 | |
| | 3 | 2.58488 | 1.52261 | 2.55792 | 1.56749 | 2.53017 | 1.61190 | 3 | |
| | 4 | 3.44651 | 2.03015 | 3.41056 | 2.08999 | 3.37356 | 2.14920 | 4 | |
| 30 | 5 | 4.30814 | 2.53769 | 4.26320 | 2.61249 | 4.21695 | 2.68650 | 5 | 30 |
| | 6 | 5.16977 | 3.04523 | 5.11584 | 3.13499 | 5.06034 | 3.22380 | 6 | |
| | 7 | 6.03140 | 3.55276 | 5.96948 | 3.65749 | 5.90373 | 3.76110 | 7 | |
| | 8 | 6.89303 | 4.06030 | 6.82112 | 4.17998 | 6.74713 | 4.29840 | 8 | |
| | 9 | 7.75466 | 4.56784 | 7.67376 | 4.70248 | 7.59052 | 4.83570 | 9 | |
| | 1 | 0.85940 | 0.51129 | 0.85035 | 0.52621 | 0.84103 | 0.54097 | 1 | |
| | 2 | 1.71881 | 1.02258 | 1.70070 | 1.05242 | 1.68207 | 1.08194 | 2 | |
| | 3 | 2.57821 | 1.53387 | 2.55105 | 1.57864 | 2.52311 | 1.62292 | 3 | |
| | 4 | 3.43762 | 2.04517 | 3.40140 | 2.10485 | 3.36415 | 2.16389 | 4 | |
| 45 | 5 | 4.29703 | 2.55646 | 4.25176 | 2.63107 | 4.20519 | 2.70487 | 5 | 15 |
| | 6 | 5.15643 | 3.06775 | 5.10211 | 3.15728 | 5.04623 | 3.24584 | 6 | |
| | 7 | 6.01584 | 3.57905 | 5.95246 | 3.68349 | 5.88827 | 3.78682 | 7 | |
| | 8 | 6.87525 | 4.09034 | 6.80281 | 4.20971 | 6.72831 | 4.32779 | 8 | |
| | 9 | 7.73465 | 4.60163 | 7.65316 | 4.73592 | 7.56935 | 4.86877 | 9 | |

| Minutes | Distance | Dep. | Lat. | Dep. | Lat. | Dep. | Lat. | Distance | Minutes |
|---|---|---|---|---|---|---|---|---|---|
| | | 59° | | 58° | | 57° | | | |

## Differences of Latitude and Departures—Continued.

| Minutes | Distance | 33° Lat. | 33° Dep. | 34° Lat. | 34° Dep. | 35° Lat. | 35° Dep. | Distance | Minutes |
|---|---|---|---|---|---|---|---|---|---|
|  | 1 | 0.83867 | 0.54463 | 0.82903 | 0.55919 | 0.81915 | 0.57357 | 1 |  |
|  | 2 | 1.67734 | 1.08927 | 1.65807 | 1.11838 | 1.63830 | 1.14715 | 2 |  |
|  | 3 | 2.51601 | 1.63391 | 2.48711 | 1.67757 | 2.45745 | 1.72072 | 3 |  |
|  | 4 | 3.35468 | 2.17855 | 3.31615 | 2.23677 | 3.27660 | 2.29430 | 4 |  |
| 0 | 5 | 4.19335 | 2.72319 | 4.14518 | 2.79596 | 4.09576 | 2.86788 | 5 | .60 |
|  | 6 | 5.03202 | 3.26783 | 4.97422 | 3.35515 | 4.91491 | 3.44145 | 6 |  |
|  | 7 | 5.87069 | 3.81247 | 5.80326 | 3.91435 | 5.73406 | 4.01503 | 7 |  |
|  | 8 | 6.70936 | 4.35711 | 6.63230 | 4.47354 | 6.55321 | 4.58861 | 8 |  |
|  | 9 | 7.54803 | 4.90175 | 7.46133 | 5.03273 | 7.37236 | 5.16218 | 9 |  |
|  | 1 | 0.83628 | 0.54829 | 0.82659 | 0.56280 | 0.81664 | 0.57714 | 1 |  |
|  | 2 | 1.67257 | 1.09658 | 1.65318 | 1.12560 | 1.63328 | 1.15429 | 2 |  |
|  | 3 | 2.50885 | 1.64487 | 2.47977 | 1.68841 | 2.44992 | 1.73143 | 3 |  |
|  | 4 | 3.34514 | 2.19317 | 3.30636 | 2.25121 | 3.26656 | 2.30858 | 4 |  |
| 15 | 5 | 4.18143 | 2.74146 | 4.13295 | 2.81402 | 4.08320 | 2.88572 | 5 | 45 |
|  | 6 | 5.01771 | 3.28975 | 4.95954 | 3.37682 | 4.89984 | 3.46287 | 6 |  |
|  | 7 | 5.85400 | 3.83805 | 5.78613 | 3.93963 | 5.71649 | 4.04001 | 7 |  |
|  | 8 | 6.69028 | 4.38634 | 6.61272 | 4.50243 | 6.53313 | 4.61716 | 8 |  |
|  | 9 | 7.52657 | 4.93463 | 7.43931 | 5.06524 | 7.34977 | 5.19430 | 9 |  |
|  | 1 | 0.83388 | 0.55193 | 0.82412 | 0.56640 | 0.81411 | 0.58070 | 1 |  |
|  | 2 | 1.66777 | 1.10387 | 1.64825 | 1.13281 | 1.62823 | 1.16140 | 2 |  |
|  | 3 | 2.50165 | 1.65581 | 2.47237 | 1.69921 | 2.44234 | 1.74210 | 3 |  |
|  | 4 | 3.33554 | 2.20774 | 3.29650 | 2.26562 | 3.25646 | 2.32281 | 4 |  |
| 30 | 5 | 4.16942 | 2.75968 | 4.12063 | 2.83203 | 4.07057 | 2.90351 | 5 | 30 |
|  | 6 | 5.00331 | 3.31162 | 4.94475 | 3.39843 | 4.88469 | 3.48421 | 6 |  |
|  | 7 | 5.83720 | 3.86355 | 5.76888 | 3.96484 | 5.69880 | 4.06492 | 7 |  |
|  | 8 | 6.67108 | 4.41549 | 6.59300 | 4.53124 | 6.51292 | 4.64562 | 8 |  |
|  | 9 | 7.50497 | 4.96743 | 7.41713 | 5.09765 | 7.32703 | 5.22632 | 9 |  |
|  | 1 | 0.83147 | 0.55557 | 0.82164 | 0.56999 | 0.81157 | 0.58425 | 1 |  |
|  | 2 | 1.66294 | 1.11114 | 1.64329 | 1.13999 | 1.62314 | 1.16850 | 2 |  |
|  | 3 | 2.49441 | 1.66671 | 2.46494 | 1.70999 | 2.43472 | 1.75275 | 3 |  |
|  | 4 | 3.32588 | 2.22228 | 3.28658 | 2.27998 | 3.24629 | 2.33700 | 4 |  |
| 45 | 5 | 4.15735 | 2.77785 | 4.10823 | 2.84998 | 4.05787 | 2.92125 | 5 | 15 |
|  | 6 | 4.98882 | 3.33342 | 4.92988 | 3.41998 | 4.86944 | 3.50550 | 6 |  |
|  | 7 | 5.82029 | 3.88899 | 5.75152 | 3.98997 | 5.68101 | 4.08975 | 7 |  |
|  | 8 | 6.65176 | 4.44456 | 6.57317 | 4.55997 | 6.49260 | 4.67400 | 8 |  |
|  | 9 | 7.48323 | 5.00013 | 7.39482 | 5.12997 | 7.30416 | 5.25825 | 9 |  |
| Minutes | Distance | Dep. | Lat. | Dep. | Lat. | Dep. | Lat. | Distance | Minutes |
|  |  | 56° | | 55° | | 54° | | | |

### *Differences of Latitude and Departures*—Continued.

| Minutes. | Distance. | 36° Lat. | 36° Dep. | 37° Lat. | 37° Dep. | 38° Lat. | 38° Dep. | Distance. | Minutes. |
|---|---|---|---|---|---|---|---|---|---|
| | 1 | 0.80901 | 0.58778 | 0.79863 | 0.60181 | 0.78801 | 0.61566 | 1 | |
| | 2 | 1.61803 | 1.17557 | 1.59727 | 1.20363 | 1.57602 | 1.23132 | 2 | |
| | 3 | 2.42705 | 1.76335 | 2.39590 | 1.80544 | 2.36403 | 1.84698 | 3 | |
| | 4 | 3.23606 | 2.35114 | 3.19454 | 2.40726 | 3.15204 | 2.46264 | 4 | |
| 0 | 5 | 4.04508 | 2.93892 | 3.99317 | 3.00907 | 3.94005 | 3.07830 | 5 | 60 |
| | 6 | 4.85410 | 3.52671 | 4.79181 | 3.61089 | 4.72806 | 3.69396 | 6 | |
| | 7 | 5.66311 | 4.11449 | 5.59044 | 4.21270 | 5.51607 | 4.30963 | 7 | |
| | 8 | 6.47213 | 4.70228 | 6.38908 | 4.81452 | 6.30408 | 4.92529 | 8 | |
| | 9 | 7.28115 | 5.29006 | 7.18771 | 5.41633 | 7.09209 | 5.54095 | 9 | |
| | 1 | 0.80644 | 0.59130 | 0.79600 | 0.60529 | 1.78531 | 0.61909 | 1 | |
| | 2 | 1.61288 | 1.18261 | 1.59200 | 1.21058 | 1.57063 | 1.23818 | 2 | |
| | 3 | 2.41933 | 1.77392 | 2.38800 | 1.81588 | 2.35595 | 1.85728 | 3 | |
| | 4 | 3.22577 | 2.36523 | 3.18400 | 2.42117 | 3.14126 | 2.47637 | 4 | |
| 15 | 5 | 4.03222 | 2.95654 | 3.98001 | 3.02647 | 3.92658 | 3.09547 | 5 | 45 |
| | 6 | 4.83866 | 3.54785 | 4.77601 | 3.63176 | 4.71190 | 3.71456 | 6 | |
| | 7 | 5.64511 | 4.13916 | 5.57201 | 4.23705 | 5.49721 | 4.33365 | 7 | |
| | 8 | 6.45155 | 4.73047 | 6.36801 | 4.84235 | 6.28253 | 4.95275 | 8 | |
| | 9 | 7.25800 | 5.32178 | 7.16401 | 5.44764 | 7.06785 | 5.57184 | 9 | |
| | 1 | 0.80385 | 0.59482 | 0.79335 | 0.60876 | 0.78260 | 0.62251 | 1 | |
| | 2 | 1.60771 | 1.18964 | 1.58670 | 1.21752 | 1.56521 | 1.24502 | 2 | |
| | 3 | 2.41157 | 1.78446 | 2.38005 | 1.82628 | 2.34782 | 1.86754 | 3 | |
| | 4 | 3.21542 | 2.37929 | 3.17341 | 2.43504 | 3.13043 | 2.49005 | 4 | |
| 30 | 5 | 4.01928 | 2.97411 | 3.96676 | 3.04380 | 3.91304 | 3.11257 | 5 | 30 |
| | 6 | 4.82314 | 3.56893 | 4.76011 | 3.65256 | 4.69564 | 3.73508 | 6 | |
| | 7 | 5.62699 | 4.16375 | 5.55347 | 4.26132 | 5.47825 | 4.35760 | 7 | |
| | 8 | 6.43085 | 4.75858 | 6.34682 | 4.87009 | 6.26086 | 4.98011 | 8 | |
| | 9 | 7.23471 | 5.35340 | 7.14017 | 5.47885 | 7.04347 | 5.60263 | 9 | |
| | 1 | 0.80125 | 0.59832 | 0.79068 | 0.61221 | 0.77988 | 0.62592 | 1 | |
| | 2 | 1.60250 | 1.19664 | 1.58137 | 1.22443 | 1.55946 | 1.25184 | 2 | |
| | 3 | 2.40376 | 1.79497 | 2.37206 | 1.83665 | 2.33965 | 1.87777 | 3 | |
| | 4 | 3.20501 | 2.39329 | 3.16275 | 2.44886 | 3.11953 | 2.50369 | 4 | |
| 45 | 5 | 4.00626 | 2.99162 | 3.95344 | 3.06108 | 3.89942 | 3.12961 | 5 | 15 |
| | 6 | 4.80752 | 3.58994 | 4.74413 | 3.67330 | 4.67930 | 3.75554 | 6 | |
| | 7 | 5.60877 | 4.18827 | 5.53482 | 4.28552 | 5.45919 | 4.38146 | 7 | |
| | 8 | 6.41003 | 4.78659 | 6.32551 | 4.89773 | 6.23907 | 5.00738 | 8 | |
| | 9 | 7.21128 | 5.38492 | 7.11620 | 5.50995 | 7.01896 | 5.63331 | 9 | |

| Minutes. | Distance. | 53° Dep. | 53° Lat. | 52° Dep. | 52° Lat. | 51° Dep. | 51° Lat. | Distance. | Minutes. |
|---|---|---|---|---|---|---|---|---|---|

## Differences of Latitude and Departures—Continued.

| Minutes. | Distance. | 39° Lat. | 39° Dep. | 40° Lat. | 40° Dep. | 41° Lat. | 41° Dep. | Distance. | Minutes. |
|---|---|---|---|---|---|---|---|---|---|
| | 1 | 0.77714 | 0.62932 | 0.76604 | 0.64278 | 0.75470 | 0.65605 | 1 | |
| | 2 | 1.55429 | 1.25864 | 1.53208 | 1.28557 | 1.50941 | 1.31211 | 2 | |
| | 3 | 2.33143 | 1.88796 | 2.29813 | 1.92836 | 2.26412 | 1.96817 | 3 | |
| | 4 | 3.10858 | 2.51728 | 3.06417 | 2.57115 | 3.01883 | 2.62423 | 4 | |
| 0 | 5 | 3.88573 | 3.14660 | 3.83022 | 3.21393 | 3.77354 | 3.28029 | 5 | 60 |
| | 6 | 4.66287 | 3.77592 | 4.59626 | 3.85672 | 4.52825 | 3.93635 | 6 | |
| | 7 | 5.44002 | 4.40524 | 5.36231 | 4.49951 | 5.28296 | 4.59241 | 7 | |
| | 8 | 6.21716 | 5.03456 | 6.12835 | 5.14230 | 6.03767 | 5.24847 | 8 | |
| | 9 | 6.99431 | 5.66388 | 6.89439 | 5.78508 | 6.79238 | 5.90453 | 9 | |
| | 1 | 0.77439 | 0.63270 | 0.76323 | 0.64612 | 0.75184 | 0.65934 | 1 | |
| | 2 | 1.54878 | 1.26541 | 1.52646 | 1.29224 | 1.50368 | 1.31869 | 2 | |
| | 3 | 2.32317 | 1.89811 | 2.28969 | 1.93837 | 2.25552 | 1.97803 | 3 | |
| | 4 | 3.09757 | 2.53082 | 3.05293 | 2.58449 | 3.00736 | 2.63738 | 4 | |
| 15 | 5 | 3.87196 | 3.16352 | 3.81616 | 3.23062 | 3.75920 | 3.29672 | 5 | 45 |
| | 6 | 4.64635 | 3.79623 | 4.57939 | 3.87674 | 4.51104 | 3.95607 | 6 | |
| | 7 | 5.42074 | 4.42893 | 5.34262 | 4.52286 | 5.26288 | 4.61542 | 7 | |
| | 8 | 6.19514 | 5.06164 | 6.10586 | 5.16899 | 6.01472 | 5.27476 | 8 | |
| | 9 | 6.96953 | 5.69434 | 6.86909 | 5.81511 | 6.76656 | 5.93411 | 9 | |
| | 1 | 0.77162 | 0.63607 | 0.76040 | 0.64944 | 0.74895 | 0.66262 | 1 | |
| | 2 | 1.54324 | 1.27215 | 1.52081 | 1.29889 | 1.49791 | 1.32524 | 2 | |
| | 3 | 2.31487 | 1.90823 | 2.28121 | 1.94834 | 2.24686 | 1.98786 | 3 | |
| | 4 | 3.08649 | 2.54431 | 3.04162 | 2.59779 | 2.99582 | 2.65048 | 4 | |
| 30 | 5 | 3.85812 | 3.18039 | 3.80203 | 3.24724 | 3.74477 | 3.31310 | 5 | 30 |
| | 6 | 4.62974 | 3.81646 | 4.56243 | 3.89668 | 4.49373 | 3.97572 | 6 | |
| | 7 | 5.40137 | 4.45254 | 5.32284 | 4.54613 | 5.24268 | 4.63834 | 7 | |
| | 8 | 6.17299 | 5.08862 | 6.08324 | 5.19558 | 5.99164 | 5.30096 | 8 | |
| | 9 | 6.94462 | 5.72470 | 6.84365 | 5.84503 | 6.74060 | 5.96358 | 9 | |
| | 1 | 0.76884 | 0.63943 | 0.75756 | 0.65276 | 0.74605 | 0.66588 | 1 | |
| | 2 | 1.53768 | 1.27887 | 1.51513 | 1.30552 | 1.49211 | 1.33176 | 2 | |
| | 3 | 2.30652 | 1.91831 | 2.27269 | 1.95828 | 2.23817 | 1.99764 | 3 | |
| | 4 | 3.07536 | 2.55775 | 3.03026 | 2.61104 | 2.98422 | 2.66352 | 4 | |
| 45 | 5 | 3.84420 | 3.19719 | 3.78782 | 3.26380 | 3.73028 | 3.32940 | 5 | 15 |
| | 6 | 4.61305 | 3.83663 | 4.54539 | 3.91656 | 4.47634 | 3.99529 | 6 | |
| | 7 | 5.38189 | 4.47607 | 5.30295 | 4.56932 | 5.22240 | 4.66117 | 7 | |
| | 8 | 6.15073 | 5.11551 | 6.06052 | 5.22208 | 5.96845 | 5.32705 | 8 | |
| | 9 | 6.91957 | 5.75495 | 6.81808 | 5.87484 | 6.71451 | 5.99293 | 9 | |
| Minutes. | Distance. | Dep. | Lat. | Dep. | Lat. | Dep. | Lat. | Distance. | Minutes. |
| | | 50° | | 49° | | 48° | | | |

## Differences of Latitude and Departures—Continued.

| Minutes. | Distance. | 42° Lat. | 42° Dep. | 43° Lat. | 43° Dep. | 44° Lat. | 44° Dep. | Distance. | Minutes. |
|---|---|---|---|---|---|---|---|---|---|
| | 1 | 0.74314 | 0.66913 | 0.73135 | 0.68199 | 0.71933 | 0.69465 | 1 | |
| | 2 | 1.48628 | 1.33826 | 1.46270 | 1.36399 | 1.43867 | 1.38931 | 2 | |
| | 3 | 2.22943 | 2.00739 | 2.19406 | 2.04599 | 2.15801 | 2.08397 | 3 | |
| | 4 | 2.97257 | 2.67652 | 2.92541 | 2.72799 | 2.87735 | 2.77863 | 4 | |
| 0 | 5 | 3.71572 | 3.34565 | 3.65676 | 3.40999 | 3.59669 | 3.47329 | 5 | 60 |
| | 6 | 4.45886 | 4.01478 | 4.38812 | 4.09199 | 4.31603 | 4.16795 | 6 | |
| | 7 | 5.20201 | 4.68391 | 5.11947 | 4.77398 | 5.03537 | 4.86260 | 7 | |
| | 8 | 5.94515 | 5.35304 | 5.85082 | 5.45598 | 5.75471 | 5.55726 | 8 | |
| | 9 | 6.68830 | 6.02217 | 6.58218 | 6.13798 | 6.47405 | 6.25192 | 9 | |
| | 1 | 0.74021 | 0.67236 | 0.72837 | 0.68518 | 0.71630 | 0.69779 | 1 | |
| | 2 | 1.48043 | 1.34473 | 1.45674 | 1.37036 | 1.43260 | 1.39558 | 2 | |
| | 3 | 2.22065 | 2.01710 | 2.18511 | 2.05554 | 2.14890 | 2.09337 | 3 | |
| | 4 | 2.96087 | 2.68946 | 2.91348 | 2.74073 | 2.86520 | 2.79116 | 4 | |
| 15 | 5 | 3.70109 | 3.36183 | 3.64185 | 3.42591 | 3.58151 | 3.48895 | 5 | 45 |
| | 6 | 4.44130 | 4.03420 | 4.37022 | 4.11109 | 4.29781 | 4.18674 | 6 | |
| | 7 | 5.18152 | 4.70656 | 5.09859 | 4.79628 | 5.01411 | 4.88453 | 7 | |
| | 8 | 5.92174 | 5.37893 | 5.82696 | 5.48146 | 5.73041 | 5.58232 | 8 | |
| | 9 | 6.66196 | 6.05130 | 6.55533 | 6.16664 | 6.44671 | 6.28011 | 9 | |
| | 1 | 0.73727 | 0.67559 | 0.72537 | 0.68835 | 0.71325 | 0.70090 | 1 | |
| | 2 | 1.47455 | 1.35118 | 1.45074 | 1.37670 | 1.42650 | 1.40181 | 2 | |
| | 3 | 2.21183 | 2.02677 | 2.17612 | 2.06506 | 2.13975 | 2.10272 | 3 | |
| | 4 | 2.94910 | 2.70236 | 2.90149 | 2.75341 | 2.85300 | 2.80363 | 4 | |
| 30 | 5 | 3.68638 | 3.37795 | 3.62687 | 3.44177 | 3.56625 | 3.50454 | 5 | 30 |
| | 6 | 4.42366 | 4.05354 | 4.35224 | 4.13012 | 4.27950 | 4.20545 | 6 | |
| | 7 | 5.16094 | 4.72913 | 5.07762 | 4.81848 | 4.99275 | 4.90636 | 7 | |
| | 8 | 5.89821 | 5.40472 | 5.80299 | 5.50683 | 5.70600 | 5.60727 | 8 | |
| | 9 | 6.63549 | 6.08031 | 6.52836 | 6.19519 | 6.41925 | 6.30818 | 9 | |
| | 1 | 0.73432 | 0.67880 | 0.72236 | 0.69151 | 0.71018 | 0.70401 | 1 | |
| | 2 | 1.46864 | 1.35760 | 1.44472 | 1.38302 | 1.42037 | 1.40802 | 2 | |
| | 3 | 2.20296 | 2.03640 | 2.16709 | 2.07453 | 2.13055 | 2.11204 | 3 | |
| | 4 | 2.93729 | 2.71520 | 2.88945 | 2.76605 | 2.84074 | 2.81605 | 4 | |
| 45 | 5 | 3.67161 | 3.39400 | 3.61182 | 3.45756 | 3.55092 | 3.52007 | 5 | 15 |
| | 6 | 4.40593 | 4.07280 | 4.33418 | 4.14907 | 4.26111 | 4.22408 | 6 | |
| | 7 | 5.14025 | 4.75160 | 5.05654 | 4.84059 | 4.97129 | 4.92810 | 7 | |
| | 8 | 5.87458 | 5.43040 | 5.77891 | 5.53210 | 5.68148 | 5.63211 | 8 | |
| | 9 | 6.60890 | 6.10920 | 6.50127 | 6.22361 | 6.39166 | 6.33613 | 9 | |
| Minutes. | Distance. | Dep. | Lat. | Dep. | Lat. | Dep. | Lat. | Distance. | Minutes. |
| | | 47° | | 46° | | 45° | | | |

## *Differences of Latitude and Departures*—Continued.

|  | 45° | | |
|---|---|---|---|
|  | Lat. | Dep. | |
| 1 | 0. 70710 | 0. 70710 | 1 |
| 2 | 1. 41421 | 1. 41421 | 2 |
| 3 | 2. 12132 | 2. 12132 | 3 |
| 4 | 2. 82842 | 2. 82842 | 4 |
| 5 | 3. 53553 | 3. 53553 | 5 |
| 6 | 4. 24264 | 4. 24264 | 6 |
| 7 | 4. 94974 | 4. 94974 | 7 |
| 8 | 5. 65685 | 5. 65685 | 8 |
| 9 | 6. 36396 | 6. 36396 | 9 |
|  | Dep. | Lat. | |
|  | 45° | | |

## *Chains, Yards, and Feet,*

### WITH THEIR RECIPROCAL EQUIVALENTS.

Link = 7. 92 inches.   Chain = 66 feet = 792 inches.

| | | CHAINS INTO FEET. | | FEET INTO LINKS. | | |
|---|---|---|---|---|---|---|
| Chains. | Links. | Yards. | Feet. | Feet. | Yards. | Links. |
| 0 | 1 | 0. 22 | 0. 66 | 0. 10 | . 033 | 0. 15 |
| 0 | 2 | 0. 44 | 1. 32 | 0. 20 | . 066 | 0. 30 |
| 0 | 3 | 0. 66 | 1. 98 | 0. 25 | . 082 | 0. 38 |
| 0 | 4 | 0. 88 | 2. 64 | 0. 30 | . 010 | 0. 45 |
| 0 | 5 | 1. 10 | 3. 30 | 0. 40 | . 133 | 0. 60 |
| 0 | 6 | 1. 32 | 3. 96 | 0. 50 | . 166 | 0. 76 |
| 0 | 7 | 1. 54 | 4. 62 | 0. 60 | . 200 | 0. 91 |
| 0 | 8 | 1. 76 | 5. 28 | 0. 70 | . 233 | 1. 06 |
| 0 | 9 | 1. 98 | 5. 94 | 0. 75 | . 250 | 1. 13 |
| 0 | 10 | 2. 20 | 6. 60 | 0. 80 | . 266 | 1. 21 |

## *Chains, Yards, and Feet*—Continued.

| | | CHAINS INTO FEET. | | FEET INTO LINKS. | | |
|---|---|---|---|---|---|---|
| Chains. | Links. | Yards. | Feet. | Feet. | Yards. | Links. |
| 0 | 20 | 4. 40 | 13. 20 | 0. 9 | . 30 | 1. 36 |
| 0 | 30 | 6. 60 | 19. 80 | 1. 0 | . 33 | 1. 51 |
| 0 | 40 | 8. 80 | 26. 40 | 2. 0 | . 66 | 3. 0 |
| 0 | 50 | 11. 00 | 33. 00 | 3. 0 | 1, 00 | 4. 5 |
| 0 | 60 | 13. 20 | 39. 60 | 4. 0 | 1. 33 | 6. 0 |
| 0 | 70 | 15. 40 | 46. 20 | 5. 0 | 1. 66 | 7. 5 |
| 0 | 80 | 17. 60 | 52. 80 | 6. 0 | 2. 00 | 9. 1 |
| 0 | 90 | 19. 80 | 59. 40 | 7. 0 | 2. 33 | 10. 6 |
| 1 | 00 | 22. 00 | 66. 00 | 8. 0 | 2. 66 | 12. 1 |
| 2 | 00 | 44. 00 | 132 | 9. 0 | 3. 00 | 13. 6 |
| | | | | | | |
| 3 | | 66 | 198 | 10 | 3. 33 | 15. 1 |
| 4 | | 88 | 264 | 15 | 5. 00 | 22. 7 |
| 5 | | 110 | 330 | 20 | 6. 66 | 30. 3 |
| 6 | | 132 | 396 | 24 | 8. 00 | 36. 3 |
| 7 | | 154 | 462 | 27 | 9. 00 | 40. 9 |
| 8 | | 176 | 528 | 30 | 10. 00 | 45. 4 |
| 9 | | 198 | 594 | 33 | 11. 00 | 50. 0 |
| 10 | | 220 | 660 | 36 | 12. 00 | 54. 5 |
| 20 | | 440 | 1320 | 39 | 13. 00 | 59. 1 |
| 30 | | 660 | 1980 | 40 | 13. 33 | 60. 6 |
| | | | | | | |
| 35 | | 770 | 2310 | 42 | 14. 00 | 63. 3 |
| 40 | | 880 | 2640 | 45 | 15. 00 | 68. 2 |
| 45 | | 990 | 2970 | 48 | 16. 00 | 72. 7 |
| 50 | | 1100 | 3300 | 50 | 16. 66 | 75. 7 |
| 55 | | 1210 | 3630 | 51 | 17. 00 | 77. 3 |
| 60 | | 1320 | 3960 | 54 | 18. 00 | 81. 8 |
| 65 | | 1430 | 4290 | 57 | 19. 00 | 86. 3 |
| 70 | | 1540 | 4620 | 60 | 20. 00 | 90. 9 |
| 75 | | 1650 | 4950 | 63 | 21. 00 | 95. 4 |
| 80 | | 1760 | 5280 | 66 | 22. 00 | 100 |

## XIX.—*To trace Railroad Curves by means of Deflections.*

### GENERAL PROPOSITIONS.

1. The angle formed by a tangent and a chord is equal to half the angle at the center of the circle subtended by the chord.

2. The angle of deflection formed by any two equal chords meeting at the circumference is equal to the angle at the center, subtended by either cord.

3. A line bisecting the angle of deflection formed by any two equal chords is a tangent to the arc at the point where the two chords meet.

4. If an arc of a circle be subdivided into any number of equal parts, and lines be drawn from the several points of subdivision so as to meet at any point in the circumference, these several lines will form equal angles at the point of meeting, and the angles thus formed will be respectively measured by one-half the subdivided arc.

### METHOD BY DEFLECTION-ANGLES.

The *degree* of a curve is determined by the angle subtended at its center by a chord of 100 feet.

The *deflection-angle* of a curve is the acute angle formed at any point between a tangent and a chord. It is, therefore, half the degree of the curve.

In order to unite two straight lines by a curve, the angle of intersection is measured, and then a radius for the curve may be assumed and the tangent calculated, or the tangent may be assumed of a certain length and the radius calculated.

Let I = angle of intersection of the two lines;

$\quad$ R = radius of circle;

$\quad$ T = length of tangent, or distance from point of intersection to point where the curvature is to commence; and

$\quad$ D = angle of deflection.

Then

$$T = R \tan \tfrac{1}{2} I$$
$$R = T \cot \tfrac{1}{2} I$$
$$\sin D = \frac{50}{R} = \frac{50 \tan \tfrac{1}{2} I}{T}$$

## XIX.—*To trace Railroad Curves, &c.*—Continued.

To lay out a curve, set the instrument at the point at which the curvature is to commence, lay off the given deflection-angle, and the *first* point in the curve will be at the end of 100 feet measured on this new direction.

Then lay off another deflection-angle equal to the first; attach the 100-foot chain to the point last found, and swing it, stretched, until its extremity intersects the new direction, which will be the *second* point; and so on. Should it be found necessary to remove the instrument from its first position, either on account of the length of the curve or of some obstruction to the sight, the first deflection at the new position of the instrument will be equal to the total deflection from the preceding position.

### METHOD BY TANGENT AND CHORD DEFLECTION.

*Tangent-deflection* is the distance between the extremities of a tangent and a chord, each 100 feet long.

*Chord-deflection* is the distance from the extremity of the first chord, produced an additional 100 feet, to the extremity of the next, and is, therefore, double the tangent-deflection.

To lay out a curve, stretch the 100-foot chain from the point of beginning in the direction of the tangent, and mark its extremity; swing the chain toward the direction of the curve, keeping the initial point fixed, until it has diverged a distance equal to the *tangent-deflection*, which will be the *first* point of the curve.

Produce the first chord an additional 100 feet, and swing the chain (round the extremity of the first chord as a pivot) until it has diverged a distance equal to the *chord-deflection*, which will be the *second* point of the curve.

Continue to lay off the chord-deflection from the preceding chord produced until the curve is finished.

#### LENGTH OF CIRCULAR ARCS IN PARTS OF RADIUS.

| ° | | | ′ | | | ″ | | |
|---|---|---|---|---|---|---|---|---|
| 1 | .01745 | 32925 | 1 | .00029 | 08882 | 1 | .00000 | 48481 |
| 2 | .03490 | 65850 | 2 | .00058 | 17764 | 2 | .00000 | 96962 |
| 3 | .05235 | 98775 | 3 | .00087 | 26646 | 3 | .00001 | 45444 |
| 4 | .06981 | 31700 | 4 | .00116 | 35528 | 4 | .00001 | 93925 |
| 5 | .08726 | 64625 | 5 | .00145 | 44410 | 5 | .00002 | 42406 |
| 6 | .10471 | 97551 | 6 | .00174 | 53292 | 6 | .00002 | 90888 |
| 7 | .12217 | 30476 | 7 | .00203 | 62174 | 7 | .00003 | 39369 |
| 8 | .13962 | 63401 | 8 | .00232 | 71056 | 8 | .00003 | 87850 |
| 9 | .15707 | 96326 | 9 | .00261 | 79938 | 9 | .00004 | 36332 |

## XIX.—*To trace Railroad Curves, &c.*—Continued.

| Degree. | | Radii. | ORDINATES. To circular arcs on a chord of 100 feet. | | | | Tangent-deflection. | Chord-deflection. |
|---|---|---|---|---|---|---|---|---|
| ° | ′ | | 12¼ | 25 | 37½ | 50 | | |
| | | *Feet.* | | | | | | |
| 0 | 5 | 68754.94 | .008 | .014 | .017 | .018 | .073 | .145 |
| | 10 | 34377.48 | .016 | .027 | .034 | .036 | .145 | .291 |
| | 15 | 22918.33 | .024 | .041 | .051 | .055 | .218 | .436 |
| | 20 | 17188.76 | .032 | .055 | .068 | .073 | .291 | .582 |
| | 25 | 13751.02 | .040 | .068 | .085 | .091 | .364 | .727 |
| | 30 | 11459.19 | .048 | .082 | .102 | .109 | .436 | .873 |
| | 35 | 9822.18 | .056 | .095 | .119 | .127 | .509 | 1.018 |
| | 40 | 8594.41 | .064 | .109 | .136 | .145 | .582 | 1.164 |
| | 45 | 7639.49 | .072 | .123 | .153 | .164 | .654 | 1.309 |
| | 50 | 6875.55 | .080 | .136 | .170 | .182 | .727 | 1.454 |
| | 55 | 6250.51 | .087 | .150 | .187 | .200 | .800 | 1.600 |
| 1 | 0 | 5729.65 | .095 | .164 | .205 | .218 | .873 | 1.745 |
| | 5 | 5288.92 | .103 | .177 | .222 | .236 | .945 | 1.891 |
| | 10 | 4911.15 | .111 | .191 | .239 | .255 | 1.018 | 2.036 |
| | 15 | 4583.75 | .119 | .205 | .256 | .273 | 1.091 | 2.182 |
| | 20 | 4297.28 | .127 | .218 | .273 | .291 | 1.164 | 2.327 |
| | 25 | 4044.51 | .135 | .232 | .290 | .309 | 1.236 | 2.472 |
| | 30 | 3819.83 | .143 | .245 | .307 | .327 | 1.309 | 2.618 |
| | 35 | 3618.80 | .151 | .259 | .324 | .345 | 1.382 | 2.763 |
| | 40 | 3437.87 | .159 | .273 | .341 | .364 | 1.454 | 2.909 |
| | 45 | 3274.17 | .167 | .286 | .358 | .382 | 1.527 | 3.054 |
| | 50 | 3125.36 | .175 | .300 | .375 | .400 | 1.600 | 3.200 |
| | 55 | 2989.48 | .183 | .314 | .392 | .418 | 1.673 | 3.345 |
| 2 | 0 | 2864.93 | .191 | .327 | .409 | .436 | 1.745 | 3.490 |
| | 5 | 2750.35 | .199 | .341 | .426 | .455 | 1.818 | 3.636 |
| | 10 | 2644.58 | .207 | .355 | .443 | .473 | 1.891 | 3.781 |
| | 15 | 2546.64 | .215 | .368 | .460 | .491 | 1.963 | 3.927 |
| | 20 | 2455.70 | .223 | .382 | .477 | .509 | 2.036 | 4.072 |
| | 25 | 2371.04 | .231 | .395 | .494 | .527 | 2.109 | 4.218 |
| | 30 | 2292.01 | .239 | .409 | .511 | .545 | 2.181 | 4.363 |
| | 35 | 2218.09 | .247 | .423 | .528 | .564 | 2.254 | 4.508 |
| | 40 | 2148.79 | .255 | .436 | .545 | .582 | 2.327 | 4.654 |
| | 45 | 2083.68 | .263 | .450 | .562 | .600 | 2.400 | 4.799 |
| | 50 | 2022.41 | .270 | .464 | .580 | .618 | 2.472 | 4.945 |
| | 55 | 1964.64 | .278 | .477 | .597 | .636 | 2.545 | 5.090 |

XIX.—*To trace Railroad Curves, &c.*—Continued.

| Degree. | | Radii. | ORDINATES. To circular arcs on a chord of 100 feet. | | | | Tangent-deflection. | Chord-deflection. |
|---|---|---|---|---|---|---|---|---|
| | | | 12½ | 25 | 37½ | 50 | | |
| 3 | 0 | 1910.08 | .286 | .491 | .614 | .655 | 2.618 | 5.235 |
| | 5 | 1858.47 | .294 | .505 | .631 | .673 | 2.690 | 5.381 |
| | 10 | 1809.57 | .302 | .518 | .648 | .691 | 2.763 | 5.526 |
| | 15 | 1763.18 | .310 | .532 | .665 | .709 | 2.836 | 5.672 |
| | 20 | 1719.12 | .318 | .545 | .682 | .727 | 2.908 | 5.817 |
| | 25 | 1677.20 | .326 | .559 | .699 | .745 | 2.981 | 5.962 |
| | 30 | 1637.28 | .334 | .573 | .716 | .764 | 3.054 | 6.108 |
| | 35 | 1599.21 | .342 | .586 | .733 | .782 | 3.127 | 6.253 |
| | 40 | 1562.88 | .350 | .600 | .750 | .800 | 3.199 | 6.398 |
| | 45 | 1528.16 | .358 | .614 | .767 | .818 | 3.272 | 6.544 |
| | 50 | 1494.95 | .366 | .627 | .784 | .836 | 3.345 | 6.689 |
| | 55 | 1463.16 | .374 | .641 | .801 | .855 | 3.417 | 6.835 |
| 4 | 0 | 1432.69 | .382 | .655 | .818 | .873 | 3.490 | 6.980 |
| | 5 | 1403.46 | .390 | .668 | .835 | .891 | 3.563 | 7.125 |
| | 10 | 1375.40 | .398 | .682 | .852 | .909 | 3.635 | 7.271 |
| | 15 | 1348.45 | .406 | .695 | .869 | .927 | 3.708 | 7.416 |
| | 20 | 1322.53 | .414 | .709 | .886 | .945 | 3.781 | 7.561 |
| | 25 | 1297.58 | .422 | .723 | .903 | .964 | 3.853 | 7.707 |
| | 30 | 1273.57 | .430 | .736 | .921 | .982 | 3.926 | 7.852 |
| | 35 | 1250.42 | .438 | .750 | .938 | 1.000 | 3.999 | 7.997 |
| | 40 | 1228.11 | .446 | .764 | .955 | 1.018 | 4.071 | 8.143 |
| | 45 | 1206.57 | .454 | .777 | .972 | 1.036 | 4.144 | 8.288 |
| | 50 | 1185.78 | .462 | .791 | .989 | 1.055 | 4.217 | 8.433 |
| | 55 | 1165.70 | .469 | .805 | 1.006 | 1.073 | 4.289 | 8.579 |
| 5 | 0 | 1146.28 | .477 | .818 | 1.023 | 1.091 | 4.362 | 8.724 |
| | 5 | 1127.50 | .485 | .832 | 1.040 | 1.109 | 4.435 | 8.869 |
| | 10 | 1109.33 | .493 | .846 | 1.057 | 1.127 | 4.507 | 9.014 |
| | 15 | 1091.73 | .501 | .859 | 1.074 | 1.146 | 4.580 | 9.160 |
| | 20 | 1074.68 | .509 | .873 | 1.091 | 1.164 | 4.653 | 9.305 |
| | 25 | 1058.16 | .517 | .887 | 1.108 | 1.182 | 4.725 | 9.450 |
| | 30 | 1042.14 | .525 | .900 | 1.125 | 1.200 | 4.798 | 9.596 |
| | 35 | 1026.60 | .533 | .914 | 1.142 | 1.218 | 4.870 | 9.741 |
| | 40 | 1011.51 | .541 | .928 | 1.159 | 1.237 | 4.943 | 9.886 |
| | 45 | 996.87 | .549 | .941 | 1.176 | 1.255 | 5.016 | 10.031 |

## XIX.—*To trace Railroad Curves, &c.*—Continued.

| Degree. | Radii. | ORDINATES. To circular arcs on a chord of 100 feet. | | | | Tangent-deflection. | Chord-deflection. |
|---|---|---|---|---|---|---|---|
| | | 12¼ | 25 | 37¼ | 50 | | |
| ° ′ | | | | | | | |
| 5  50 | 982. 64 | . 557 | . 955 | 1. 193 | 1. 273 | 5. 088 | 10. 177 |
| 55 | 968. 81 | . 565 | . 968 | 1. 210 | 1. 291 | 5. 161 | 10. 322 |
| 6  0 | 955.37 | . 573 | . 982 | 1. 228 | 1. 309 | 5. 234 | 10. 467 |
| 5 | 942. 29 | . 581 | . 996 | 1. 245 | 1. 327 | 5. 306 | 10. 612 |
| 10 | 929. 57 | . 589 | 1. 009. | 1. 262 | 1. 346 | 5. 379 | 10. 758 |
| 15 | 917. 19 | . 597 | 1. 023 | 1. 279 | 1. 364 | 5. 451 | 10. 903 |
| 20 | 905. 13 | . 605 | 1. 037 | 1. 296 | 1. 382 | 5. 524 | 11. 048 |
| 25 | 893. 39 | . 613 | 1. 050 | 1. 313 | 1. 400 | 5. 597 | 11. 193 |
| 30 | 881. 95 | . 621 | 1. 064 | 1. 330 | 1. 418 | 5. 669 | 11. 339 |
| 35 | 870. 79 | . 629 | 1. 078 | 1. 347 | 1. 437 | 5. 742 | 11. 484 |
| 40 | 859. 92 | . 637 | 1. 091 | 1. 364 | 1. 455 | 5. 814 | 11. 629 |
| 45 | 849. 32 | . 645 | 1. 105 | 1. 381 | 1. 473 | 5. 887 | 11. 774 |
| 50 | 838. 97 | . 653 | 1. 118 | 1. 398 | 1. 491 | 5. 960 | 11. 919 |
| 55 | 828. 88 | . 661 | 1. 132 | 1. 415 | 1. 510 | 6. 032 | 12. 065 |
| 7  0 | 819. 02 | . 669 | 1. 146 | 1. 432 | 1. 528 | 6. 105 | 12. 210 |
| 5 | 809. 40 | . 677 | 1. 159 | 1. 449 | 1. 546 | 6. 177 | 12. 355 |
| 10 | 800. 00 | . 685 | 1. 173 | 1. 466 | 1. 564 | 6. 250 | 12. 500 |
| 15 | 790. 81 | . 693 | 1. 187 | 1. 483 | 1. 582 | 6. 323 | 12. 645 |
| 20 | 781. 84 | . 701 | 1. 200 | 1. 501 | 1. 600 | 6. 395 | 12. 790 |
| 25 | 773. 07 | . 709 | 1. 214 | 1. 517 | 1. 619 | 6. 468 | 12. 936 |
| 30 | 764. 49 | . 717 | 1. 228 | 1. 535 | 1. 637 | 6. 540 | 13. 081 |
| 35 | 756. 10 | . 725 | 1. 242 | 1. 552 | 1. 655 | 6. 613 | 13. 226 |
| 40 | 747. 89 | . 733 | 1. 255 | 1. 569 | 1. 673 | 6. 685 | 13. 371 |
| 45 | 739. 86 | . 740 | 1. 269 | 1. 586 | 1. 691 | 6. 758 | 13. 516 |
| 50 | 732. 01 | . 748 | 1. 283 | 1. 603 | 1. 710 | 6. 831 | 13. 661 |
| 55 | 724. 31 | . 756 | 1. 296 | 1. 620 | 1. 728 | 6. 903 | 13. 806 |
| 8  0 | 716. 78 | . 764 | 1. 310 | 1. 637 | 1. 746 | 6. 976 | 13. 951 |
| 5 | 709. 40 | . 772 | 1. 324 | 1. 654 | 1. 764 | 7. 048 | 14. 096 |
| 10 | 702. 18 | . 780 | 1. 337 | 1. 671 | 1. 782 | 7. 121 | 14. 241 |
| 15 | 695. 09 | . 788 | 1. 351 | 1. 688 | 1. 801 | 7. 193 | 14. 387 |
| 20 | 688. 16 | . 796 | 1. 365 | 1. 705 | 1. 819 | 7. 266 | 14. 532 |
| 25 | 681. 35 | . 804 | 1. 378 | 1. 722 | 1. 837 | 7. 338 | 14. 677 |
| 30 | 674. 69 | . 812 | 1. 392 | 1. 739 | 1. 855 | 7. 411 | 14. 822 |
| 35 | 668. 15 | . 820 | 1. 406 | 1. 757 | 1. 873 | 7. 483 | 14. 967 |

## XIX.—*To trace Railroad Curves, &c.*—Continued.

| Degree. | Radii. | ORDINATES. To circular arcs on a chord of 100 feet. | | | | Tangent-deflection. | Chord-deflection. |
|---|---|---|---|---|---|---|---|
| | | 12½ | 25 | 37½ | 50 | | |
| 8  40 | 661.74 | .828 | 1.419 | 1.774 | 1.892 | 7.556 | 15.112 |
| 45 | 655.45 | .836 | 1.433 | 1.791 | 1.910 | 7.628 | 15.257 |
| 50 | 649.27 | .844 | 1.447 | 1.808 | 1.928 | 7.701 | 15.402 |
| 55 | 643.22 | .852 | 1.460 | 1.825 | 1.946 | 7.773 | 15.547 |
| 9  0 | 637.27 | .860 | 1.474 | 1.842 | 1.965 | 7.846 | 15.692 |
| 5 | 631.44 | .868 | 1.488 | 1.859 | 1.983 | 7.918 | 15.837 |
| 10 | 625.71 | .876 | 1.501 | 1.876 | 2.001 | 7.991 | 15.982 |
| 15 | 620.09 | .884 | 1.515 | 1.893 | 2.019 | 8.063 | 16.127 |
| 20 | 614.56 | .892 | 1.529 | 1.910 | 2.037 | 8.136 | 16.272 |
| 25 | 609.14 | .900 | 1.542 | 1.927 | 2.056 | 8.208 | 16.417 |
| 30 | 603.80 | .908 | 1.556 | 1.944 | 2.074 | 8.281 | 16.562 |
| 35 | 598.57 | .916 | 1.570 | 1.961 | 2.092 | 8.353 | 16.707 |
| 40 | 593.42 | .924 | 1.583 | 1.979 | 2.110 | 8.426 | 16.852 |
| 45 | 588.36 | .932 | 1.597 | 1.996 | 2.128 | 8.498 | 16.996 |
| 50 | 583.38 | .940 | 1.611 | 2.013 | 2.147 | 8.571 | 17.141 |
| 55 | 578.49 | .948 | 1.624 | 2.030 | 2.165 | 8.643 | 17.286 |
| 10  0 | 573.69 | .956 | 1.638 | 2.047 | 2.183 | 8.716 | 17.431 |
| 10 | 564.31 | .972 | 1.665 | 2.081 | 2.219 | 8.860 | 17.721 |
| 20 | 555.23 | .988 | 1.693 | 2.115 | 2.256 | 9.005 | 18.011 |
| 30 | 546.44 | 1.004 | 1.720 | 2.149 | 2.292 | 9.150 | 18.300 |
| 40 | 537.92 | 1.020 | 1.748 | 2.184 | 2.329 | 9.295 | 18.590 |
| 50 | 529.67 | 1.036 | 1.775 | 2.218 | 2.365 | 9.440 | 18.880 |
| 11  0 | 521.67 | 1.052 | 1.802 | 2.252 | 2.402 | 9.585 | 19.169 |
| 10 | 513.91 | 1.068 | 1.830 | 2.286 | 2.438 | 9.729 | 19.459 |
| 20 | 506.38 | 1.084 | 1.857 | 2.320 | 2.475 | 9.874 | 19.748 |
| 30 | 499.06 | 1.100 | 1.884 | 2.354 | 2.511 | 10.019 | 20.038 |
| 40 | 491.96 | 1.116 | 1.912 | 2.389 | 2.547 | 10.164 | 20.327 |
| 50 | 485.05 | 1.132 | 1.938 | 2.423 | 2.584 | 10.308 | 20.616 |
| 12  0 | 478.34 | 1.148 | 1.967 | 2.457 | 2.620 | 10.453 | 20.906 |
| 10 | 471.81 | 1.164 | 1.994 | 2.491 | 2.657 | 10.597 | 21.195 |
| 20 | 465.46 | 1.180 | 2.021 | 2.525 | 2.693 | 10.742 | 21.484 |
| 30 | 459.28 | 1.196 | 2.049 | 2.560 | 2.730 | 10.887 | 21.773 |
| 40 | 453.26 | 1.212 | 2.076 | 2.594 | 2.766 | 11.031 | 22.063 |
| 50 | 447.40 | 1.228 | 2.104 | 2.628 | 2.803 | 11.176 | 22.352 |

## XIX.—*To trace Railroad Curves, &c.*—Continued.

| Degree. | Radii. | ORDINATES. To circular arcs on a chord of 100 feet. | | | | Tangent-deflection. | Chord-deflection. |
|---|---|---|---|---|---|---|---|
| | | 12½ | 25 | 37½ | 50 | | |
| 13  0 | 441.68 | 1.244 | 2.131 | 2.662 | 2.839 | 11.320 | 22.641 |
| 10 | 436.12 | 1.260 | 2.159 | 2.697 | 2.876 | 11.465 | 22.930 |
| 20 | 430.69 | 1.277 | 2.186 | 2.731 | 2.912 | 11.609 | 23.219 |
| 30 | 425.40 | 1.293 | 2.213 | 2.765 | 2.949 | 11.754 | 23.507 |
| 40 | 420.23 | 1.309 | 2.241 | 2.799 | 2.985 | 11.898 | 23.796 |
| 50 | 415.19 | 1.325 | 2.268 | 2.833 | 3.022 | 12.043 | 24.085 |
| 14  0 | 410.28 | 1.341 | 2.296 | 2.868 | 3.058 | 12.187 | 24.374 |
| 10 | 405.47 | 1.357 | 2.323 | 2.902 | 3.095 | 12.331 | 24.663 |
| 20 | 400.78 | 1.373 | 2.351 | 2.936 | 3.131 | 12.476 | 24.951 |
| 30 | 396.20 | 1.389 | 2.378 | 2.970 | 3.168 | 12.620 | 25.240 |
| 40 | 391.72 | 1.405 | 2.406 | 3.005 | 3.204 | 12.764 | 25.528 |
| 50 | 387.34 | 1.421 | 2.433 | 3.039 | 3.241 | 12.908 | 25.817 |
| 15  0 | 383.06 | 1.437 | 2.461 | 3.073 | 3.277 | 13.053 | 26.105 |
| 10 | 378.88 | 1.453 | 2.488 | 3.107 | 3.314 | 13.197 | 26.394 |
| 20 | 374.79 | 1.469 | 2.515 | 3.142 | 3.350 | 13.341 | 26.682 |
| 30 | 370.78 | 1.486 | 2.543 | 3.176 | 3.387 | 13.485 | 26.970 |
| 40 | 366.86 | 1.502 | 2.570 | 3.210 | 3.423 | 13.629 | 27.258 |
| 50 | 363.02 | 1.518 | 2.598 | 3.245 | 3.460 | 13.773 | 27.547 |
| 16  0 | 359.26 | 1.534 | 2.625 | 3.279 | 3.496 | 13.917 | 27.835 |
| 10 | 355.59 | 1.550 | 2.653 | 3.313 | 3.533 | 14.061 | 28.123 |
| 20 | 351.98 | 1.566 | 2.680 | 3.347 | 3.569 | 14.205 | 28.411 |
| 30 | 348.45 | 1.582 | 2.708 | 3.382 | 3.606 | 14.349 | 28.699 |
| 40 | 344.99 | 1.598 | 2.736 | 3.416 | 3.643 | 14.493 | 28.986 |
| 50 | 341.60 | 1.615 | 2.763 | 3.450 | 3.679 | 14.637 | 29.274 |
| 17  0 | 338.27 | 1.631 | 2.791 | 3.485 | 3.716 | 14.781 | 29.562 |
| 10 | 335.01 | 1.647 | 2.818 | 3.519 | 3.752 | 14.925 | 29.850 |
| 20 | 331.82 | 1.663 | 2.846 | 3.553 | 3.789 | 15.069 | 30.137 |
| 30 | 328.68 | 1.679 | 2.873 | 3.588 | 3.825 | 15.212 | 30.425 |
| 40 | 325.60 | 1.695 | 2.901 | 3.622 | 3.862 | 15.356 | 30.712 |
| 50 | 322.59 | 1.711 | 2.928 | 3.656 | 3.898 | 15.500 | 31.000 |
| 18  0 | 319.62 | 1.728 | 2.956 | 3.691 | 3.935 | 15.643 | 31.287 |
| 10 | 316.71 | 1.744 | 2.983 | 3.725 | 3.972 | 15.787 | 31.574 |
| 20 | 313.86 | 1.760 | 3.011 | 3.759 | 4.008 | 15.931 | 31.861 |

## XIX.—*To trace Railroad Curves, &c.*—Continued.

| Degree. | Radii. | ORDINATES. To circular arcs on a chord of 100 feet. | | | | Tangent-deflection. | Chord-deflection. |
|---|---|---|---|---|---|---|---|
| ° ′ | | 12½ | 25 | 37½ | 50 | | |
| 18 30 | 311.06 | 1.776 | 3.039 | 3.794 | 4.045 | 16.074 | 32.149 |
| 40 | 308.30 | 1.792 | 3.066 | 3.828 | 4.081 | 16.218 | 32.436 |
| 50 | 305.60 | 1.809 | 3.094 | 3.862 | 4.118 | 16.361 | 32.723 |
| 19 0 | 302.94 | 1.825 | 3.121 | 3.897 | 4.155 | 16.505 | 33.010 |
| 10 | 300.33 | 1.841 | 3.149 | 3.931 | 4.191 | 16.648 | 33.296 |
| 20 | 297.77 | 1.857 | 3.177 | 3.965 | 4.228 | 16.792 | 33.583 |
| 30 | 295.25 | 1.873 | 3.204 | 4.000 | 4.265 | 16.935 | 33.870 |
| 40 | 292.77 | 1.890 | 3.232 | 4.034 | 4.301 | 17.078 | 34.157 |
| 50 | 290.33 | 1.906 | 3.259 | 4.069 | 4.338 | 17.222 | 34.443 |

LONG CHORDS.

| Degree of curve. | 2 stations. | 3 stations. | 4 stations. | 5 stations. | 6 stations. |
|---|---|---|---|---|---|
| ° ′ | | | | | |
| 0 10 | 200.000 | 299.999 | 399.998 | 499.996 | 599.993 |
| 20 | 199.999 | .997 | .992 | .983 | .970 |
| 30 | .998 | .992 | .981 | .962 | .933 |
| 40 | .997 | .986 | .966 | .932 | .882 |
| 50 | .995 | .979 | .947 | .894 | .815 |
| 1 0 | 199.992 | 299.970 | 399.924 | 499.848 | 599.733 |
| 10 | .990 | .959 | .896 | .793 | .637 |
| 20 | .986 | .946 | .865 | .729 | .526 |
| 30 | .983 | .932 | .829 | .657 | .401 |
| 40 | .979 | .915 | .789 | .577 | .260 |
| 50 | .974 | .898 | .744 | .488 | .105 |
| 2 0 | 199.970 | 299.878 | 399.695 | 499.391 | 598.934 |
| 10 | .964 | .857 | .643 | .285 | .750 |
| 20 | .959 | .834 | .586 | .171 | .550 |
| 30 | .952 | .810 | .524 | .049 | .336 |
| 40 | .946 | .783 | .459 | 498.918 | .106 |
| 50 | .939 | .756 | .389 | .778 | 597.862 |

## XIX.—*To trace Railroad Curves, &c.*—Continued.

### LONG CHORDS—Continued.

| Degree of curve. | 2 stations. | 3 stations. | 4 stations. | 5 stations. | 6 stations. |
|---|---|---|---|---|---|
| ° ′ | | | | | |
| 3　0 | 199.931 | 299.726 | 399.315 | 498.630 | 597.604 |
| 10 | .924 | .695 | .237 | .474 | .331 |
| 20 | .915 | .662 | .154 | .309 | .043 |
| 30 | .907 | .627 | .068 | .136 | 596.740 |
| 40 | .898 | .591 | 398.977 | 497.955 | .423 |
| 50 | .888 | .553 | .882 | .765 | .091 |
| | | | | | |
| 4　0 | 199.878 | 299.513 | 398.782 | 497.566 | 595.744 |
| 10 | .868 | .471 | .679 | .360 | .383 |
| 20 | .857 | .428 | .571 | .145 | .007 |
| 30 | .846 | .383 | .459 | 496.921 | 594.617 |
| 40 | .834 | .337 | .343 | .689 | .212 |
| 50 | .822 | .289 | .223 | .449 | 593.792 |
| | | | | | |
| 5　0 | 199.810 | 299.239 | 398.099 | 496.200 | 593.358 |
| 10 | .797 | .187 | 397.970 | 495.944 | 592.909 |
| 20 | .783 | .134 | .837 | .678 | .446 |
| 30 | .770 | .079 | .700 | .405 | 591.968 |
| 40 | .756 | .023 | .559 | .123 | .476 |
| 50 | .741 | 298.964 | .413 | 494.832 | 590.970 |
| | | | | | |
| 6　0 | 199.726 | 298.904 | 397.264 | 494.534 | 590.449 |
| 10 | .710 | .843 | .110. | .227 | 589.913 |
| 20 | .695 | .779 | 396.952 | 493.912 | .364 |
| 30 | .678 | .714 | .790 | .588 | 588.800 |
| 40 | .662 | .648 | .623 | .257 | .221 |
| 50 | .644 | .579 | .453 | 492.917 | 587.628 |
| | | | | | |
| 7　0 | 199.627 | 298.509 | 396.278 | 492.568 | 587.021 |
| 10 | .609 | .438 | .099 | .212 | 586.400 |
| 20 | .591 | .364 | 395.916 | 491.847 | 585.765 |
| 30 | .572 | .289 | .729 | .474 | .115 |
| 40 | .553 | .212 | .538 | .093 | 584.451 |
| 50 | .533 | .134 | .342 | 490.704 | 583.773 |
| | | | | | |
| 8　0 | 199.513 | 298.054 | 395.142 | 490.306 | 583.081 |

### XX.—*To ascertain the Discharge of Water in any Stream.*

1. For practically gauging large rivers a locality is selected in a straight portion of the stream where the water flows smoothly and without obstruction. A base-line about 200 feet long is laid out parallel to the current, and the exact cross-section in front of this base is determined by careful sounding.

To obtain the discharge, two theodolites are established, and the angular distance from, and the times of transit past, each end of the base, of numerous floats, well distributed between the banks, are noted.

The floats should be made double, the surface-float being a minute tin ellipsoid, a piece of cork, or some other small light body, bearing a small flag. The lower float may be a large box or keg without top or bottom, kept upright by lead ballasting; or better, because lighter, two sheets of tin bent at right angles, and soldered together at the bend, so as to make all the angles between the four faces right angles; the essential conditions being that the lower float shall so greatly preponderate in area over the upper, and shall be connected by so fine a wire or cord, that its rate of movement will govern the whole combination.

The center of the lower float should be placed at the *mid-depth of the stream*, in each vertical plane of transit, because the rate of movement will then be unaffected by wind.

As it is sometimes troublesome to adjust to mid-depth in the different planes of transit, when there is a tolerably uniform and symmetrical cross-section, the average mid-depth of the river may be adopted for all the floats without sensible error.

If floats passing near the surface are used, errors in the computed discharge may be caused by an ordinary breeze, and as these errors are positive or negative according to the direction of the wind, discrepancies may result in the measurements of different days when there is no real variation in the discharge. All this uncertainty is avoided by using mid-depth floats.

The exact level of the water-surface on a permanent gauge-rod should be carefully noted when the observations begin and terminate.

XX.—*To ascertain the Discharge of Water, &c.*—Continued.

Upon a sheet of section-paper the base-line and the two perpendiculars across which the times of transit were noted are then
laid down, and, from the recorded angles and a table of natural
tangents, the distances from the base-line to the points at which
each float passed both lines are plotted. These points, being
connected, indicate the paths of the floats. Upon each path the
difference between the two recorded times of transit is written in
seconds. These seconds of transit are next examined, and the
total width of the river is marked off into as many "divisions" as
it seems proper to assume are traversed by water moving with
sensibly unvarying velocity, say, for instance, that each division
is about $\frac{1}{10}$ of the width of the stream. A mean of the seconds
of transit of all the floats in each "division" is next taken, and,
when reduced to velocity in feet per second, is adopted as the
mid-depth velocity in that "division."

A mean of all these mean mid-depth velocities—interpolations
being made if any are missing—closely approximates the mean
velocity of the river, provided the "divisions" are equal in width.

This method involves two errors, which nearly balance each
other, viz: the inequality in area of the divisions, and the difference between the mid-depth velocity and the mean velocity in
any vertical plane, giving a resulting mean velocity of about 0.95
times its true value.

The mean velocity in each plane is obtained from the mid-
depth velocities, $V \frac{1}{2} D$, by the formula—

$$v = 1.075 \, V \tfrac{1}{2} D + 0.004 \, b - 0.093 \, (V \tfrac{1}{2} D \cdot b)^{\frac{1}{2}}$$

and—

$$b = \frac{1.69}{(D + 1.5)^{\frac{1}{2}}}$$

where—

    $D$ = the depth of the stream at any point of the surface.

If $a$ is the area of cross-section, and $a'$, $a''$, etc., the partial
division areas, the discharge may be found by—

$$Q = v \, a = \left[ a' \, (V \tfrac{1}{2} D) - \tfrac{1}{12} \, (b \cdot v)^{\frac{1}{2}} \right]$$

where $\left[ \; \right]$ denotes the sum of similar quantities.

XX.—*To ascertain the Discharge of Water, &c.*—Continued.

2. Determination of the mean velocity in terms of the dimensions of the cross-section and the slope.

### *Humphreys-Abbot Formula.*

(Not applicable to water flowing in smooth artificial channels.)

$$v = \left( \sqrt{0.0081\, b + (225\, r_1\, s^{\frac{1}{2}})^{\frac{1}{2}}} - 0.09\, (b)^{\frac{1}{2}} \right)^2 - \frac{2.4\,(v')^{\frac{1}{2}}}{1 + p}$$

where the symbols have the following signification, all expressed in English feet:

$v$ = mean velocity = $\dfrac{Q}{a}$;

$a$ = area of cross-section;

$W$ = width;

$r$ = mean radius, or $\dfrac{a}{p}$;

$v'$ = value of first term in expression for $v$;

$p$ = wetted perimeter;

$Q$ = discharge in cubic feet per second;

$r_1 = \dfrac{a}{p + W}$;

$s$ = sine of slope of water surface corrected for bends;

$b$ = function of the depth, for small streams = $\dfrac{1.69}{(r + 1.5)^{\frac{1}{2}}}$

For rivers whose mean radius exceeds 12 or 15 feet, $b$ may be assumed to be 0.1856, which will make the numerical value of the term involving $b$ so small that it may be generally neglected, reducing the above equation to—

$$v = ([\, 225\, r_1\, s^{\frac{1}{2}}\, ]^{\frac{1}{2}} - 0.0388)^2$$

The following formulæ give the value of each variable in terms of the others and known quantities:

$$z = 0.93\, v + 0.167\, (b\, v)^{\frac{1}{2}};$$

and when $p$ is not known by measurement it may, for ordinary natural channels, be assumed to be 1.015 $W$.

$$s = \left( \frac{(p + W)\, z^2}{195\, a} \right)^2$$

$$a = \frac{(p + W)\, z^2}{195\, (s)^{\frac{1}{2}}}$$

$$p + W = \frac{195\, a\, (s)^{\frac{1}{2}}}{z^2}$$

XX.—*To ascertain the Discharge of Water, &c.*—Continued.

APPLICATION.—The variables which enter these formulæ require a knowledge of the mean cross-section of the stream, and a map of the course of the channel between two selected points of the water-surface, whose difference of level should be exactly known.

Whenever practicable, the two points should be located on a straight and regular portion of the river to eliminate the effects of bends. As this is not always possible, the general case is considered in the above formulæ, and bends are assumed to exist between the points selected.

The field-operations consist in a survey of the channel, with numerous soundings between permanent bench-marks placed near the water, and in running a line of levels between those marks, so as to give their relative level with the most extreme accuracy.

These points should be located with care, as far apart as practicable, distant from any eddy, and placed where the current on the banks flows with *equal velocity*. This latter condition is necessary, because, as water in motion exerts less pressure than when at rest, if it moved rapidly past one bench-mark and was nearly stationary at the other, a difference of level, which has nothing to do with the motive power of the stream, would vitiate the observation.

In determining the mean dimensions of cross-sections, care must be taken to extend the soundings throughout the entire distance between the bench-marks, and it must be borne in mind that measured fall in water-surface between two stations corresponds to *the mean channel between them.*

When the soundings are made, the water-level should be referred to the bench-marks in order to determine the area corresponding to any subsequent stand of the river.

These soundings completed, frequent gauging of the river can be made by referring at any time, by accurate levels, the water-surface at the two points to their respective bench-marks, thus determining the fall and corresponding cross-section of the river from which the discharge is computed.

The observations must be simultaneous in order to avoid the effect of any oscillation in the river, and *calm days* should be selected because waves render it difficult to determine the exact level of the water-surface; and, also, changes of level result from the general piling up or lowering of the water under the influence of winds.

XX.—*To ascertain the Discharge of Water, &c.*—Continued.

*Correction for Bends.*—A line following the mid-channel is drawn on the map, composed of straight lines with angular changes, wherever necessary, of 30°. A mean velocity is assumed, to be corrected subsequently if required, and the value of $h$ is computed in the following formulæ, in which $N$ represents the number of deflections:

$$h = \frac{v^2 . N \sin^2 30°}{134}$$

The deduced value of $h$ is next subtracted from the total fall in the water-surface between the two stations; the remainder divided by the distance in feet between these stations, measured on the middle line of the river, is the true value of $s$ in the formula for mean velocity.

If any material error has been made in assuming $v$, the computation should be repeated until the requisite approximation has been made.

By expressing the formula in the form—

$$v = \left( \sqrt{M} + \left( \frac{225\, a\sqrt{s}}{p + W} \right)^2 - \sqrt{M} \right)^2 - M' \sqrt{v'}$$

The following table will facilitate its application:

| $v$ | M | $\sqrt{M}$ | $p$ | M' | Log M' |
|---|---|---|---|---|---|
| 1 | 0.0087 | 0.0930 | 5 | 0.400 | 9.602060 |
| 2 | 73 | 855 | 6 | .343 | 9.535294 |
| 3 | 65 | 803 | 7 | .300 | 9.477121 |
| 4 | 58 | 764 | 8 | .267 | 9.426511 |
| 5 | 54 | 733 | 9 | .240 | 9.380211 |
| 6 | 50 | 707 | 10 | .218 | 9.338456 |
| 7 | 47 | 685 | 12 | .185 | 9.267172 |
| 8 | 44 | 666 | 14 | .160 | 9.204120 |
| 9 | 42 | 649 | 16 | .141 | 9.149219 |
| 10 | 40 | 634 | 18 | .126 | 9.100371 |
| 12 | 37 | 610 | 20 | .114 | 9.056905 |
| 14 | 35 | 590 | 22 | .104 | 9.017033 |
| 16 | 33 | 573 | 24 | .096 | 8.982271 |
| 18 | 31 | 558 | 26 | .089 | 8.949390 |
| 20 | 29 | 544 | 28 | .083 | 8.919078 |
| 30 | 24 | 494 | 30 | .078 | 8.892095 |
| 50 | 0.0019 | 0.0437 | 50 | 0.047 | 8.672098 |

For streams larger than 50 or 100 feet in cross-section the term involving M' may be dropped, and for larger rivers, exceeding 12 or 20 feet in mean radius, M, but not $\sqrt{M}$, may be neglected.

## XX.—To ascertain the Discharge of Water, &c.—Continued

### EXAMPLES OF RESULTING MEAN VELOCITIES FROM GIVEN CROSS-SECTIONS AND SLOPES.

| Stream. | Locality. | Date. | Dimensions of cross-section. | | | | Slope. | | | Mean velocity. | | Authority. |
| --- | --- | --- | --- | --- | --- | --- | --- | --- | --- | --- | --- | --- |
| | | | Area. | Width. | Perimeter. | Max. Depth. | $s$ | $\log s$ | $\log s^{\frac{1}{2}}$ | Observed. | Computed. | |
| | | | Sq. feet. | Feet. | Feet. | Feet. | | | | | | |
| 1. Mississippi River........ | Carrolton........ | H. W. of 1851 | 193968 | 2658 | 2693 | 136 | 0.00002051 | 5.3119657 | 7.6559828 | 5.9288 | 5.8903 | Delta survey. |
| 2. Bayou Plaquemine. | Near upper mouth | Jan. 16, 1859 | 4259 | 268 | 278 | 24 | 0.0014372 | 6.1575172 | 8.0787586 | 3.9589 | 4.3460 | Do. |
| 3. Chesapeake & Ohio Canal feeder. | Near Georgetown, D. C. | Nov. 26, 1859 | 121 | 23 | 32.7 | 7.6 | 0.0069851 | 6.8441726 | 8.4220863 | 3.0323 | 3.1032 | Do. |

XX.—*To ascertain the Discharge of Water, &c.*—Continued.

3. Formulæ for the mean velocity, from other authorities:

Chezy...
{
Downing's and others' co-efficient .............. $v = 100.0 \ (rs)^{\frac{1}{2}}$
Eytelwein's co-efficient ...................... $v = 93.4 \ (rs)^{\frac{1}{2}}$
Young's co-efficient ...................... $v = 84.3 \ (rs)^{\frac{1}{2}}$
}

De Prony
{
For canals .............. $v = (0.0556 \ + 10593 \ rs)^{\frac{1}{2}} - 0.2357$
For canals and pipes ........ $v = (0.0237 \ + 9966 \ rs)^{\frac{1}{2}} - 0.1542$
Eytelwein's co-efficient ..... $v = (0.0119 \ + 8963 \ rs)^{\frac{1}{2}} - 0.1089$
Weisbach's co-efficient ..... $v = (0.00024 + 8675 \ rs)^{\frac{1}{2}} - 0.0154$
}

Darcy-Bazin .............................. $v = r \left( \dfrac{1000 \ s}{0.08534 \ r + 0.35} \right)^{\frac{1}{2}}$

### XXI.—*Motion of Water in Conduit-Pipes.*

Discharge through pipes of uniform dimensions, and having no sudden changes of direction:

For ordinary cases:

$$Q = 38.436 \ \sqrt{\frac{H \ D^5}{L}} - 0.070862 \ D^2$$

In great velocities:

$$Q = 36.769 \ \sqrt{\frac{H \ D^5}{L}} \ .$$

If the velocity be required, divide the discharge by the area of the section ($0.7854 \ D^2$).

To find the diameter of a conduit-pipe for a given discharge under a given head:

$$D = 0.2323 \ \sqrt[5]{\frac{L \ Q^2}{H}}$$

Where $Q$ = discharge in cubic feet per second; H, the head, and D and L the diameter and length of the pipe in feet.

The resistance of curves is proportional to the square of the velocity of the fluid, to the number of angles of reflextion, and to the square of their sine,

or, in function of Q, $= 0.006079 \ \dfrac{Q^2}{D^4} \cdot s^2$

$s^2$ being the sum of the squares of all the sines of the angles of reflexion.

## XXII.—*Logarithms of Numbers.*

| Nat. Nos. | 0 | 1 | 2 | 3 | 4 | 5 | 6 | 7 | 8 | 9 | Proportional parts. | | | | | | | | |
|---|---|---|---|---|---|---|---|---|---|---|---|---|---|---|---|---|---|---|---|
| | | | | | | | | | | | 1 | 2 | 3 | 4 | 5 | 6 | 7 | 8 | 9 |
| 10 | .0000 | .0043 | .0086 | .0128 | .0170 | .0212 | .0253 | .0294 | .0334 | .0374 | 4 | 8 | 12 | 17 | 21 | 25 | 29 | 33 | 37 |
| 11 | .0414 | .0453 | .0492 | .0531 | .0569 | .0607 | .0645 | .0682 | .0719 | .0755 | 4 | 8 | 11 | 15 | 19 | 23 | 26 | 30 | 34 |
| 12 | .0792 | .0828 | .0864 | .0899 | .0934 | .0969 | .1004 | .1038 | .1072 | .1106 | 3 | 7 | 10 | 14 | 17 | 21 | 24 | 28 | 31 |
| 13 | .1139 | .1173 | .1206 | .1239 | .1271 | .1303 | .1335 | .1367 | .1399 | .1430 | 3 | 6 | 10 | 13 | 16 | 19 | 23 | 26 | 29 |
| 14 | .1461 | .1492 | .1523 | .1553 | .1584 | .1614 | .1644 | .1673 | .1703 | .1732 | 3 | 6 | 9 | 12 | 15 | 18 | 21 | 24 | 27 |
| 15 | .1761 | .1790 | .1818 | .1847 | .1875 | .1903 | .1931 | .1959 | .1987 | .2014 | 3 | 6 | 8 | 11 | 14 | 17 | 20 | 22 | 25 |
| 16 | .2041 | .2068 | .2095 | .2122 | .2148 | .2175 | .2201 | .2227 | .2253 | .2279 | 3 | 5 | 8 | 11 | 13 | 16 | 18 | 21 | 24 |
| 17 | .2304 | .2330 | .2355 | .2380 | .2405 | .2430 | .2455 | .2480 | .2504 | .2529 | 2 | 5 | 7 | 10 | 12 | 15 | 17 | 20 | 22 |
| 18 | .2553 | .2577 | .2601 | .2625 | .2648 | .2672 | .2695 | .2718 | .2742 | .2765 | 2 | 5 | 7 | 9 | 12 | 14 | 16 | 19 | 21 |
| 19 | .2788 | .2810 | .2833 | .2856 | .2878 | .2900 | .2923 | .2945 | .2967 | .2989 | 2 | 4 | 7 | 9 | 11 | 13 | 16 | 18 | 20 |
| 20 | .3010 | .3032 | .3054 | .3075 | .3096 | .3118 | .3139 | .3160 | .3181 | .3201 | 2 | 4 | 6 | 8 | 11 | 13 | 15 | 17 | 19 |
| 21 | .3222 | .3243 | .3263 | .3284 | .3304 | .3324 | .3345 | .3365 | .3385 | .3404 | 2 | 4 | 6 | 8 | 10 | 12 | 14 | 16 | 18 |
| 22 | .3424 | .3444 | .3464 | .3483 | .3502 | .3522 | .3541 | .3560 | .3579 | .3598 | 2 | 4 | 6 | 8 | 10 | 12 | 14 | 15 | 17 |
| 23 | .3617 | .3636 | .3655 | .3674 | .3692 | .3711 | .3729 | .3747 | .3766 | .3784 | 2 | 4 | 6 | 7 | 9 | 11 | 13 | 15 | 17 |
| 24 | .3802 | .3820 | .3838 | .3856 | .3874 | .3892 | .3909 | .3927 | .3945 | .3962 | 2 | 4 | 5 | 7 | 9 | 11 | 12 | 14 | 16 |
| 25 | .3979 | .3997 | .4014 | .4031 | .4048 | .4065 | .4082 | .4099 | .4116 | .4133 | 2 | 3 | 5 | 7 | 9 | 10 | 12 | 14 | 15 |
| 26 | .4150 | .4166 | .4183 | .4200 | .4216 | .4232 | .4249 | .4265 | .4281 | .4298 | 2 | 3 | 5 | 7 | 8 | 10 | 11 | 13 | 15 |
| 27 | .4314 | .4330 | .4346 | .4362 | .4378 | .4393 | .4409 | .4425 | .4440 | .4456 | 2 | 3 | 5 | 6 | 8 | 9 | 11 | 13 | 14 |
| 28 | .4472 | .4487 | .4502 | .4518 | .4533 | .4548 | .4564 | .4579 | .4594 | .4609 | 2 | 3 | 5 | 6 | 8 | 9 | 11 | 12 | 14 |
| 29 | .4624 | .4639 | .4654 | .4669 | .4683 | .4698 | .4713 | .4728 | .4742 | .4757 | 1 | 3 | 4 | 6 | 7 | 9 | 10 | 12 | 13 |
| 30 | .4771 | .4786 | .4800 | .4814 | .4829 | .4843 | .4857 | .4871 | .4886 | .4900 | 1 | 3 | 4 | 6 | 7 | 9 | 10 | 11 | 13 |
| 31 | .4914 | .4928 | .4942 | .4955 | .4969 | .4983 | .4997 | .5011 | .5024 | .5038 | 1 | 3 | 4 | 6 | 7 | 8 | 10 | 11 | 12 |
| 32 | .5051 | .5065 | .5079 | .5092 | .5105 | .5119 | .5132 | .5145 | .5159 | .5172 | 1 | 3 | 4 | 5 | 7 | 8 | 9 | 11 | 12 |
| 33 | .5185 | .5198 | .5211 | .5224 | .5237 | .5250 | .5263 | .5276 | .5289 | .5302 | 1 | 3 | 4 | 5 | 6 | 8 | 9 | 10 | 12 |
| 34 | .5315 | .5328 | .5340 | .5353 | .5366 | .5378 | .5391 | .5403 | .5416 | .5428 | 1 | 3 | 4 | 5 | 6 | 8 | 9 | 10 | 11 |
| 35 | .5441 | .5453 | .5465 | .5478 | .5490 | .5502 | .5514 | .5527 | .5539 | .5551 | 1 | 2 | 4 | 5 | 6 | 7 | 9 | 10 | 11 |
| 36 | .5563 | .5575 | .5587 | .5599 | .5611 | .5623 | .5635 | .5647 | .5658 | .5670 | 1 | 2 | 4 | 5 | 6 | 7 | 8 | 10 | 11 |
| 37 | .5682 | .5694 | .5705 | .5717 | .5729 | .5740 | .5752 | .5763 | .5775 | .5786 | 1 | 2 | 3 | 5 | 6 | 7 | 8 | 9 | 10 |
| 38 | .5798 | .5809 | .5821 | .5832 | .5843 | .5855 | .5866 | .5877 | .5888 | .5899 | 1 | 2 | 3 | 5 | 6 | 7 | 8 | 9 | 10 |
| 39 | .5911 | .5922 | .5933 | .5944 | .5955 | .5966 | .5977 | .5988 | .5999 | .6010 | 1 | 2 | 3 | 4 | 5 | 7 | 8 | 9 | 10 |
| 40 | .6021 | .6031 | .6042 | .6053 | .6064 | .6075 | .6085 | .6096 | .6107 | .6117 | 1 | 2 | 3 | 4 | 5 | 6 | 8 | 9 | 10 |
| 41 | .6128 | .6138 | .6149 | .6160 | .6170 | .6180 | .6191 | .6201 | .6212 | .6222 | 1 | 2 | 3 | 4 | 5 | 6 | 7 | 8 | 9 |
| 42 | .6232 | .6243 | .6253 | .6263 | .6274 | .6284 | .6294 | .6304 | .6314 | .6325 | 1 | 2 | 3 | 4 | 5 | 6 | 7 | 8 | 9 |
| 43 | .6335 | .6345 | .6355 | .6365 | .6375 | .6385 | .6395 | .6405 | .6415 | .6425 | 1 | 2 | 3 | 4 | 5 | 6 | 7 | 8 | 9 |
| 44 | .6435 | .6444 | .6454 | .6464 | .6474 | .6484 | .6493 | .6503 | .6513 | .6522 | 1 | 2 | 3 | 4 | 5 | 6 | 7 | 8 | 9 |
| 45 | .6532 | .6542 | .6551 | .6561 | .6571 | .6580 | .6590 | .6599 | .6609 | .6618 | 1 | 2 | 3 | 4 | 5 | 6 | 7 | 8 | 9 |
| 46 | .6628 | .6637 | .6646 | .6656 | .6665 | .6675 | .6684 | .6693 | .6702 | .6712 | 1 | 2 | 3 | 4 | 5 | 6 | 7 | 7 | 8 |
| 47 | .6721 | .6730 | .6739 | .6749 | .6758 | .6767 | .6776 | .6785 | .6794 | .6803 | 1 | 2 | 3 | 4 | 5 | 5 | 6 | 7 | 8 |
| 48 | .6812 | .6821 | .6830 | .6839 | .6848 | .6857 | .6866 | .6875 | .6884 | .6893 | 1 | 2 | 3 | 4 | 4 | 5 | 6 | 7 | 8 |
| 49 | .6902 | .6911 | .6920 | .6928 | .6937 | .6946 | .6955 | .6964 | .6972 | .6981 | 1 | 2 | 3 | 4 | 4 | 5 | 6 | 7 | 8 |
| 50 | .6990 | .6998 | .7007 | .7016 | .7024 | .7033 | .7042 | .7050 | .7059 | .7067 | 1 | 2 | 3 | 3 | 4 | 5 | 6 | 7 | 8 |
| 51 | .7076 | .7084 | .7093 | .7101 | .7110 | .7118 | .7126 | .7135 | .7143 | .7152 | 1 | 2 | 3 | 3 | 4 | 5 | 6 | 7 | 8 |
| 52 | .7160 | .7168 | .7177 | .7185 | .7193 | .7202 | .7210 | .7218 | .7226 | .7235 | 1 | 2 | 2 | 3 | 4 | 5 | 6 | 7 | 7 |
| 53 | .7243 | .7251 | .7259 | .7267 | .7275 | .7284 | .7292 | .7300 | .7308 | .7316 | 1 | 2 | 2 | 3 | 4 | 5 | 6 | 6 | 7 |
| 54 | .7324 | .7332 | .7340 | .7348 | .7356 | .7364 | .7372 | .7380 | .7388 | .7396 | 1 | 2 | 2 | 3 | 4 | 5 | 6 | 6 | 7 |

## XXII.—*Logarithms of Numbers*—Continued.

| Nat. Nos. | 0 | 1 | 2 | 3 | 4 | 5 | 6 | 7 | 8 | 9 | Proportional parts. 1 2 3 4 5 6 7 8 9 |
|---|---|---|---|---|---|---|---|---|---|---|---|
| 55 | .7404 | .7412 | .7419 | .7427 | .7435 | .7443 | .7451 | .7459 | .7466 | .7474 | 1 2 2 3 4 5 5 6 7 |
| 56 | .7482 | .7490 | .7497 | .7505 | .7513 | .7520 | .7528 | .7536 | .7543 | .7551 | 1 2 2 3 4 5 5 6 7 |
| 57 | .7559 | .7566 | .7574 | .7582 | .7589 | .7597 | .7604 | .7612 | .7619 | .7627 | 1 2 2 3 4 5 5 6 7 |
| 58 | .7634 | .7642 | .7649 | .7657 | .7664 | .7672 | .7679 | .7686 | .7694 | .7701 | 1 1 2 3 4 4 5 6 7 |
| 59 | .7709 | .7716 | .7723 | .7731 | .7738 | .7745 | .7752 | .7760 | .7767 | .7774 | 1 1 2 3 4 4 5 6 7 |
| 60 | .7782 | .7789 | .7796 | .7803 | .7810 | .7818 | .7825 | .7832 | .7839 | .7846 | 1 1 2 3 4 4 5 6 6 |
| 61 | .7853 | .7860 | .7868 | .7875 | .7882 | .7889 | .7896 | .7903 | .7910 | .7917 | 1 1 2 3 4 4 5 6 6 |
| 62 | .7924 | .7931 | .7938 | .7945 | .7952 | .7959 | .7966 | .7973 | .7980 | .7987 | 1 1 2 3 4 4 5 6 6 |
| 63 | .7993 | .8000 | .8007 | .8014 | .8021 | .8028 | .8035 | .8041 | .8048 | .8055 | 1 1 2 3 4 5 5 6 |
| 64 | .8062 | .8069 | .8075 | .8082 | .8089 | .8096 | .8102 | .8109 | .8116 | .8122 | 1 1 2 3 3 4 5 5 6 |
| 65 | .8129 | .8136 | .8142 | .8149 | .8156 | .8162 | .8169 | .8176 | .8182 | .8189 | 1 1 2 3 4 5 5 6 |
| 66 | .8195 | .8202 | .8209 | .8215 | .8222 | .8228 | .8235 | .8241 | .8248 | .8254 | 1 1 2 3 3 4 5 5 6 |
| 67 | .8261 | .8267 | .8274 | .8280 | .8287 | .8293 | .8299 | .8306 | .8312 | .8319 | 1 1 2 3 3 4 5 5 6 |
| 68 | .8325 | .8331 | .8338 | .8344 | .8351 | .8357 | .8363 | .8370 | .8376 | .8382 | 1 1 2 3 3 4 4 5 6 |
| 69 | .8388 | .8395 | .8401 | .8407 | .8414 | .8420 | .8426 | .8432 | .8439 | .8445 | 1 1 2 3 4 4 5 6 |
| 70 | .8451 | .8457 | .8463 | .8470 | .8476 | .8482 | .8488 | .8494 | .8500 | .8506 | 1 1 2 2 3 4 4 5 6 |
| 71 | .8513 | .8519 | .8525 | .8531 | .8537 | .8543 | .8549 | .8555 | .8561 | .8567 | 1 1 2 2 3 4 4 5 5 |
| 72 | .8573 | .8579 | .8585 | .8591 | .8597 | .8603 | .8609 | .8615 | .8621 | .8627 | 1 1 2 2 3 4 4 5 5 |
| 73 | .8633 | .8639 | .8645 | .8651 | .8657 | .8663 | .8669 | .8675 | .8681 | .8686 | 1 1 2 2 3 4 4 5 5 |
| 74 | .8692 | .8698 | .8704 | .8710 | .8716 | .8722 | .8727 | .8733 | .8739 | .8745 | 1 1 2 2 3 4 4 5 5 |
| 75 | .8751 | .8756 | .8762 | .8768 | .8774 | .8779 | .8785 | .8791 | .8797 | .8802 | 1 1 2 2 3 3 4 5 5 |
| 76 | .8808 | .8814 | .8820 | .8825 | .8831 | .8837 | .8842 | .8848 | .8854 | .8859 | 1 1 2 2 3 3 4 5 5 |
| 77 | .8865 | .8871 | .8876 | .8882 | .8887 | .8893 | .8899 | .8904 | .8910 | .8915 | 1 1 2 2 3 3 4 4 5 |
| 78 | .8921 | .8927 | .8932 | .8938 | .8943 | .8949 | .8954 | .8960 | .8965 | .8971 | 1 1 2 2 3 3 4 4 5 |
| 79 | .8976 | .8982 | .8987 | .8993 | .8998 | .9004 | .9009 | .9015 | .9020 | .9025 | 1 1 2 2 3 3 4 4 5 |
| 80 | .9031 | .9036 | .9042 | .9047 | .9053 | .9058 | .9063 | .9069 | .9074 | .9079 | 1 1 2 3 3 4 4 5 |
| 81 | .9085 | .9090 | .9096 | .9101 | .9106 | .9112 | .9117 | .9122 | .9128 | .9133 | 1 1 2 2 3 3 4 4 5 |
| 82 | .9138 | .9143 | .9149 | .9154 | .9159 | .9165 | .9170 | .9175 | .9180 | .9186 | 1 1 2 2 3 3 4 4 5 |
| 83 | .9191 | .9196 | .9201 | .9206 | .9212 | .9217 | .9222 | .9227 | .9232 | .9238 | 1 1 2 2 3 3 4 4 5 |
| 84 | .9243 | .9248 | .9253 | .9258 | .9263 | .9269 | .9274 | .9279 | .9284 | .9289 | 1 1 2 2 3 3 4 4 5 |
| 85 | .9294 | .9299 | .9304 | .9309 | .9315 | .9320 | .9325 | .9330 | .9335 | .9340 | 1 1 2 2 3 3 4 4 5 |
| 86 | .9345 | .9350 | .9355 | .9360 | .9365 | .9370 | .9375 | .9380 | .9385 | .9390 | 1 1 2 2 3 3 4 4 5 |
| 87 | .9395 | .9400 | .9405 | .9410 | .9415 | .9420 | .9425 | .9430 | .9435 | .9440 | 0 1 1 2 2 3 3 4 4 |
| 88 | .9445 | .9450 | .9455 | .9460 | .9465 | .9469 | .9474 | .9479 | .9484 | .9489 | 0 1 1 2 2 3 3 4 4 |
| 89 | .9494 | .9499 | .9504 | .9509 | .9513 | .9518 | .9523 | .9528 | .9533 | .9538 | 0 1 1 2 2 3 3 4 4 |
| 90 | .9542 | .9547 | .9552 | .9557 | .9562 | .9566 | .9571 | .9576 | .9581 | .9586 | 0 1 1 2 2 3 3 4 4 |
| 91 | .9590 | .9595 | .9600 | .9605 | .9609 | .9614 | .9619 | .9624 | .9628 | .9633 | 0 1 1 2 2 3 3 4 4 |
| 92 | .9638 | .9643 | .9647 | .9652 | .9657 | .9661 | .9666 | .9671 | .9675 | .9680 | 0 1 1 2 2 3 3 4 4 |
| 93 | .9685 | .9689 | .9694 | .9699 | .9703 | .9708 | .9713 | .9717 | .9722 | .9727 | 0 1 1 2 2 3 3 4 4 |
| 94 | .9731 | .9736 | .9741 | .9745 | .9750 | .9754 | .9759 | .9763 | .9768 | .9773 | 0 1 1 2 2 3 3 4 4 |
| 95 | .9777 | .9782 | .9786 | .9791 | .9795 | .9800 | .9805 | .9809 | .9814 | .9818 | 0 1 1 2 2 3 3 4 4 |
| 96 | .9823 | .9827 | .9832 | .9836 | .9841 | .9845 | .9850 | .9854 | .9859 | .9863 | 0 1 1 2 2 3 3 4 4 |
| 97 | .9868 | .9872 | .9877 | .9881 | .9886 | .9890 | .9894 | .9899 | .9903 | .9908 | 0 1 1 2 2 3 3 4 4 |
| 98 | .9912 | .9917 | .9921 | .9926 | .9930 | .9934 | .9939 | .9943 | .9948 | .9952 | 0 1 1 2 2 3 3 4 4 |
| 99 | .9956 | .9961 | .9965 | .9969 | .9974 | .9978 | .9983 | .9987 | .9991 | .9996 | 0 1 1 2 2 3 3 3 4 |

## XXIII.—*Logarithms of Sines and Tangents.*

| | 0° | | | | 1° | | | | |
|---|---|---|---|---|---|---|---|---|---|
| | Sin. | Cos. | Tan. | Cot. | Sin. | Cos. | Tan. | Cot. | |
| 0′ | | 0.0000 | | | 8.2419 | 9.9999 | 8.2419 | 1.7581 | 60′ |
| 1 | 6.4637 | .0000 | 6.4637 | 3.5363 | .2490 | .9999 | .2491 | .7509 | 59 |
| 2 | .7648 | .0000 | .7648 | .2352 | .2561 | .9999 | .2562 | .7438 | 58 |
| 3 | 6.9408 | .0000 | 6.9408 | 3.0592 | .2630 | .9999 | .2631 | .7369 | 57 |
| 4 | 7.0658 | .0000 | 7.0658 | 2.9342 | .2699 | .9999 | .2700 | .7300 | 56 |
| 5 | .1627 | .0000 | .1627 | .8373 | .2766 | .9999 | .2767 | .7233 | 55 |
| 6 | .2419 | .0000 | .2419 | .7581 | .2832 | .9999 | .2833 | .7167 | 54 |
| 7 | .3088 | .0000 | .3088 | .6912 | .2898 | .9999 | .2899 | .7101 | 53 |
| 8 | .3668 | .0000 | .3668 | .6332 | .2962 | .9999 | .2963 | .7037 | 52 |
| 9 | .4180 | .0000 | .4180 | .5820 | .3025 | .9999 | .3026 | .6974 | 51 |
| 10 | .4637 | .0000 | .4637 | .5363 | .3088 | .9999 | .3089 | .6911 | 50 |
| 11 | .5051 | .0000 | .5051 | .4949 | .3150 | .9999 | .3150 | .6850 | 49 |
| 12 | .5429 | .0000 | .5429 | .4571 | .3210 | .9999 | .3211 | .6789 | 48 |
| 13 | .5777 | .0000 | .5777 | .4223 | .3270 | .9999 | .3271 | .6729 | 47 |
| 14 | .6099 | .0000 | .6099 | .3901 | .3329 | .9999 | .3330 | .6670 | 46 |
| 15 | .6398 | .0000 | .6398 | .3602 | .3388 | .9999 | .3389 | .6611 | 45 |
| 16 | .6678 | .0000 | .6678 | .3322 | .3445 | .9999 | .3446 | .6554 | 44 |
| 17 | .6942 | .0000 | .6942 | .3058 | .3502 | .9999 | .3503 | .6497 | 43 |
| 18 | .7190 | .0000 | .7190 | .2810 | .3558 | .9999 | .3559 | .6441 | 42 |
| 19 | .7425 | .0000 | .7425 | .2575 | .3613 | .9999 | .3614 | .6386 | 41 |
| 20 | .7648 | .0000 | .7648 | .2352 | .3668 | .9999 | .3669 | .6331 | 40 |
| 21 | .7859 | .0000 | .7860 | .2140 | .3722 | .9999 | .3723 | .6277 | 39 |
| 22 | .8061 | .0000 | .8062 | .1938 | .3775 | .9999 | .3776 | .6224 | 38 |
| 23 | .8255 | .0000 | .8255 | .1745 | .3828 | .9999 | .3829 | .6171 | 37 |
| 24 | .8439 | .0000 | .8439 | .1561 | .3880 | .9999 | .3881 | .6119 | 36 |
| 25 | .8617 | .0000 | .8617 | .1383 | .3931 | .9999 | .3932 | .6068 | 35 |
| 26 | .8787 | .0000 | .8787 | .1213 | .3982 | .9999 | .3983 | .6017 | 34 |
| 27 | .8951 | .0000 | .8951 | .1049 | .4032 | .9999 | .4033 | .5967 | 33 |
| 28 | .9109 | .0000 | .9109 | .0891 | .4082 | .9999 | .4083 | .5917 | 32 |
| 29 | .9261 | .0000 | .9261 | .0739 | .4131 | .9999 | .4132 | .5868 | 31 |
| 30 | .9408 | .0000 | .9409 | .0591 | .4179 | .9999 | .4181 | .5819 | 30 |
| 31 | .9551 | .0000 | .9551 | .0449 | .4227 | .9998 | .4229 | .5771 | 29 |
| 32 | .9689 | .0000 | .9689 | .0311 | .4275 | .9998 | .4276 | .5724 | 28 |
| 33 | .9822 | .0000 | .9823 | .0177 | .4322 | .9998 | .4323 | .5677 | 27 |
| 34 | 7.9952 | .0000 | 7.9952 | 2.0048 | .4368 | .9998 | .4370 | .5630 | 26 |
| 35 | 8.0078 | .0000 | 8.0078 | 1.9922 | .4414 | .9998 | .4416 | .5584 | 25 |
| 36 | .0200 | .0000 | .0200 | .9800 | .4459 | .9998 | .4461 | .5539 | 24 |
| 37 | .0319 | .0000 | .0319 | .9681 | .4504 | .9998 | .4506 | .5494 | 23 |
| 38 | .0435 | .0000 | .0435 | .9565 | .4549 | .9998 | .4551 | .5449 | 22 |
| 39 | .0548 | .0000 | .0548 | .9452 | .4593 | .9998 | .4595 | .5405 | 21 |
| 40 | .0658 | .0000 | .0658 | .9342 | .4637 | .9998 | .4638 | .5362 | 20 |
| 41 | .0765 | .0000 | .0765 | .9235 | .4680 | .9998 | .4682 | .5318 | 19 |
| 42 | .0870 | .0000 | .0870 | .9130 | .4723 | .9998 | .4725 | .5275 | 18 |
| 43 | .0972 | .0000 | .0972 | .9028 | .4765 | .9998 | .4767 | .5233 | 17 |
| 44 | .1072 | .0000 | .1072 | .8928 | .4807 | .9998 | .4809 | .5191 | 16 |
| 45 | .1169 | .0000 | .1170 | .8830 | .4848 | .9998 | .4851 | .5149 | 15 |
| 46 | .1265 | .0000 | .1265 | .8735 | .4890 | .9998 | .4892 | .5108 | 14 |
| 47 | .1358 | .0000 | .1359 | .8641 | .4930 | .9998 | .4933 | .5067 | 13 |
| 48 | .1450 | .0000 | .1450 | .8550 | .4971 | .9998 | .4973 | .5027 | 12 |
| 49 | .1539 | .0000 | .1540 | .8460 | .5011 | .9998 | .5013 | .4987 | 11 |
| 50 | .1627 | .0000 | .1627 | .8373 | .5050 | .9998 | .5053 | .4947 | 10 |
| 51 | .1713 | .0000 | .1713 | .8287 | .5090 | .9998 | .5092 | .4908 | 9 |
| 52 | .1797 | .0000 | .1798 | .8202 | .5129 | .9998 | .5131 | .4869 | 8 |
| 53 | .1880 | 9.9999 | .1880 | .8120 | .5167 | .9998 | .5170 | .4830 | 7 |
| 54 | .1961 | .9999 | .1962 | .8038 | .5206 | .9998 | .5208 | .4792 | 6 |
| 55 | .2041 | .9999 | .2041 | .7959 | .5243 | .9998 | .5246 | .4754 | 5 |
| 56 | .2119 | .9999 | .2120 | .7880 | .5281 | .9998 | .5283 | .4717 | 4 |
| 57 | .2196 | .9999 | .2196 | .7804 | .5318 | .9997 | .5321 | .4679 | 3 |
| 58 | .2271 | .9999 | .2272 | .7728 | .5355 | .9997 | .5358 | .4642 | 2 |
| 59 | .2346 | .9999 | .2346 | .7654 | .5392 | .9997 | .5394 | .4606 | 1 |
| 60 | 8.2419 | 9.9999 | 8.2419 | 1.7581 | 8.5428 | 9.9997 | 8.5431 | 1.4569 | 0 |
| | Cos. | Sin. | Cot. | Tan. | Cos. | Sin. | Cot. | Tan. | |
| | | 89° | | | | 88° | | | |

## XXIII.—*Logarithms of Sines and Tangents*—Continued.

| | 2° | | | | 3° | | | | 4° | | | | |
|---|---|---|---|---|---|---|---|---|---|---|---|---|---|
| | Sin. | Cos. | Tan. | Cot. | Sin. | Cos. | Tan. | Cot. | Sin. | Cos. | Tan. | Cot. | |
| 0' | 8.5428 | 9.9997 | 8.5431 | 1.4569 | 8.7188 | 9.9994 | 8.7194 | 1.2806 | 8.8436 | 9.9989 | 8.8446 | 1.1554 | 60' |
| 1 | .5464 | .9997 | .5467 | .4533 | .7212 | .9994 | .7218 | .2782 | .8454 | .9989 | .8465 | .1535 | 59 |
| 2 | .5500 | .9997 | .5503 | .4497 | .7236 | .9994 | .7242 | .2758 | .8472 | .9989 | .8483 | .1517 | 58 |
| 3 | .5535 | .9997 | .5538 | .4462 | .7260 | .9994 | .7266 | .2734 | .8490 | .9989 | .8501 | .1499 | 57 |
| 4 | .5571 | .9997 | .5573 | .4427 | .7283 | .9994 | .7290 | .2710 | .8508 | .9989 | .8518 | .1482 | 56 |
| 5 | .5605 | .9997 | .5608 | .4392 | .7307 | .9994 | .7313 | .2687 | .8525 | .9989 | .8536 | .1464 | 55 |
| 6 | .5640 | .9997 | .5643 | .4357 | .7330 | .9994 | .7337 | .2663 | .8543 | .9989 | .8554 | .1446 | 54 |
| 7 | .5674 | .9997 | .5677 | .4323 | .7354 | .9994 | .7360 | .2640 | .8560 | .9989 | .8572 | .1428 | 53 |
| 8 | .5708 | .9997 | .5711 | .4289 | .7377 | .9994 | .7383 | .2617 | .8578 | .9989 | .8589 | .1411 | 52 |
| 9 | .5742 | .9997 | .5745 | .4255 | .7400 | .9993 | .7406 | .2594 | .8595 | .9989 | .8607 | .1393 | 51 |
| 10 | .5776 | .9997 | .5779 | .4221 | .7423 | .9993 | .7429 | .2571 | .8613 | .9989 | .8624 | .1376 | 50 |
| 11 | .5809 | .9997 | .5812 | .4188 | .7445 | .9993 | .7452 | .2548 | .8630 | .9988 | .8642 | .1358 | 49 |
| 12 | .5842 | .9997 | .5845 | .4155 | .7468 | .9993 | .7475 | .2525 | .8647 | .9988 | .8659 | .1341 | 48 |
| 13 | .5875 | .9997 | .5878 | .4122 | .7491 | .9993 | .7497 | .2503 | .8665 | .9988 | .8676 | .1324 | 47 |
| 14 | .5907 | .9997 | .5911 | .4089 | .7513 | .9993 | .7520 | .2480 | .8682 | .9988 | .8694 | .1306 | 46 |
| 15 | .5939 | .9997 | .5943 | .4057 | .7535 | .9993 | .7542 | .2458 | .8699 | .9988 | .8711 | .1289 | 45 |
| 16 | .5972 | .9997 | .5975 | .4025 | .7557 | .9993 | .7565 | .2435 | .8716 | .9988 | .8728 | .1272 | 44 |
| 17 | .6003 | .9997 | .6007 | .3993 | .7580 | .9993 | .7587 | .2413 | .8733 | .9988 | .8745 | .1255 | 43 |
| 18 | .6035 | .9996 | .6038 | .3962 | .7602 | .9993 | .7609 | .2391 | .8749 | .9988 | .8762 | .1238 | 42 |
| 19 | .6066 | .9996 | .6070 | .3930 | .7623 | .9993 | .7631 | .2369 | .8766 | .9988 | .8778 | .1222 | 41 |
| 20 | .6097 | .9996 | .6101 | .3899 | .7645 | .9993 | .7652 | .2348 | .8783 | .9988 | .8795 | .1205 | 40 |
| 21 | .6128 | .9996 | .6132 | .3868 | .7667 | .9993 | .7674 | .2326 | .8799 | .9987 | .8812 | .1188 | 39 |
| 22 | .6159 | .9996 | .6163 | .3837 | .7688 | .9992 | .7696 | .2304 | .8816 | .9987 | .8829 | .1171 | 38 |
| 23 | .6189 | .9996 | .6193 | .3807 | .7710 | .9992 | .7717 | .2283 | .8833 | .9987 | .8845 | .1155 | 37 |
| 24 | .6220 | .9996 | .6223 | .3777 | .7731 | .9992 | .7739 | .2261 | .8849 | .9987 | .8862 | .1138 | 36 |
| 25 | .6250 | .9996 | .6254 | .3746 | .7752 | .9992 | .7760 | .2240 | .8865 | .9987 | .8878 | .1122 | 35 |
| 26 | .6279 | .9996 | .6283 | .3717 | .7773 | .9992 | .7781 | .2219 | .8882 | .9987 | .8895 | .1105 | 34 |
| 27 | .6309 | .9996 | .6313 | .3687 | .7794 | .9992 | .7802 | .2198 | .8898 | .9987 | .8911 | .1089 | 33 |
| 28 | .6339 | .9996 | .6343 | .3657 | .7815 | .9992 | .7823 | .2177 | .8914 | .9987 | .8927 | .1073 | 32 |
| 29 | .6368 | .9996 | .6372 | .3628 | .7836 | .9992 | .7844 | .2156 | .8930 | .9987 | .8944 | .1056 | 31 |
| 30 | .6397 | .9996 | .6401 | .3599 | .7857 | .9992 | .7865 | .2135 | .8946 | .9987 | .8960 | .1040 | 30 |
| 31 | .6426 | .9996 | .6430 | .3570 | .7877 | .9992 | .7886 | .2114 | .8962 | .9986 | .8976 | .1024 | 29 |
| 32 | .6454 | .9996 | .6459 | .3541 | .7898 | .9992 | .7906 | .2094 | .8978 | .9986 | .8992 | .1008 | 28 |
| 33 | .6483 | .9996 | .6487 | .3513 | .7918 | .9992 | .7927 | .2073 | .8994 | .9986 | .9008 | .0992 | 27 |
| 34 | .6511 | .9996 | .6515 | .3485 | .7939 | .9992 | .7947 | .2053 | .9010 | .9986 | .9024 | .0976 | 26 |
| 35 | .6539 | .9996 | .6544 | .3456 | .7959 | .9992 | .7967 | .2033 | .9026 | .9986 | .9040 | .0960 | 25 |
| 36 | .6567 | .9996 | .6571 | .3429 | .7979 | .9991 | .7988 | .2012 | .9342 | .9986 | .9056 | .0944 | 24 |
| 37 | .6595 | .9995 | .6599 | .3401 | .7999 | .9991 | .8008 | .1992 | .9057 | .9986 | .9071 | .0929 | 23 |
| 38 | .6622 | .9995 | .6627 | .3373 | .8019 | .9991 | .8028 | .1972 | .9073 | .9986 | .9087 | .0913 | 22 |
| 39 | .6650 | .9995 | .6654 | .3346 | .8039 | .9991 | .8048 | .1952 | .9089 | .9986 | .9103 | .0897 | 21 |
| 40 | .6677 | .9995 | .6682 | .3318 | .8059 | .9991 | .8067 | .1933 | .9104 | .9986 | .9118 | .0882 | 20 |
| 41 | .6704 | .9995 | .6709 | .3291 | .8078 | .9991 | .8087 | .1913 | .9119 | .9985 | .9134 | .0866 | 19 |
| 42 | .6731 | .9995 | .6736 | .3264 | .8098 | .9991 | .8107 | .1893 | .9135 | .9985 | .9150 | .0850 | 18 |
| 43 | .6758 | .9995 | .6762 | .3238 | .8117 | .9991 | .8126 | .1874 | .9150 | .9985 | .9165 | .0835 | 17 |
| 44 | .6784 | .9995 | .6789 | .3211 | .8137 | .9991 | .8146 | .1854 | .9166 | .9985 | .9180 | .0820 | 16 |
| 45 | .6810 | .9995 | .6815 | .3185 | .8156 | .9991 | .8165 | .1835 | .9181 | .9985 | .9196 | .0804 | 15 |
| 46 | .6837 | .9995 | .6842 | .3158 | .8175 | .9991 | .8185 | .1815 | .9196 | .9985 | .9211 | .0789 | 14 |
| 47 | .6863 | .9995 | .6868 | .3132 | .8194 | .9991 | .8204 | .1796 | .9211 | .9985 | .9226 | .0774 | 13 |
| 48 | .6889 | .9995 | .6894 | .3106 | .8213 | .9990 | .8223 | .1777 | .9226 | .9985 | .9241 | .0759 | 12 |
| 49 | .6914 | .9995 | .6920 | .3080 | .8232 | .9990 | .8242 | .1758 | .9241 | .9985 | .9256 | .0744 | 11 |
| 50 | .6940 | .9995 | .6945 | .3055 | .8251 | .9990 | .8261 | .1739 | .9256 | .9985 | .9272 | .0728 | 10 |
| 51 | .6965 | .9995 | .6971 | .3029 | .8270 | .9990 | .8280 | .1720 | .9271 | .9984 | .9287 | .0713 | 9 |
| 52 | .6991 | .9995 | .6996 | .3004 | .8289 | .9990 | .8299 | .1701 | .9286 | .9984 | .9302 | .0698 | 8 |
| 53 | .7016 | .9994 | .7021 | .2979 | .8307 | .9990 | .8317 | .1683 | .9301 | .9984 | .9316 | .0684 | 7 |
| 54 | .7041 | .9994 | .7046 | .2954 | .8326 | .9990 | .8336 | .1664 | .9315 | .9984 | .9331 | .0669 | 6 |
| 55 | .7066 | .9994 | .7071 | .2929 | .8345 | .9990 | .8355 | .1645 | .9330 | .9984 | .9346 | .0654 | 5 |
| 56 | .7090 | .9994 | .7096 | .2904 | .8363 | .9990 | .8373 | .1627 | .9345 | .9984 | .9361 | .0639 | 4 |
| 57 | .7115 | .9994 | .7121 | .2879 | .8381 | .9990 | .8391 | .1608 | .9359 | .9984 | .9376 | .0624 | 3 |
| 58 | .7140 | .9994 | .7145 | .2855 | .8400 | .9990 | .8410 | .1590 | .9374 | .9984 | .9390 | .0610 | 2 |
| 59 | .7164 | .9994 | .7170 | .2830 | .8418 | .9989 | .8428 | .1572 | .9388 | .9983 | .9405 | .0595 | 1 |
| 60 | 8.7188 | 9.9994 | 8.7194 | 1.2806 | 8.8436 | 9.9989 | 8.8446 | 1.1554 | 8.9403 | 9.9983 | 8.9420 | 1.0580 | 0 |
| | Cos. | Sin. | Cot. | Tan. | Cos. | Sin. | Cot. | Tan. | Cos. | Sin. | Cot. | Tan. | |
| | | 87° | | | | 86° | | | | 85° | | | |

## XXIII.—*Logarithms of Sines and Tangents*—Continued.

| Arc | Sin. | Df. | Cos. | Df. | Tan. | Df. | Cot. | Arc | Arc | Sin. | Df. | Cos. | Df. | Tan. | Df. | Cot. | Arc |
|---|---|---|---|---|---|---|---|---|---|---|---|---|---|---|---|---|---|
| °  ′ | | | | | | | | °  ′ | °  ′ | | | | | | | | °  ′ |
| 5 0 | 8.9403 | 142 | 9.9983 | 1 | 8.9420 | 143 | 1.0580 | 85 0 | 15 0 | 9.4130 | 47 | 9.9849 | 3 | 9.4281 | 50 | 0.5719 | 75 0 |
| 10 | .9545 | 137 | .9982 | 1 | .9563 | 138 | .0437 | 50 | 10 | .4177 | 46 | .9846 | 3 | .4331 | 50 | .5669 | 50 |
| 20 | .9682 | 134 | .9981 | 1 | .9701 | 135 | .0299 | 40 | 20 | .4223 | 46 | .9843 | 4 | .4381 | 49 | .5619 | 40 |
| 30 | 8.9816 | 129 | .9980 | 1 | .9836 | 130 | .0164 | 30 | 30 | .4269 | 45 | .9839 | 3 | .4430 | 49 | .5570 | 30 |
| 40 | 8.9945 | 125 | .9979 | 2 | 8.9966 | 127 | 1.0034 | 20 | 40 | .4314 | 45 | .9836 | 4 | .4479 | 48 | .5521 | 20 |
| 50 | 9.0070 | 122 | .9977 | 1 | 9.0093 | 123 | 0.9907 | 10 | 50 | .4359 | 44 | .9832 | 4 | .4527 | 48 | .5473 | 10 |
| 6 0 | .0192 | 119 | .9976 | 1 | .0216 | 120 | .9784 | 84 0 | 16 0 | .4403 | 44 | .9828 | 3 | .4575 | 47 | .5425 | 74 0 |
| 10 | .0311 | 115 | .9975 | 2 | .0336 | 117 | .9664 | 50 | 10 | .4447 | 44 | .9825 | 4 | .4622 | 47 | .5378 | 50 |
| 20 | .0426 | 113 | .9973 | 1 | .0453 | 114 | .9547 | 40 | 20 | .4491 | 42 | .9821 | 4 | .4669 | 47 | .5331 | 40 |
| 30 | .0539 | 109 | .9972 | 1 | .0567 | 111 | .9433 | 30 | 30 | .4533 | 43 | .9817 | 3 | .4716 | 46 | .5284 | 30 |
| 40 | .0648 | 107 | .9971 | 2 | .0678 | 108 | .9322 | 20 | 40 | .4576 | 42 | .9814 | 4 | .4762 | 46 | .5238 | 20 |
| 50 | .0755 | 104 | .9969 | 1 | .0786 | 105 | .9214 | 10 | 50 | .4618 | 41 | .9810 | 4 | .4808 | 45 | .5192 | 10 |
| 7 0 | .0859 | 102 | .9968 | 2 | .0891 | 104 | .9109 | 83 0 | 17 0 | .4659 | 41 | .9806 | 4 | .4853 | 45 | .5147 | 73 0 |
| 10 | .0961 | 99 | .9966 | 2 | .0995 | 101 | .9005 | 50 | 10 | .4700 | 41 | .9802 | 4 | .4898 | 45 | .5102 | 50 |
| 20 | .1060 | 97 | .9964 | 1 | .1096 | 98 | .8904 | 40 | 20 | .4741 | 40 | .9798 | 4 | .4943 | 44 | .5057 | 40 |
| 30 | .1157 | 95 | .9963 | 2 | .1194 | 97 | .8806 | 30 | 30 | .4781 | 40 | .9794 | 4 | .4987 | 44 | .5013 | 30 |
| 40 | .1252 | 93 | .9961 | 2 | .1291 | 94 | .8709 | 20 | 40 | .4821 | 40 | .9790 | 4 | .5031 | 44 | .4969 | 20 |
| 50 | .1345 | 91 | .9959 | 1 | .1385 | 93 | .8615 | 10 | 50 | .4861 | 39 | .9786 | 4 | .5075 | 43 | .4925 | 10 |
| 8 0 | .1436 | 89 | .9958 | 2 | .1478 | 91 | .8522 | 82 0 | 18 0 | .4900 | 39 | .9782 | 4 | .5118 | 43 | .4882 | 72 0 |
| 10 | .1525 | 87 | .9956 | 2 | .1569 | 89 | .8431 | 50 | 10 | .4939 | 38 | .9778 | 4 | .5161 | 42 | .4839 | 50 |
| 20 | .1612 | 85 | .9954 | 2 | .1658 | 87 | .8342 | 40 | 20 | .4977 | 38 | .9774 | 4 | .5203 | 42 | .4797 | 40 |
| 30 | .1697 | 84 | .9952 | 2 | .1745 | 86 | .8255 | 30 | 30 | .5015 | 37 | .9770 | 5 | .5245 | 42 | .4755 | 30 |
| 40 | .1781 | 82 | .9950 | 2 | .1831 | 84 | .8169 | 20 | 40 | .5052 | 38 | .9765 | 4 | .5287 | 42 | .4713 | 20 |
| 50 | .1863 | 80 | .9948 | 2 | .1915 | 82 | .8085 | 10 | 50 | .5090 | 36 | .9761 | 4 | .5329 | 41 | .4671 | 10 |
| 9 0 | .1943 | 79 | .9946 | 2 | .1997 | 81 | .8003 | 81 0 | 19 0 | .5126 | 37 | .9757 | 5 | .5370 | 41 | .4630 | 71 0 |
| 10 | .2022 | 78 | .9944 | 2 | .2078 | 80 | .7922 | 50 | 10 | .5163 | 36 | .9752 | 4 | .5411 | 40 | .4589 | 50 |
| 20 | .2100 | 76 | .9942 | 2 | .2158 | 78 | .7842 | 40 | 20 | .5199 | 36 | .9748 | 5 | .5451 | 40 | .4549 | 40 |
| 30 | .2176 | 75 | .9940 | 2 | .2236 | 77 | .7764 | 30 | 30 | .5235 | 35 | .9743 | 4 | .5491 | 40 | .4509 | 30 |
| 40 | .2251 | 73 | .9938 | 2 | .2313 | 76 | .7687 | 20 | 40 | .5270 | 36 | .9739 | 5 | .5531 | 40 | .4469 | 20 |
| 50 | .2324 | 73 | .9936 | 2 | .2389 | 74 | .7611 | 10 | 50 | .5306 | 35 | .9734 | 4 | .5571 | 40 | .4429 | 10 |
| 10 0 | .2397 | 71 | .9934 | 3 | .2463 | 73 | .7537 | 80 0 | 20 0 | .5341 | 34 | .9730 | 5 | .5611 | 39 | .4389 | 70 0 |
| 10 | .2468 | 70 | .9931 | 2 | .2536 | 73 | .7464 | 50 | 10 | .5375 | 34 | .9725 | 5 | .5650 | 39 | .4350 | 50 |
| 20 | .2538 | 68 | .9929 | 2 | .2609 | 71 | .7391 | 40 | 20 | .5409 | 34 | .9721 | 5 | .5689 | 38 | .4311 | 40 |
| 30 | .2606 | 68 | .9927 | 3 | .2680 | 70 | .7320 | 30 | 30 | .5443 | 34 | .9716 | 5 | .5727 | 39 | .4273 | 30 |
| 40 | .2674 | 66 | .9924 | 2 | .2750 | 69 | .7250 | 20 | 40 | .5477 | 33 | .9711 | 5 | .5766 | 38 | .4234 | 20 |
| 50 | .2740 | 66 | .9922 | 3 | .2819 | 68 | .7181 | 10 | 50 | .5510 | 33 | .9706 | 4 | .5804 | 38 | .4196 | 10 |
| 11 0 | .2806 | 64 | .9919 | 2 | .2887 | 66 | .7113 | 79 0 | 21 0 | .5543 | 33 | .9702 | 5 | .5842 | 37 | .4158 | 69 0 |
| 10 | .2870 | 64 | .9917 | 3 | .2953 | 67 | .7047 | 50 | 10 | .5576 | 33 | .9697 | 5 | .5879 | 38 | .4121 | 50 |
| 20 | .2934 | 63 | .9914 | 2 | .3020 | 65 | .6980 | 40 | 20 | .5609 | 32 | .9692 | 5 | .5917 | 37 | .4083 | 40 |
| 30 | .2997 | 61 | .9912 | 3 | .3085 | 64 | .6915 | 30 | 30 | .5641 | 32 | .9687 | 5 | .5954 | 37 | .4046 | 30 |
| 40 | .3058 | 61 | .9909 | 2 | .3149 | 63 | .6851 | 20 | 40 | .5673 | 31 | .9682 | 5 | .5991 | 37 | .4009 | 20 |
| 50 | .3119 | 60 | .9907 | 3 | .3212 | 63 | .6788 | 10 | 50 | .5704 | 32 | .9677 | 5 | .6028 | 36 | .3972 | 10 |
| 12 0 | .3179 | 59 | .9904 | 3 | .3275 | 61 | .6725 | 78 0 | 22 0 | .5736 | 31 | .9672 | 5 | .6064 | 36 | .3936 | 68 0 |
| 10 | .3238 | 58 | .9901 | 3 | .3336 | 61 | .6664 | 50 | 10 | .5767 | 31 | .9667 | 6 | .6100 | 36 | .3900 | 50 |
| 20 | .3296 | 57 | .9899 | 3 | .3397 | 61 | .6603 | 40 | 20 | .5798 | 30 | .9661 | 5 | .6136 | 36 | .3864 | 40 |
| 30 | .3353 | 57 | .9896 | 3 | .3458 | 59 | .6542 | 30 | 30 | .5828 | 31 | .9656 | 5 | .6172 | 36 | .3828 | 30 |
| 40 | .3410 | 56 | .9893 | 3 | .3517 | 59 | .6483 | 20 | 40 | .5859 | 30 | .9651 | 5 | .6208 | 35 | .3792 | 20 |
| 50 | .3466 | 55 | .9890 | 3 | .3576 | 58 | .6424 | 10 | 50 | .5889 | 30 | .9646 | 6 | .6243 | 36 | .3757 | 10 |
| 13 0 | .3521 | 54 | .9887 | 3 | .3634 | 57 | .6366 | 77 0 | 23 0 | .5919 | 29 | .9640 | 5 | .6279 | 35 | .3721 | 67 0 |
| 10 | .3575 | 54 | .9884 | 3 | .3691 | 57 | .6309 | 50 | 10 | .5948 | 30 | .9635 | 6 | .6314 | 34 | .3686 | 50 |
| 20 | .3629 | 53 | .9881 | 3 | .3748 | 56 | .6252 | 40 | 20 | .5978 | 29 | .9629 | 5 | .6348 | 35 | .3652 | 40 |
| 30 | .3682 | 52 | .9878 | 3 | .3804 | 55 | .6196 | 30 | 30 | .6007 | 29 | .9624 | 6 | .6383 | 34 | .3617 | 30 |
| 40 | .3734 | 52 | .9875 | 3 | .3859 | 55 | .6141 | 20 | 40 | .6036 | 29 | .9618 | 5 | .6417 | 35 | .3583 | 20 |
| 50 | .3786 | 51 | .9872 | 3 | .3914 | 54 | .6086 | 10 | 50 | .6065 | 28 | .9613 | 6 | .6452 | 34 | .3548 | 10 |
| 14 0 | .3837 | 50 | .9869 | 3 | .3968 | 53 | .6032 | 76 0 | 24 0 | .6093 | 28 | .9607 | 5 | .6486 | 34 | .3514 | 66 0 |
| 10 | .3887 | 50 | .9866 | 3 | .4021 | 53 | .5979 | ·50 | 10 | .6121 | 28 | .9602 | 6 | .6520 | 33 | .3480 | 50 |
| 20 | .3937 | 49 | .9863 | 4 | .4074 | 53 | .5926 | 40 | 20 | .6149 | 28 | .9596 | 6 | .6553 | 34 | .3447 | 40 |
| 30 | .3986 | 49 | .9859 | 3 | .4127 | 52 | .5873 | 30 | 30 | .6177 | 28 | .9590 | 6 | .6587 | 33 | .3413 | 30 |
| 40 | .4035 | 48 | .9856 | 3 | .4178 | 51 | .5822 | 20 | 40 | .6205 | 27 | .9584 | 5 | .6620 | 34 | .3380 | 20 |
| 50 | .4083 | 47 | .9853 | 4 | .4230 | 51 | .5770 | 10 | 50 | .6232 | 27 | .9579 | 6 | .6654 | 33 | .3346 | 10 |
| 15 0 | 9.4130 | 47 | 9.9849 | 3 | 9.4281 | 50 | 0.5719 | 75 0 | 25 0 | 0.6259 | 27 | 9.9573 | 7 | 9.6687 | 33 | 0.3313 | 65 0 |
| Arc | Cos. | Df. | Sin. | Df. | Cot. | Df. | Tan. | Arc | Arc | Cos. | Df. | Sin. | Df. | Cot. | Df. | Tan. | Arc |

## XXIII.—*Logarithms of Sines and Tangents*—Continued.

| Arc | Sin. | Df. | Cos. | Df. | Tan. | Df. | Cot. | Arc | Arc | Sin. | Df. | Cos. | Df. | Tan. | Df. | Cot. | Arc |
|---|---|---|---|---|---|---|---|---|---|---|---|---|---|---|---|---|---|
| ° ′ | | | | | | | | ° ′ | ° ′ | | | | | | | | ° ′ |
| 25 0 | 9.6259 | 27 | 9.9573 | 6 | 9.6687 | 33 | 0.3313 | 65 0 | 35 0 | 9.7586 | 18 | 9.9134 | 9 | 9.8452 | 27 | 0.1548 | 55 0 |
| 10 | .6286 | 27 | .9567 | 6 | .6720 | 32 | .3280 | 50 | 10 | .7604 | 18 | .9125 | 9 | .8479 | 27 | .1521 | 50 |
| 20 | .6313 | 27 | .9561 | 6 | .6752 | 33 | .3248 | 40 | 20 | .7622 | 18 | .9116 | 9 | .8506 | 27 | .1494 | 40 |
| 30 | .6340 | 26 | .9555 | 6 | .6785 | 32 | .3215 | 30 | 30 | .7640 | 17 | .9107 | 9 | .8533 | 26 | .1467 | 30 |
| 40 | .6366 | 26 | .9549 | 6 | .6817 | 33 | .3183 | 20 | 40 | .7657 | 18 | .9098 | 9 | .8559 | 27 | .1441 | 20 |
| 50 | .6392 | 26 | .9543 | 6 | .6850 | 32 | .3150 | 10 | 50 | .7675 | 17 | .9089 | 9 | .8586 | 27 | .1414 | 10 |
| 26 0 | .6418 | 26 | .9537 | 7 | .6882 | 32 | .3118 | 64 0 | 36 0 | .7692 | 18 | .9080 | 10 | .8613 | 26 | .1387 | 54 0 |
| 10 | .6444 | 26 | .9530 | 6 | .6914 | 32 | .3086 | 50 | 10 | .7710 | 17 | .9070 | 9 | .8639 | 27 | .1361 | 50 |
| 20 | .6470 | 25 | .9524 | 6 | .6946 | 31 | .3054 | 40 | 20 | .7727 | 17 | .9061 | 9 | .8666 | 26 | .1334 | 40 |
| 30 | .6495 | 26 | .9518 | 6 | .6977 | 32 | .3023 | 30 | 30 | .7744 | 17 | .9052 | 10 | .8692 | 26 | .1308 | 30 |
| 40 | .6521 | 25 | .9512 | 7 | .7009 | 31 | .2991 | 20 | 40 | .7761 | 17 | .9042 | 9 | .8718 | 27 | .1282 | 20 |
| 50 | .6546 | 24 | .9505 | 6 | .7040 | 32 | .2960 | 10 | 50 | .7778 | 17 | .9033 | 10 | .8745 | 26 | .1255 | 10 |
| 27 0 | .6570 | 25 | .9499 | 7 | .7072 | 31 | .2928 | 63 0 | 37 0 | .7795 | 16 | .9023 | 9 | .8771 | 26 | .1229 | 53 0 |
| 10 | .6595 | 25 | .9492 | 6 | .7103 | 31 | .2897 | 50 | 10 | .7811 | 17 | .9014 | 10 | .8797 | 27 | .1203 | 50 |
| 20 | .6620 | 24 | .9486 | 7 | .7134 | 31 | .2866 | 40 | 20 | .7828 | 16 | .9004 | 9 | .8824 | 26 | .1176 | 40 |
| 30 | .6644 | 24 | .9479 | 6 | .7165 | 31 | .2835 | 30 | 30 | .7844 | 17 | .8995 | 10 | .8850 | 26 | .1150 | 30 |
| 40 | .6668 | 24 | .9473 | 7 | .7196 | 30 | .2804 | 20 | 40 | .7861 | 16 | .8985 | 10 | .8876 | 26 | .1124 | 20 |
| 50 | .6692 | 24 | .9466 | 7 | .7226 | 31 | .2774 | 10 | 50 | .7877 | 16 | .8975 | 10 | .8902 | 26 | .1098 | 10 |
| 28 0 | .6716 | 24 | .9459 | 6 | .7257 | 30 | .2743 | 62 0 | 38 0 | .7893 | 17 | .8965 | 10 | .8928 | 26 | .1072 | 52 0 |
| 10 | .6740 | 23 | .9453 | 7 | .7287 | 30 | .2713 | 50 | 10 | .7910 | 16 | .8955 | 10 | .8954 | 26 | .1046 | 50 |
| 20 | .6763 | 24 | .9446 | 7 | .7317 | 31 | .2683 | 40 | 20 | .7926 | 15 | .8945 | 10 | .8980 | 26 | .1020 | 40 |
| 30 | .6787 | 23 | .9439 | 7 | .7348 | 30 | .2652 | 30 | 30 | .7941 | 16 | .8935 | 10 | .9006 | 26 | .0994 | 30 |
| 40 | .6810 | 23 | .9432 | 7 | .7378 | 30 | .2622 | 20 | 40 | .7957 | 16 | .8925 | 10 | .9032 | 26 | .0968 | 20 |
| 50 | .6833 | 23 | .9425 | 7 | .7408 | 30 | .2592 | 10 | 50 | .7973 | 16 | .8915 | 10 | .9058 | 26 | .0942 | 10 |
| 29 0 | .6856 | 22 | .9418 | 7 | .7438 | 29 | .2562 | 61 0 | 39 0 | .7989 | 15 | .8905 | 10 | .9084 | 26 | .0916 | 51 0 |
| 10 | .6878 | 23 | .9411 | 7 | .7467 | 30 | .2533 | 50 | 10 | .8004 | 16 | .8895 | 11 | .9110 | 25 | .0890 | 50 |
| 20 | .6901 | 22 | .9404 | 7 | .7497 | 29 | .2503 | 40 | 20 | .8020 | 15 | .8884 | 10 | .9135 | 26 | .0865 | 40 |
| 30 | .6923 | 23 | .9397 | 7 | .7526 | 30 | .2474 | 30 | 30 | .8035 | 15 | .8874 | 10 | .9161 | 26 | .0839 | 30 |
| 40 | .6946 | 22 | .9390 | 7 | .7556 | 29 | .2444 | 20 | 40 | .8050 | 16 | .8864 | 11 | .9187 | 25 | .0813 | 20 |
| 50 | .6968 | 22 | .9383 | 8 | .7585 | 29 | .2415 | 10 | 50 | .8066 | 15 | .8853 | 10 | .9212 | 26 | .0788 | 10 |
| 30 0 | .6990 | 22 | .9375 | 7 | .7614 | 30 | .2386 | 60 0 | 40 0 | .8081 | 15 | .8843 | 11 | .9238 | 26 | .0762 | 50 0 |
| 10 | .7012 | 21 | .9368 | 7 | .7644 | 29 | .2356 | 50 | 10 | .8096 | 15 | .8832 | 11 | .9264 | 25 | .0736 | 50 |
| 20 | .7033 | 22 | .9361 | 8 | .7673 | 28 | .2327 | 40 | 20 | .8111 | 14 | .8821 | 11 | .9289 | 26 | .0711 | 40 |
| 30 | .7055 | 21 | .9353 | 7 | .7701 | 29 | .2299 | 30 | 30 | .8125 | 15 | .8810 | 10 | .9315 | 26 | .0685 | 30 |
| 40 | .7076 | 21 | .9346 | 8 | .7730 | 29 | .2270 | 20 | 40 | .8140 | 15 | .8800 | 11 | .9341 | 25 | .0659 | 20 |
| 50 | .7097 | 22 | .9338 | 7 | .7759 | 29 | .2241 | 10 | 50 | .8155 | 14 | .8789 | 11 | .9366 | 26 | .0634 | 10 |
| 31 0 | .7118 | 21 | .9331 | 8 | .7788 | 28 | .2212 | 59 0 | 41 0 | .8169 | 15 | .8778 | 11 | .9392 | 25 | .0608 | 49 0 |
| 10 | .7139 | 21 | .9323 | 8 | .7816 | 29 | .2184 | 50 | 10 | .8184 | 14 | .8767 | 11 | .9417 | 26 | .0583 | 50 |
| 20 | .7160 | 21 | .9315 | 7 | .7845 | 28 | .2155 | 40 | 20 | .8198 | 15 | .8756 | 11 | .9443 | 25 | .0557 | 40 |
| 30 | .7181 | 20 | .9308 | 8 | .7873 | 29 | .2127 | 30 | 30 | .8213 | 14 | .8745 | 12 | .9468 | 26 | .0532 | 30 |
| 40 | .7201 | 21 | .9300 | 8 | .7902 | 28 | .2098 | 20 | 40 | .8227 | 14 | .8733 | 11 | .9494 | 25 | .0506 | 20 |
| 50 | .7222 | 20 | .9292 | 8 | .7930 | 28 | .2070 | 10 | 50 | .8241 | 14 | .8722 | 11 | .9519 | 25 | .0481 | 10 |
| 32 0 | .7242 | 20 | .9284 | 8 | .7958 | 28 | .2042 | 58 0 | 42 0 | .8255 | 14 | .8711 | 12 | .9544 | 26 | .0456 | 48 0 |
| 10 | .7262 | 20 | .9276 | 8 | .7986 | 28 | .2014 | 50 | 10 | .8269 | 14 | .8699 | 11 | .9570 | 25 | .0430 | 50 |
| 20 | .7282 | 20 | .9268 | 8 | .8014 | 28 | .1986 | 40 | 20 | .8283 | 14 | .8688 | 12 | .9595 | 26 | .0405 | 40 |
| 30 | .7302 | 20 | .9260 | 8 | .8042 | 28 | .1958 | 30 | 30 | .8297 | 14 | .8676 | 11 | .9621 | 25 | .0379 | 30 |
| 40 | .7322 | 20 | .9252 | 8 | .8070 | 27 | .1930 | 20 | 40 | .8311 | 13 | .8665 | 12 | .9646 | 25 | .0354 | 20 |
| 50 | .7342 | 19 | .9244 | 8 | .8097 | 28 | .1903 | 10 | 50 | .8324 | 14 | .8653 | 12 | .9671 | 26 | .0329 | 10 |
| 33 0 | .7361 | 19 | .9236 | 8 | .8125 | 28 | .1875 | 57 0 | 43 0 | .8338 | 13 | .8641 | 12 | .9697 | 25 | .0303 | 47 0 |
| 10 | .7380 | 20 | .9228 | 9 | .8153 | 27 | .1847 | 50 | 10 | .8351 | 14 | .8629 | 11 | .9722 | 25 | .0278 | 50 |
| 20 | .7400 | 19 | .9219 | 8 | .8180 | 28 | .1820 | 40 | 20 | .8365 | 13 | .8618 | 12 | .9747 | 25 | .0253 | 40 |
| 30 | .7419 | 19 | .9211 | 8 | .8208 | 27 | .1792 | 30 | 30 | .8378 | 13 | .8606 | 12 | .9772 | 26 | .0228 | 30 |
| 40 | .7438 | 19 | .9203 | 9 | .8235 | 28 | .1765 | 20 | 40 | .8391 | 13 | .8594 | 12 | .9798 | 25 | .0202 | 20 |
| 50 | .7457 | 19 | .9194 | 8 | .8263 | 27 | .1737 | 10 | 50 | .8405 | 13 | .8582 | 13 | .9823 | 25 | .0177 | 10 |
| 34 0 | .7476 | 18 | .9186 | 9 | .8290 | 27 | .1710 | 56 0 | 44 0 | .8418 | 13 | .8569 | 12 | .9848 | 26 | .0152 | 46 0 |
| 10 | .7494 | 19 | .9177 | 8 | .8317 | 27 | .1683 | 50 | 10 | .8431 | 13 | .8557 | 12 | .9874 | 25 | .0126 | 50 |
| 20 | .7513 | 18 | .9169 | 9 | .8344 | 27 | .1656 | 40 | 20 | .8444 | 13 | .8545 | 13 | .9899 | 25 | .0101 | 40 |
| 30 | .7531 | 19 | .9160 | 9 | .8371 | 27 | .1629 | 30 | 30 | .8457 | 12 | .8532 | 12 | .9924 | 25 | .0076 | 30 |
| 40 | .7550 | 18 | .9151 | 9 | .8398 | 27 | .1602 | 20 | 40 | .8469 | 13 | .8520 | 13 | .9949 | 26 | .0051 | 20 |
| 50 | .7568 | 18 | .9142 | 8 | .8425 | 27 | .1575 | 10 | 50 | .8482 | 13 | .8507 | 12 | .9975 | 25 | .0025 | 10 |
| 35 0 | 9.7586 | 18 | 9.9134 | 9 | 9.8452 | 27 | 0.1548 | 55 0 | 45 0 | 9.8495 | | 9.8495 | | 0.0000 | | 0.0000 | 45 0 |
| Arc | Cos. | Df. | Sin. | Df. | Cot. | Df. | Tan. | Arc | Arc | Cos. | Df. | Sin. | Df. | Cot. | Df. | Tan. | Arc |

## XXIV.—*Squares and Square Roots.*

| No. | Square. | Square root. | No. | Square. | Square root. |
|---|---|---|---|---|---|
| 1 | 1 | 1.000 | 51 | 2601 | 7.141 |
| 2 | 4 | 1.414 | 52 | 2704 | 7.211 |
| 3 | 9 | 1.732 | 53 | 2809 | 7.280 |
| 4 | 16 | 2.000 | 54 | 2916 | 7.348 |
| 5 | 25 | 2.236 | 55 | 3025 | 7.416 |
| 6 | 36 | 2.449 | 56 | 3136 | 7.483 |
| 7 | 49 | 2.646 | 57 | 3249 | 7.550 |
| 8 | 64 | 2.828 | 58 | 3364 | 7.616 |
| 9 | 81 | 3.000 | 59 | 3481 | 7.681 |
| 10 | 100 | 3.162 | 60 | 3600 | 7.746 |
| 11 | 121 | 3.317 | 61 | 3721 | 7.810 |
| 12 | 144 | 3.464 | 62 | 3844 | 7.874 |
| 13 | 169 | 3.606 | 63 | 3969 | 7.937 |
| 14 | 196 | 3.742 | 64 | 4096 | 8.000 |
| 15 | 225 | 3.873 | 65 | 4225 | 8.062 |
| 16 | 256 | 4.000 | 66 | 4356 | 8.124 |
| 17 | 289 | 4.123 | 67 | 4489 | 8.185 |
| 18 | 324 | 4.243 | 68 | 4624 | 8.246 |
| 19 | 361 | 4.359 | 69 | 4761 | 8.307 |
| 20 | 400 | 4.472 | 70 | 4900 | 8.367 |
| 21 | 441 | 4.583 | 71 | 5041 | 8.426 |
| 22 | 484 | 4.690 | 72 | 5184 | 8.485 |
| 23 | 529 | 4.796 | 73 | 5329 | 8.544 |
| 24 | 576 | 4.899 | 74 | 5476 | 8.602 |
| 25 | 625 | 5.000 | 75 | 5625 | 8.660 |
| 26 | 676 | 5.099 | 76 | 5776 | 8.718 |
| 27 | 729 | 5.196 | 77 | 5929 | 8.775 |
| 28 | 784 | 5.292 | 78 | 6084 | 8.832 |
| 29 | 841 | 5.385 | 79 | 6241 | 8.888 |
| 30 | 900 | 5.477 | 80 | 6400 | 8.944 |
| 31 | 961 | 5.568 | 81 | 6561 | 9.000 |
| 32 | 1024 | 5.657 | 82 | 6724 | 9.055 |
| 33 | 1089 | 5.745 | 83 | 6889 | 9.110 |
| 34 | 1156 | 5.831 | 84 | 7056 | 9.165 |
| 35 | 1225 | 5.916 | 85 | 7225 | 9.220 |
| 36 | 1296 | 6.000 | 86 | 7396 | 9.274 |
| 37 | 1369 | 6.083 | 87 | 7569 | 9.327 |
| 38 | 1444 | 6.164 | 88 | 7744 | 9.381 |
| 39 | 1521 | 6.245 | 89 | 7921 | 9.434 |
| 40 | 1600 | 6.325 | 90 | 8100 | 9.487 |
| 41 | 1681 | 6.403 | 91 | 8281 | 9.539 |
| 42 | 1764 | 6.481 | 92 | 8464 | 9.592 |
| 43 | 1849 | 6.557 | 93 | 8649 | 9.644 |
| 44 | 1936 | 6.633 | 94 | 8836 | 9.695 |
| 45 | 2025 | 6.708 | 95 | 9025 | 9.747 |
| 46 | 2116 | 6.782 | 96 | 9216 | 9.798 |
| 47 | 2209 | 6.856 | 97 | 9409 | 9.849 |
| 48 | 2304 | 6.928 | 98 | 9604 | 9.899 |
| 49 | 2401 | 7.000 | 99 | 9801 | 9.950 |
| 50 | 2500 | 7.071 | 100 | 10000 | 10.000 |

## XXIV.—*Squares and Square Roots*—Continued.

| No. | Square. | Square root. | No. | Square. | Square root. |
|---|---|---|---|---|---|
| 101 | 10201 | 10.050 | 151 | 22801 | 12.288 |
| 102 | 10404 | 10.100 | 152 | 23104 | 12.329 |
| 103 | 10609 | 10.149 | 153 | 23409 | 12.369 |
| 104 | 10816 | 10.198 | 154 | 23716 | 12.410 |
| 105 | 11025 | 10.247 | 155 | 24025 | 12.450 |
| 106 | 11236 | 10.296 | 156 | 24336 | 12.490 |
| 107 | 11449 | 10.344 | 157 | 24649 | 12.530 |
| 108 | 11664 | 10.392 | 158 | 24964 | 12.570 |
| 109 | 11881 | 10.440 | 159 | 25281 | 12.610 |
| 110 | 12100 | 10.488 | 160 | 25600 | 12.650 |
| 111 | 12321 | 10.536 | 161 | 25921 | 12.689 |
| 112 | 12544 | 10.583 | 162 | 26244 | 12.728 |
| 113 | 12769 | 10.630 | 163 | 26569 | 12.767 |
| 114 | 12996 | 10.677 | 164 | 26896 | 12.806 |
| 115 | 13225 | 10.724 | 165 | 27225 | 12.845 |
| 116 | 13456 | 10.771 | 166 | 27556 | 12.884 |
| 117 | 13689 | 10.817 | 167 | 27889 | 12.923 |
| 118 | 13924 | 10.863 | 168 | 28224 | 12.961 |
| 119 | 14161 | 10.909 | 169 | 28561 | 13.000 |
| 120 | 14400 | 10.954 | 170 | 28900 | 13.038 |
| 121 | 14641 | 11.000 | 171 | 29241 | 13.077 |
| 122 | 14884 | 11.045 | 172 | 29584 | 13.115 |
| 123 | 15129 | 11.091 | 173 | 29929 | 13.153 |
| 124 | 15376 | 11.136 | 174 | 30276 | 13.191 |
| 125 | 15625 | 11.180 | 175 | 30625 | 13.229 |
| 126 | 15876 | 11.225 | 176 | 30976 | 13.266 |
| 127 | 16129 | 11.269 | 177 | 31329 | 13.304 |
| 128 | 16384 | 11.314 | 178 | 31684 | 13.342 |
| 129 | 16641 | 11.358 | 179 | 32041 | 13.379 |
| 130 | 16900 | 11.402 | 180 | 32400 | 13.416 |
| 131 | 17161 | 11.446 | 181 | 32761 | 13.454 |
| 132 | 17424 | 11.489 | 182 | 33124 | 13.491 |
| 133 | 17689 | 11.533 | 183 | 33489 | 13.528 |
| 134 | 17956 | 11.576 | 184 | 33856 | 13.565 |
| 135 | 18225 | 11.619 | 185 | 34225 | 13.601 |
| 136 | 18496 | 11.662 | 186 | 34596 | 13.638 |
| 137 | 18769 | 11.705 | 187 | 34969 | 13.675 |
| 138 | 19044 | 11.747 | 188 | 35344 | 13.711 |
| 139 | 19321 | 11.790 | 189 | 35721 | 13.748 |
| 140 | 19600 | 11.832 | 190 | 36100 | 13.784 |
| 141 | 19881 | 11.874 | 191 | 36481 | 13.820 |
| 142 | 20164 | 11.916 | 192 | 36864 | 13.856 |
| 143 | 20449 | 11.958 | 193 | 37249 | 13.892 |
| 144 | 20736 | 12.000 | 194 | 37636 | 13.928 |
| 145 | 21025 | 12.042 | 195 | 38025 | 13.964 |
| 146 | 21316 | 12.083 | 196 | 38416 | 14.000 |
| 147 | 21609 | 12.124 | 197 | 38809 | 14.036 |
| 148 | 21904 | 12.166 | 198 | 39204 | 14.071 |
| 149 | 22201 | 12.207 | 199 | 39601 | 14.107 |
| 150 | 22500 | 12.247 | 200 | 40000 | 14.142 |

## XXIV.—*Squares and Square Roots*—Continued.

| No. | Square. | Square root. | No. | Square. | Square root. |
|-----|---------|--------------|-----|---------|--------------|
| 201 | 40401 | 14.177 | 251 | 63001 | 15.843 |
| 202 | 40804 | 14.213 | 252 | 63504 | 15.875 |
| 203 | 41209 | 14.248 | 253 | 64009 | 15.906 |
| 204 | 41616 | 14.283 | 254 | 64516 | 15.937 |
| 205 | 42025 | 14.318 | 255 | 65025 | 15.969 |
| 206 | 42436 | 14.353 | 256 | 65536 | 16.000 |
| 207 | 42849 | 14.387 | 257 | 66049 | 16.031 |
| 208 | 43264 | 14.422 | 258 | 66564 | 16.062 |
| 209 | 43681 | 14.457 | 259 | 67081 | 16.093 |
| 210 | 44100 | 14.491 | 260 | 67600 | 16.125 |
| 211 | 44521 | 14.526 | 261 | 68121 | 16.155 |
| 212 | 44944 | 14.560 | 262 | 68644 | 16.186 |
| 213 | 45369 | 14.595 | 263 | 69169 | 16.217 |
| 214 | 45796 | 14.629 | 264 | 69696 | 16.248 |
| 215 | 46225 | 14.663 | 265 | 70225 | 16.279 |
| 216 | 46656 | 14.697 | 266 | 70756 | 16.310 |
| 217 | 47089 | 14.731 | 267 | 71289 | 16.340 |
| 218 | 47524 | 14.765 | 268 | 71824 | 16.371 |
| 219 | 47961 | 14.799 | 269 | 72361 | 16.401 |
| 220 | 48400 | 14.832 | 270 | 72900 | 16.432 |
| 221 | 48841 | 14.866 | 271 | 73441 | 16.462 |
| 222 | 49284 | 14.900 | 272 | 73984 | 16.492 |
| 223 | 49729 | 14.933 | 273 | 74529 | 16.523 |
| 224 | 50176 | 14.967 | 274 | 75076 | 16.553 |
| 225 | 50625 | 15.000 | 275 | 75625 | 16.583 |
| 226 | 51076 | 15.033 | 276 | 76176 | 16.613 |
| 227 | 51529 | 15.067 | 277 | 76729 | 16.643 |
| 228 | 51984 | 15.100 | 278 | 77284 | 16.673 |
| 229 | 52441 | 15.133 | 279 | 77841 | 16.703 |
| 230 | 52900 | 15.166 | 280 | 78400 | 16.733 |
| 231 | 53361 | 15.199 | 281 | 78961 | 16.763 |
| 232 | 53824 | 15.232 | 282 | 79524 | 16.793 |
| 233 | 54289 | 15.264 | 283 | 80089 | 16.823 |
| 234 | 54756 | 15.297 | 284 | 80656 | 16.852 |
| 235 | 55225 | 15.330 | 285 | 81225 | 16.882 |
| 236 | 55696 | 15.362 | 286 | 81796 | 16.912 |
| 237 | 56169 | 15.395 | 287 | 82369 | 16.941 |
| 238 | 56644 | 15.427 | 288 | 82944 | 16.971 |
| 239 | 57121 | 15.460 | 289 | 83521 | 17.000 |
| 240 | 57600 | 15.492 | 290 | 84100 | 17.029 |
| 241 | 58081 | 15.524 | 291 | 84681 | 17.059 |
| 242 | 58564 | 15.556 | 292 | 85264 | 17.088 |
| 243 | 59049 | 15.588 | 293 | 85849 | 17.117 |
| 244 | 59536 | 15.620 | 294 | 86436 | 17.146 |
| 245 | 60025 | 15.652 | 295 | 87025 | 17.176 |
| 246 | 60516 | 15.684 | 296 | 87616 | 17.205 |
| 247 | 61009 | 15.716 | 297 | 88209 | 17.234 |
| 248 | 61504 | 15.748 | 298 | 88804 | 17.263 |
| 249 | 62001 | 15.780 | 299 | 89401 | 17.292 |
| 250 | 62500 | 15.811 | 300 | 90000 | 17.321 |

## XXIV.—*Squares and Square Roots*—Continued.

| No. | Square. | Square root. | No. | Square. | Square root. |
|---|---|---|---|---|---|
| 301 | 90601 | 17.349 | 351 | 123201 | 18.735 |
| 302 | 91204 | 17.378 | 352 | 123904 | 18.762 |
| 303 | 91809 | 17.407 | 353 | 124609 | 18.788 |
| 304 | 92416 | 17.436 | 354 | 125316 | 18.815 |
| 305 | 93025 | 17.464 | 355 | 126025 | 18.841 |
| 306 | 93636 | 17.493 | 356 | 126736 | 18.868 |
| 307 | 94249 | 17.521 | 357 | 127449 | 18.894 |
| 308 | 94864 | 17.550 | 358 | 128164 | 18.921 |
| 309 | 95481 | 17.578 | 359 | 128881 | 18.947 |
| 310 | 96100 | 17.607 | 360 | 129600 | 18.974 |
| 311 | 96721 | 17.635 | 361 | 130321 | 19.000 |
| 312 | 97344 | 17.664 | 362 | 131044 | 19.026 |
| 313 | 97969 | 17.692 | 363 | 131769 | 19.053 |
| 314 | 98596 | 17.720 | 364 | 132496 | 19.079 |
| 315 | 99225 | 17.748 | 365 | 133225 | 19.105 |
| 316 | 99856 | 17.776 | 366 | 133956 | 19.131 |
| 317 | 100489 | 17.804 | 367 | 134689 | 19.157 |
| 318 | 101124 | 17.833 | 368 | 135424 | 19.183 |
| 319 | 101761 | 17.861 | 369 | 136161 | 19.209 |
| 320 | 102400 | 17.889 | 370 | 136900 | 19.235 |
| 321 | 103041 | 17.916 | 371 | 137641 | 19.261 |
| 322 | 103684 | 17.944 | 372 | 138384 | 19.287 |
| 323 | 104329 | 17.972 | 373 | 139129 | 19.313 |
| 324 | 104976 | 18.000 | 374 | 139876 | 19.339 |
| 325 | 105625 | 18.028 | 375 | 140625 | 19.365 |
| 326 | 106276 | 18.055 | 376 | 141376 | 19.391 |
| 327 | 106929 | 18.083 | 377 | 142129 | 19.416 |
| 328 | 107584 | 18.111 | 378 | 142884 | 19.442 |
| 329 | 108241 | 18.138 | 379 | 143641 | 19.468 |
| 330 | 108900 | 18.166 | 380 | 144400 | 19.494 |
| 331 | 109561 | 18.193 | 381 | 145161 | 19.519 |
| 332 | 110224 | 18.221 | 382 | 145924 | 19.545 |
| 333 | 110889 | 18.248 | 383 | 146689 | 19.570 |
| 334 | 111556 | 18.276 | 384 | 147456 | 19.596 |
| 335 | 112225 | 18.303 | 385 | 148225 | 19.621 |
| 336 | 112896 | 18.330 | 386 | 148996 | 19.647 |
| 337 | 113569 | 18.358 | 387 | 149769 | 19.672 |
| 338 | 114244 | 18.385 | 388 | 150544 | 19.698 |
| 339 | 114921 | 18.412 | 389 | 151321 | 19.723 |
| 340 | 115600 | 18.439 | 390 | 152100 | 19.748 |
| 341 | 116281 | 18.466 | 391 | 152881 | 19.774 |
| 342 | 116964 | 18.493 | 392 | 153664 | 19.799 |
| 343 | 117649 | 18.520 | 393 | 154449 | 19.824 |
| 344 | 118336 | 18.547 | 394 | 155236 | 19.849 |
| 345 | 119025 | 18.574 | 395 | 156025 | 19.875 |
| 346 | 119716 | 18.601 | 396 | 156816 | 19.900 |
| 347 | 120409 | 18.628 | 397 | 157609 | 19.925 |
| 348 | 121104 | 18.655 | 398 | 158404 | 19.950 |
| 349 | 121801 | 18.682 | 399 | 159201 | 19.975 |
| 350 | 122500 | 18.703 | 400 | 160000 | 20.000 |

## XXIV.—*Squares and Square Roots*—Continued.

| No. | Square. | Square root. | No. | Square. | Square root. |
|---|---|---|---|---|---|
| 401 | 160801 | 20.025 | 451 | 203401 | 21.237 |
| 402 | 161604 | 20.050 | 452 | 204304 | 21.260 |
| 403 | 162409 | 20.075 | 453 | 205209 | 21.284 |
| 404 | 163216 | 20.100 | 454 | 206116 | 21.307 |
| 405 | 164025 | 20.125 | 455 | 207025 | 21.331 |
| 406 | 164836 | 20.149 | 456 | 207936 | 21.354 |
| 407 | 165649 | 20.174 | 457 | 208849 | 21.378 |
| 408 | 166464 | 20.199 | 458 | 209764 | 21.401 |
| 409 | 167281 | 20.224 | 459 | 210681 | 21.424 |
| 410 | 168100 | 20.248 | 460 | 211600 | 21.448 |
| 411 | 168921 | 20.273 | 461 | 212521 | 21.471 |
| 412 | 169744 | 20.298 | 462 | 213444 | 21.494 |
| 413 | 170569 | 20.322 | 463 | 214369 | 21.517 |
| 414 | 171396 | 20.347 | 464 | 215296 | 21.541 |
| 415 | 172225 | 20.372 | 465 | 216225 | 21.564 |
| 416 | 173056 | 20.396 | 466 | 217156 | 21.587 |
| 417 | 173889 | 20.421 | 467 | 218089 | 21.610 |
| 418 | 174724 | 20.445 | 468 | 219024 | 21.633 |
| 419 | 175561 | 20.469 | 469 | 219961 | 21.656 |
| 420 | 176400 | 20.494 | 470 | 220900 | 21.679 |
| 421 | 177241 | 20.518 | 471 | 221841 | 21.703 |
| 422 | 178084 | 20.543 | 472 | 222784 | 21.726 |
| 423 | 178929 | 20.567 | 473 | 223729 | 21.749 |
| 424 | 179776 | 20.591 | 474 | 224676 | 21.772 |
| 425 | 180625 | 20.616 | 475 | 225625 | 21.794 |
| 426 | 181476 | 20.640 | 476 | 226576 | 21.817 |
| 427 | 182329 | 20.664 | 477 | 227529 | 21.840 |
| 428 | 183184 | 20.688 | 478 | 228484 | 21.863 |
| 429 | 184041 | 20.712 | 479 | 229441 | 21.886 |
| 430 | 184900 | 20.736 | 480 | 230400 | 21.909 |
| 431 | 185761 | 20.761 | 481 | 231361 | 21.932 |
| 432 | 186624 | 20.785 | 482 | 232324 | 21.954 |
| 433 | 187489 | 20.809 | 483 | 233289 | 21.977 |
| 434 | 188356 | 20.833 | 484 | 234256 | 22.000 |
| 435 | 189225 | 20.857 | 485 | 235225 | 22.023 |
| 436 | 190096 | 20.881 | 486 | 236196 | 22.045 |
| 437 | 190969 | 20.905 | 487 | 237169 | 22.068 |
| 438 | 191844 | 20.928 | 488 | 238144 | 22.091 |
| 439 | 192721 | 20.952 | 489 | 239121 | 22.113 |
| 440 | 193600 | 20.976 | 490 | 240100 | 22.136 |
| 441 | 194481 | 21.000 | 491 | 241081 | 22.159 |
| 442 | 195364 | 21.024 | 492 | 242064 | 22.181 |
| 443 | 196249 | 21.048 | 493 | 243049 | 22.204 |
| 444 | 197136 | 21.071 | 494 | 244036 | 22.226 |
| 445 | 198025 | 21.095 | 495 | 245025 | 22.249 |
| 446 | 198916 | 21.119 | 496 | 246016 | 22.271 |
| 447 | 199809 | 21.142 | 497 | 247009 | 22.293 |
| 448 | 200704 | 21.166 | 498 | 248004 | 22.316 |
| 449 | 201601 | 21.190 | 499 | 249001 | 22.338 |
| 450 | 202500 | 21.213 | 500 | 250000 | 22.361 |

XXIV.—*Squares and Square Roots*—Continued.

| No. | Square. | Square root. | No. | Square. | Square root. |
|---|---|---|---|---|---|
| 501 | 251001 | 22.383 | 551 | 303601 | 23.473 |
| 502 | 252004 | 22.405 | 552 | 304704 | 23.495 |
| 503 | 253009 | 22.428 | 553 | 305809 | 23.516 |
| 504 | 254016 | 22.450 | 554 | 306916 | 23.537 |
| 505 | 255025 | 22.472 | 555 | 308025 | 23.558 |
| 506 | 256036 | 22.494 | 556 | 309136 | 23.580 |
| 507 | 257049 | 22.517 | 557 | 310249 | 23.601 |
| 508 | 258064 | 22.539 | 558 | 311364 | 23.622 |
| 509 | 259081 | 22.561 | 559 | 312481 | 23.643 |
| 510 | 260100 | 22.583 | 560 | 313600 | 23.664 |
| 511 | 261121 | 22.605 | 561 | 314721 | 23.685 |
| 512 | 262144 | 22.627 | 562 | 315844 | 23.707 |
| 513 | 263169 | 22.650 | 563 | 316969 | 23.728 |
| 514 | 264196 | 22.672 | 564 | 318096 | 23.749 |
| 515 | 265225 | 22.694 | 565 | 319225 | 23.770 |
| 516 | 266256 | 22.716 | 566 | 320356 | 23.791 |
| 517 | 267289 | 22.738 | 567 | 321489 | 23.812 |
| 518 | 268324 | 22.760 | 568 | 322624 | 23.833 |
| 519 | 269361 | 22.782 | 569 | 323761 | 23.854 |
| 520 | 270400 | 22.804 | 570 | 324900 | 23.875 |
| 521 | 271441 | 22.825 | 571 | 326041 | 23.896 |
| 522 | 272484 | 22.847 | 572 | 327184 | 23.917 |
| 523 | 273529 | 22.869 | 573 | 328329 | 23.937 |
| 524 | 274576 | 22.891 | 574 | 329476 | 23.958 |
| 525 | 275625 | 22.913 | 575 | 330625 | 23.979 |
| 526 | 276676 | 22.935 | 576 | 331776 | 24.000 |
| 527 | 277729 | 22.956 | 577 | 332929 | 24.021 |
| 528 | 278784 | 22.978 | 578 | 334084 | 24.042 |
| 529 | 279841 | 23.000 | 579 | 335241 | 24.062 |
| 530 | 280900 | 23.022 | 580 | 336400 | 24.083 |
| 531 | 281961 | 23.043 | 581 | 337561 | 24.104 |
| 532 | 283024 | 23.065 | 582 | 338724 | 24.125 |
| 533 | 284089 | 23.087 | 583 | 339889 | 24.145 |
| 534 | 285156 | 23.108 | 584 | 341056 | 24.166 |
| 535 | 286225 | 23.130 | 585 | 342225 | 24.187 |
| 536 | 287296 | 23.152 | 586 | 343396 | 24.207 |
| 537 | 288369 | 23.173 | 587 | 344569 | 24.228 |
| 538 | 289444 | 23.195 | 588 | 345744 | 24.249 |
| 539 | 290521 | 23.216 | 589 | 346921 | 24.269 |
| 540 | 291600 | 23.238 | 590 | 348100 | 24.290 |
| 541 | 292681 | 23.259 | 591 | 349281 | 24.310 |
| 542 | 293764 | 23.281 | 592 | 350464 | 24.331 |
| 543 | 294849 | 23.302 | 593 | 351649 | 24.352 |
| 544 | 295936 | 23.324 | 594 | 352836 | 24.372 |
| 545 | 297025 | 23.345 | 595 | 354025 | 24.393 |
| 546 | 298116 | 23.367 | 596 | 355216 | 24.413 |
| 547 | 299209 | 23.388 | 597 | 356409 | 24.434 |
| 548 | 300304 | 23.409 | 598 | 357604 | 24.454 |
| 549 | 301401 | 23.431 | 599 | 358801 | 24.474 |
| 550 | 302500 | 23.452 | 600 | 360000 | 24.495 |

## XXIV.—*Squares and Square Roots*—Continued.

| No. | Square. | Square root. | No. | Square. | Square root. |
|-----|---------|--------------|-----|---------|--------------|
| 601 | 361201 | 24.515 | 651 | 423801 | 25.515 |
| 602 | 362404 | 24.536 | 652 | 425104 | 25.534 |
| 603 | 363609 | 24.556 | 653 | 426409 | 25.554 |
| 604 | 364816 | 24.576 | 654 | 427716 | 25.573 |
| 605 | 366025 | 24.597 | 655 | 429025 | 25.593 |
| 606 | 367236 | 24.617 | 656 | 430336 | 25.612 |
| 607 | 368449 | 24.637 | 657 | 431649 | 25.632 |
| 608 | 369664 | 24.658 | 658 | 432964 | 25.652 |
| 609 | 370881 | 24.678 | 659 | 434281 | 25.671 |
| 610 | 372100 | 24.698 | 660 | 435600 | 25.690 |
| 611 | 373321 | 24.718 | 661 | 436921 | 25.710 |
| 612 | 374544 | 24.739 | 662 | 438244 | 25.720 |
| 613 | 375769 | 24.759 | 663 | 439569 | 25.749 |
| 614 | 376996 | 24.779 | 664 | 440896 | 25.768 |
| 615 | 378225 | 24.799 | 665 | 442225 | 25.788 |
| 616 | 379456 | 24.819 | 666 | 443556 | 25.807 |
| 617 | 380689 | 24.839 | 667 | 444889 | 25.826 |
| 618 | 381924 | 24.860 | 668 | 446224 | 25.846 |
| 619 | 383161 | 24.880 | 669 | 447561 | 25.865 |
| 620 | 384400 | 24.900 | 670 | 448900 | 25.884 |
| 621 | 385641 | 24.920 | 671 | 450241 | 25.904 |
| 622 | 386884 | 24.940 | 672 | 451584 | 25.923 |
| 623 | 388129 | 24.960 | 673 | 452929 | 25.942 |
| 624 | 389376 | 24.980 | 674 | 454276 | 25.962 |
| 625 | 390625 | 25.000 | 675 | 455625 | 25.981 |
| 626 | 391876 | 25.020 | 676 | 456976 | 26.000 |
| 627 | 393129 | 25.040 | 677 | 458329 | 26.019 |
| 628 | 394384 | 25.060 | 678 | 459684 | 26.038 |
| 629 | 395641 | 25.080 | 679 | 461041 | 26.058 |
| 630 | 396900 | 25.100 | 680 | 462400 | 26.077 |
| 631 | 398161 | 25.120 | 681 | 463761 | 26.096 |
| 632 | 399424 | 25.140 | 682 | 465124 | 26.115 |
| 633 | 400689 | 25.160 | 683 | 466489 | 26.134 |
| 634 | 401956 | 25.180 | 684 | 467856 | 26.153 |
| 635 | 403225 | 25.200 | 685 | 469225 | 26.173 |
| 636 | 404496 | 25.220 | 686 | 470596 | 26.192 |
| 637 | 405769 | 25.239 | 687 | 471969 | 26.211 |
| 638 | 407044 | 25.259 | 688 | 473344 | 26.230 |
| 639 | 408321 | 25.278 | 689 | 474721 | 26.249 |
| 640 | 409600 | 25.298 | 690 | 476100 | 26.268 |
| 641 | 410881 | 25.318 | 691 | 477481 | 26.287 |
| 642 | 412164 | 25.338 | 692 | 478864 | 26.306 |
| 643 | 413449 | 25.357 | 693 | 480249 | 26.325 |
| 644 | 414736 | 25.377 | 694 | 481636 | 26.344 |
| 645 | 416025 | 25.397 | 695 | 483025 | 26.363 |
| 646 | 417316 | 25.417 | 696 | 484416 | 26.382 |
| 647 | 418609 | 25.436 | 697 | 485809 | 26.401 |
| 648 | 419904 | 25.456 | 698 | 487204 | 26.420 |
| 649 | 421201 | 25.475 | 699 | 488601 | 26.439 |
| 650 | 422500 | 25.495 | 700 | 490000 | 26.458 |

## XXIV.—*Squares and Square Roots*—Continued.

| No. | Square. | Square root. | No. | Square. | Square root. |
|---|---|---|---|---|---|
| 701 | 491401 | 26.476 | 751 | 564001 | 27.404 |
| 702 | 492804 | 26.495 | 752 | 565504 | 27.423 |
| 703 | 494209 | 26.514 | 753 | 567009 | 27.441 |
| 704 | 495616 | 26.532 | 754 | 568516 | 27.459 |
| 705 | 497025 | 26.552 | 755 | 570025 | 27.477 |
| 706 | 498436 | 26.571 | 756 | 571536 | 27.495 |
| 707 | 499849 | 26.589 | 757 | 573049 | 27.514 |
| 708 | 501264 | 26.608 | 758 | 574564 | 27.532 |
| 709 | 502681 | 26.627 | 759 | 576081 | 27.550 |
| 710 | 504100 | 26.646 | 760 | 577600 | 27.568 |
| 711 | 505521 | 26.665 | 761 | 579121 | 27.586 |
| 712 | 506944 | 26.683 | 762 | 580644 | 27.604 |
| 713 | 508369 | 26.702 | 763 | 582169 | 27.622 |
| 714 | 509796 | 26.721 | 764 | 583696 | 27.641 |
| 715 | 511225 | 26.739 | 765 | 585225 | 27.659 |
| 716 | 512656 | 26.758 | 766 | 586756 | 27.677 |
| 717 | 514089 | 26.777 | 767 | 588289 | 27.695 |
| 718 | 515524 | 26.796 | 768 | 289824 | 27.713 |
| 719 | 516961 | 26.814 | 769 | 591361 | 27.731 |
| 720 | 518400 | 26.833 | 770 | 592900 | 27.749 |
| 721 | 519841 | 26.851 | 771 | 594441 | 27.767 |
| 722 | 521284 | 26.870 | 772 | 595984 | 27.785 |
| 723 | 522729 | 26.889 | 773 | 597529 | 27.803 |
| 724 | 524176 | 26.907 | 774 | 599076 | 27.821 |
| 725 | 525625 | 26.926 | 775 | 600625 | 27.839 |
| 726 | 527076 | 26.944 | 776 | 602176 | 27.857 |
| 727 | 528529 | 26.963 | 777 | 603729 | 27.875 |
| 728 | 529984 | 26.981 | 778 | 605284 | 27.893 |
| 729 | 531441 | 27.000 | 779 | 606841 | 27.911 |
| 730 | 532900 | 27.019 | 780 | 608400 | 27.928 |
| 731 | 534361 | 27.037 | 781 | 609961 | 27.946 |
| 732 | 535824 | 27.055 | 782 | 611524 | 27.964 |
| 733 | 537289 | 27.074 | 783 | 613089 | 27.982 |
| 734 | 538756 | 27.092 | 784 | 614656 | 28.000 |
| 735 | 540225 | 27.111 | 785 | 616225 | 28.018 |
| 736 | 541696 | 27.129 | 786 | 617796 | 28.036 |
| 737 | 543169 | 27.148 | 787 | 619369 | 28.054 |
| 738 | 544644 | 27.166 | 788 | 620944 | 28.071 |
| 739 | 546121 | 27.185 | 789 | 622521 | 28.089 |
| 740 | 547600 | 27.203 | 790 | 624100 | 28.107 |
| 741 | 549081 | 27.221 | 791 | 625681 | 28.125 |
| 742 | 550564 | 27.240 | 792 | 627264 | 28.142 |
| 743 | 552049 | 27.258 | 793 | 628849 | 28.160 |
| 744 | 553536 | 27.276 | 794 | 630436 | 28.178 |
| 745 | 555025 | 27.295 | 795 | 632025 | 28.196 |
| 746 | 556516 | 27.313 | 796 | 633616 | 28.213 |
| 747 | 558009 | 27.331 | 797 | 635209 | 28.231 |
| 748 | 559504 | 27.350 | 798 | 636804 | 28.249 |
| 749 | 561001 | 27.368 | 799 | 638401 | 28.267 |
| 750 | 562500 | 27.386 | 800 | 640000 | 28.284 |

### XXIV.—*Squares and Square Roots*—Continued.

| No. | Square. | Square root. | No. | Square. | Square root. |
|-----|---------|--------------|-----|---------|--------------|
| 801 | 641601 | 28.302 | 851 | 724201 | 29.172 |
| 802 | 643204 | 28.320 | 852 | 725904 | 29.189 |
| 803 | 644809 | 28.337 | 853 | 727609 | 29.206 |
| 804 | 646416 | 28.555 | 854 | 729316 | 29.223 |
| 805 | 648025 | 28.373 | 855 | 731025 | 29.240 |
| 806 | 649636 | 28.390 | 856 | 732736 | 29.257 |
| 807 | 651249 | 28.408 | 857 | 734449 | 29.275 |
| 808 | 652864 | 28.425 | 858 | 736164 | 29.292 |
| 809 | 654481 | 28.443 | 859 | 737881 | 29.309 |
| 810 | 656100 | 28.460 | 860 | 739600 | 29.326 |
| 811 | 657721 | 28.478 | 861 | 741321 | 29.343 |
| 812 | 659344 | 28.496 | 862 | 743044 | 29.360 |
| 813 | 660969 | 28.513 | 863 | 744769 | 29.377 |
| 814 | 662596 | 28.531 | 864 | 746496 | 29.394 |
| 815 | 664225 | 28.548 | 865 | 748225 | 29.411 |
| 816 | 665856 | 28.566 | 866 | 749956 | 29.428 |
| 817 | 667489 | 28.583 | 867 | 751689 | 29.445 |
| 818 | 669124 | 28.601 | 868 | 753424 | 29.462 |
| 819 | 670761 | 28.618 | 869 | 755161 | 29.479 |
| 820 | 672400 | 28.636 | 870 | 756900 | 29.496 |
| 821 | 674041 | 28.653 | 871 | 758641 | 29.513 |
| 822 | 675684 | 28.671 | 872 | 760384 | 29.530 |
| 823 | 677329 | 28.688 | 873 | 762129 | 29.547 |
| 824 | 678976 | 28.705 | 874 | 763876 | 29.563 |
| 825 | 680625 | 28.723 | 875 | 765625 | 29.580 |
| 826 | 682276 | 28.740 | 876 | 767376 | 29.597 |
| 827 | 683929 | 28.758 | 877 | 769129 | 29.614 |
| 828 | 685584 | 28.775 | 878 | 770884 | 29.631 |
| 829 | 687241 | 28.792 | 879 | 772641 | 29.648 |
| 830 | 688900 | 28.810 | 880 | 774400 | 29.665 |
| 831 | 690561 | 28.827 | 881 | 776161 | 29.682 |
| 832 | 692224 | 28.844 | 882 | 777924 | 29.698 |
| 833 | 693889 | 28.862 | 883 | 779689 | 29.715 |
| 834 | 695556 | 28.879 | 884 | 781456 | 29.732 |
| 835 | 697225 | 28.896 | 885 | 783225 | 29.749 |
| 836 | 698896 | 28.914 | 886 | 784996 | 29.766 |
| 837 | 700569 | 28.931 | 887 | 786769 | 29.783 |
| 838 | 702244 | 28.948 | 888 | 788544 | 29.799 |
| 839 | 703921 | 28.965 | 889 | 790321 | 29.816 |
| 840 | 705600 | 28.983 | 890 | 792100 | 29.833 |
| 841 | 707281 | 29.000 | 891 | 793881 | 29.850 |
| 842 | 708964 | 29.017 | 892 | 795664 | 29.866 |
| 843 | 710649 | 29.034 | 893 | 797449 | 29.883 |
| 844 | 712336 | 29.052 | 894 | 799236 | 29.900 |
| 845 | 714025 | 29.069 | 895 | 801025 | 29.917 |
| 846 | 715716 | 29.086 | 896 | 802816 | 29.933 |
| 847 | 717409 | 29.103 | 897 | 804609 | 29.950 |
| 848 | 719104 | 29.120 | 898 | 806404 | 29.967 |
| 849 | 720801 | 29.138 | 899 | 808201 | 29.983 |
| 850 | 722500 | 29.155 | 900 | 810000 | 30.000 |

## XXIV.—*Squares and Square Roots*—Continued.

| No. | Square. | Square root. | No. | Square. | Square root. |
|-----|---------|--------------|-----|---------|--------------|
| 901 | 811801 | 30.017 | 951 | 904401 | 30.838 |
| 902 | 813604 | 30.033 | 952 | 906304 | 30.854 |
| 903 | 815409 | 30.050 | 953 | 908209 | 30.871 |
| 904 | 817216 | 30.067 | 954 | 910116 | 30.887 |
| 905 | 819025 | 30.083 | 955 | 912025 | 30.903 |
| 906 | 820836 | 30.100 | 956 | 913936 | 30.919 |
| 907 | 822649 | 30.116 | 957 | 915849 | 30.935 |
| 908 | 824464 | 30.133 | 958 | 917764 | 30.952 |
| 909 | 826281 | 30.150 | 959 | 919681 | 30.968 |
| 910 | 828100 | 30.166 | 960 | 921600 | 30.984 |
| 911 | 829921 | 30.183 | 961 | 923521 | 31.000 |
| 912 | 831744 | 30.199 | 962 | 925444 | 31.016 |
| 913 | 833569 | 30.216 | 963 | 927369 | 31.032 |
| 914 | 835396 | 30.232 | 964 | 929296 | 31.048 |
| 915 | 837225 | 30.249 | 965 | 931225 | 31.064 |
| 916 | 839056 | 30.265 | 966 | 933156 | 31.081 |
| 917 | 840889 | 30.282 | 967 | 935089 | 31.097 |
| 918 | 842724 | 30.299 | 968 | 937024 | 31.113 |
| 919 | 844561 | 30.315 | 969 | 938961 | 31.129 |
| 920 | 846400 | 30.332 | 970 | 940900 | 31.145 |
| 921 | 848241 | 30.348 | 971 | 942841 | 31.161 |
| 922 | 850084 | 30.364 | 972 | 944784 | 31.177 |
| 923 | 851929 | 30.381 | 973 | 946729 | 31.193 |
| 924 | 853776 | 30.397 | 974 | 948676 | 31.209 |
| 925 | 855625 | 30.414 | 975 | 950625 | 31.225 |
| 926 | 857476 | 30.430 | 976 | 952576 | 31.241 |
| 927 | 859329 | 30.447 | 977 | 954529 | 31.257 |
| 928 | 861184 | 30.463 | 978 | 956484 | 31.273 |
| 929 | 863041 | 30.480 | 979 | 958441 | 31.289 |
| 930 | 864900 | 30.496 | 980 | 960400 | 31.305 |
| 931 | 866761 | 30.512 | 981 | 962361 | 31.321 |
| 932 | 868624 | 30.529 | 982 | 964324 | 31.337 |
| 933 | 870489 | 30.545 | 983 | 966289 | 31.353 |
| 934 | 872356 | 30.561 | 984 | 968256 | 31.369 |
| 935 | 874225 | 30.578 | 985 | 970225 | 31.385 |
| 936 | 876096 | 30.594 | 986 | 972196 | 31.401 |
| 937 | 877969 | 30.610 | 987 | 974169 | 31.417 |
| 938 | 879844 | 30.627 | 988 | 976144 | 31.432 |
| 939 | 881721 | 30.643 | 989 | 978121 | 31.448 |
| 940 | 883600 | 30.659 | 990 | 980100 | 31.464 |
| 941 | 885481 | 30.676 | 991 | 982081 | 31.480 |
| 942 | 887364 | 30.692 | 992 | 984064 | 31.496 |
| 943 | 889249 | 30.708 | 993 | 986049 | 31.512 |
| 944 | 891136 | 30.725 | 994 | 988036 | 31.528 |
| 945 | 893025 | 30.741 | 995 | 990025 | 31.544 |
| 946 | 894916 | 30.757 | 996 | 992016 | 31.560 |
| 947 | 896809 | 30.773 | 997 | 994009 | 31.575 |
| 948 | 898704 | 30.790 | 998 | 996004 | 31.591 |
| 949 | 900601 | 30.806 | 999 | 998001 | 31.607 |
| 950 | 902500 | 30.822 | 1000 | 1000000 | 31.623 |

### Bessel's Co-efficients.

| Parts of the unit of time. | 2d diff. $t \cdot \frac{t-1}{2}$ | 3d diff. $t \cdot \frac{t-1}{2} \cdot \frac{t-\frac{1}{2}}{3}$ | 4th diff. $t \ldots \frac{t-2}{4}$ | Parts of the unit of time. | 2d diff. $t \cdot \frac{t-1}{2}$ | 3d diff. $t \cdot \frac{t-1}{2} \cdot \frac{t-\frac{1}{2}}{3}$ | 4th diff. $t \ldots \frac{t-2}{4}$ |
|---|---|---|---|---|---|---|---|
| 0.01 | −.00495 | .00081 | .00083 | 0.51 | −.12495 | −.00042 | .02343 |
| .02 | .00980 | .00157 | .00165 | .52 | .12480 | .00083 | .02340 |
| .03 | .01455 | .00228 | .00246 | .53 | .12455 | .00125 | .02334 |
| .04 | .01920 | .00294 | .00326 | .54 | .12420 | .00166 | .02327 |
| .05 | .02375 | .00356 | .00405 | .55 | .12375 | .00206 | .02318 |
| .06 | .02820 | .00414 | .00483 | .56 | .12320 | .00246 | .02306 |
| .07 | .03255 | .00467 | .00560 | .57 | .12255 | .00286 | .02293 |
| .08 | .03680 | .00515 | .00636 | .58 | .12180 | .00325 | .02278 |
| .09 | .04095 | .00560 | .00711 | .59 | .12095 | .00363 | .02260 |
| .10 | .04500 | .00600 | .00784 | .60 | .12000 | .00400 | .02240 |
| .11 | .04895 | .00636 | .00856 | .61 | .11895 | .00436 | .02218 |
| .12 | .05280 | .00669 | .00927 | .62 | .11780 | .00471 | .02194 |
| .13 | .05655 | .00697 | .00996 | .63 | .11655 | .00505 | .02169 |
| .14 | .06020 | .00723 | .01064 | .64 | .11520 | .00538 | .02141 |
| .15 | .06375 | .00744 | .01130 | .65 | .11375 | .00569 | .02111 |
| .16 | .06720 | .00762 | .01195 | .66 | .11220 | .00598 | .02080 |
| .17 | .07055 | .00776 | .01259 | .67 | .11055 | .00626 | .02046 |
| .18 | .07380 | .00787 | .01321 | .68 | .10880 | .00653 | .02010 |
| .19 | .07695 | .00795 | .01381 | .69 | .10695 | .00677 | .01973 |
| .20 | .08000 | .00800 | .01440 | .70 | .10500 | .00700 | .01934 |
| .21 | .08295 | .00802 | .01497 | .71 | .10295 | .00721 | .01893 |
| .22 | .08580 | .00801 | .01553 | .72 | .10080 | .00739 | .01850 |
| .23 | .08855 | .00797 | .01606 | .73 | .09855 | .00756 | .01805 |
| .24 | .09120 | .00790 | .01658 | .74 | .09620 | .00770 | .01758 |
| .25 | .09375 | .00781 | .01709 | .75 | .09375 | .00781 | .01709 |
| .26 | .09620 | .00770 | .01758 | .76 | .09120 | .00790 | .01658 |
| .27 | .09855 | .00756 | .01805 | .77 | .08855 | .00797 | .01606 |
| .28 | .10080 | .00739 | .01850 | .78 | .08580 | .00801 | .01553 |
| .29 | .10295 | .00721 | .01893 | .79 | .08295 | .00802 | .01497 |
| .30 | .10500 | .00700 | .01934 | .80 | .08000 | .00800 | .01440 |
| .31 | .10695 | .00677 | .01973 | .81 | .07695 | .00795 | .01381 |
| .32 | .10880 | .00653 | .02010 | .82 | .07380 | .00787 | .01321 |
| .33 | .11055 | .00626 | .02046 | .83 | .07055 | .00776 | .01259 |
| .34 | .11220 | .00598 | .02080 | .84 | .06720 | .00762 | .01195 |
| .35 | .11375 | .00569 | .02111 | .85 | .06375 | .00744 | .01130 |
| .36 | .11520 | .00538 | .02141 | .86 | .06020 | .00723 | .01064 |
| .37 | .11655 | .00505 | .02169 | .87 | .05655 | .00697 | .00996 |
| .38 | .11780 | .00471 | .02194 | .88 | .05280 | .00669 | .00927 |
| .39 | .11895 | .00436 | .02218 | .89 | .04895 | .00636 | .00856 |
| .40 | .12000 | .00400 | .02240 | .90 | .04500 | .00600 | .00784 |
| .41 | .12095 | .00363 | .02260 | .91 | .04095 | .00560 | .00711 |
| .42 | .12180 | .00325 | .02278 | .92 | .03680 | .00515 | .00636 |
| .43 | .12255 | .00286 | .02293 | .93 | .03255 | .00467 | .00560 |
| .44 | .12320 | .00246 | .02306 | .94 | .02820 | .00414 | .00483 |
| .45 | .12375 | .00206 | .02318 | .95 | .02375 | .00356 | .00405 |
| .46 | .12420 | .00166 | .02327 | .96 | .01920 | .00294 | .00326 |
| .47 | .12455 | .00125 | .02334 | .97 | .01455 | .00228 | .00246 |
| .48 | .12480 | .00083 | .02340 | .98 | .00980 | .00157 | .00165 |
| .49 | .12495 | .00042 | .02343 | .99 | .00495 | .00081 | .00083 |
| .50 | −.12500 | .00000 | .02344 | 1.00 | −.00000 | −.00000 | .00000 |

# TABLES AND FORMULÆ.

## PART II.

# GEODESY.

### XXV.—*Reduction to Center of Station.*

Call—

P the place of the instrument;

C the center of the station;

O the angle at P, between two objects, A and B;

*y* the angle at P, between C and the *left*-hand object, B;

*r* the distance, C P;

C the unknown angle at C;

D the distance A C; and

G the distance B C,

then—

$$C = O + \frac{r \sin (O + y)}{D \sin 1''} - \frac{r \sin y}{G \sin 1''}$$

In the use of this formula proper attention should be paid to the signs of sin (O + *y*) and sin *y*; for the first term will be *positive* when (O + *y*) is less than 180°, (the reverse with sin *y*;) D being the distance of the *right*-hand object, the graduation of the instrument running from left to right.

*r* being small, the lengths of D and G are computed with the angle O.

### XXVI.—*Reduction to Center of Signal Observed, or Correction for Phase in Tin Cones Used as Signals.*

$$\text{Correctio} = \pm \frac{r \cos^2 \frac{1}{2} Z}{D \sin 1''}$$

where—

*r* = radius of the signal;

Z = angle at the point of observation between the sun and the signal; and

D = the distance.

### XXVII.—*Spherical Excess.*

$$E = \frac{S}{r^2 \sin 1''} = \frac{a\,b \sin C}{2 r^2 \sin 1''}$$

S being the area of the triangle; $r$, the radius of the earth.

$$S = \frac{a\,b \sin c}{2}$$

$$= \sqrt{s\,(s-a)\,(s-b)\,(s-c)}$$

$s$ being $= \dfrac{a+b+c}{2}$

Between latitudes 45° and 25° the spherical excess amounts to about $1''$ for an area of 75.5 square miles.

Hence, if the area in square miles be known, a close approximation to the spherical excess will be had by dividing the area by 75.5.

log mean radius of the earth in yards = 6.8427917

If the three angles of a triangle are assumed to have been equally well determined, the previous determination of the spherical excess is not necessary for the calculation of the sides, though it will be required for estimating the relative accuracy of the observations; for the sides of a spherical triangle may be computed as if they were rectilineal when one-third the excess of the sum of the three angles above 180° is deducted from each of the three observed angles.   Then—

$$\text{side } b = \text{side } a \, \sin (B - \tfrac{1}{3} E) \div \sin (A - \tfrac{1}{3} E)$$

For large triangles :

$$E = \frac{a\,b \sin C\,(1 + e^2 \cos 2\,L)}{2\,A^2 \sin 1''}$$

A being the equatorial radius, and L the mean latitude of the three stations.

XXVIII.—*To Reduce the Length of an Inclined Base to Horizontal Measure.*

Let—

B be the length of the base on the inclined plane;

$b$ that reduced to the horizontal plane; and

$\theta$ the inclination,

then—

$$b = \text{B} \cos \theta$$

But as $\theta$ is generally a small angle, and need not be known with extreme precision, it is better to compute the excess of B above $b$; and, supposing $\theta$ to be given in minutes,

$$\text{B} - b = \text{B}\,(1 - \cos \theta) = 2\,\text{B} \sin^2 \frac{\theta}{2} = \tfrac{1}{2}\,\text{B}\,\theta^2 \sin^2 1' = \frac{\sin^2 1'}{2}\,\theta^2\,\text{B}$$

or, $\text{B} - b = 0.00000004231\,\theta^2\,\text{B}$

or, by logarithms,

$$\log\,(\text{B} - b) = \text{const}\,\log\,2.626422 + 2\,\log\,\theta + \log\,\text{B}$$

XXIX.—*To Reduce a Broken Base to a Straight Line.*

Let—

$a$ and $b$ be the given sides; and

C the contained angle, very nearly 180°.

Make $\text{C} = 180° - \theta$; $\theta$ being small, and $\cos \theta = 1 - \tfrac{1}{2}\theta^2$,

then—

$$\text{side } c = a + b - \frac{\sin^2 1'}{2} \cdot \frac{a\,b\,\theta^2}{a + b}$$

$$= a + b - 0.00000004231 \times \frac{a\,b\,\theta}{a + b}$$

$\theta$ being expressed in minutes.

$$\log\,0.00000004231 = 2.6264222$$

XXX.—*To Find the Length*, B D = *x*, *of a Portion of a Straight Line*, A H, *Knowing the Two Other Portions*, A B = *a*, D H = *b*, *and also the Angles a, β, γ, from any Exterior Station*, C, *between B and A, D and A, and H and A*.

The problem being intended to supply by observation any portion of a base which cannot be directly measured—

$$\tan^2 \varphi = \frac{4\,a\,b}{(a-b)^2} \times \frac{\sin \beta \sin (\gamma - a)}{\sin a \sin (\gamma - \beta)}$$

$$x = -\frac{a+b}{2} \pm \frac{a-b}{2 \cos \varphi}$$

XXXI.—*To Reduce a Measured Base to the Level of the Sea.*

Let—

　　*r* represent the radius of the earth (or better, the normal, N,) corresponding to the base *b* at the level of the sea; and

　　*r + a* the radius referred to the level of the measured base B,

then—

$$r + a : r :: B : b = B \times \frac{r}{r+a}$$

and—

$$B - b = B - B \frac{r}{r+a} = B \times \left( \frac{a}{r} - \frac{a^2}{r^2} + \text{etc.} \right)$$

But the radius of the earth being very great in comparison to the difference of level, *a*, we have the correction δ sufficiently accurate by retaining only the first term; hence—

$$\delta = \frac{B\,a}{r}.$$

### XXXII.—*Correction for Temperature in Metallic Rods.*

Let—

$e$ = the linear expansion for $1°$ of Fahrenheit ;

$l$ = the length of the rod before expansion ;

$l'$ = the length of the rod after expansion ;

$t$ = the number of degrees Fahrenheit,

then—

$$\text{total expansion} = e\,t$$

and—

$$l' = l\,(1 + e\,t)$$

The following expansions were adopted by Mr. Hassler in his comparisons of weights and measures, (report of 1832 :)

| Expansion for $1°$ F. $= e$ | For 1 in a yard's length. | |
|---|---|---|
| Platinum ..0.0000051344 | ..0.0001848384 | English inches. |
| Brass bar..0.00001050903 | ...0.00037832508 | " |
| Iron bar...0.000006963535 | ..0.000240687260 | " |

Other authorities :

| Expansion for $1°$ F. $= e$ | For 1 in a yard's length. | | |
|---|---|---|---|
| Brass bar..0.000010480 | .....0.0003772800 | Eng. in. | Bailey. |
| Brass rod..0.0000105155 | ....0.0003785580 | " | Roy. |
| Brass rod..0.0000106666 | ....0.0003839976 | " | Troughton. |
| Brass wire..0.0000107407 | ....0.0003866652 | " | Smeaton. |
| Iron bar...0.0000069907 | ....0.0002516652 | " | Smeaton. |
| Steel rod...0.0000063596 | ....0.0002289456 | " | Roy. |
| Glass barom- eter-tubes .0.0000043119 | ....0.0001552284 | " | Roy. |
| White Nor- way pine..0.0000022685 | ....0.0000816660 | " | Kater. |

### XXXIII.—*Measurement of Distances by Sound.*

The *velocity* of sound, in one second of time, at 32° Fahrenheit, is about 1090 English feet. For any higher temperature,

$$v = 1089^{ft}.42 \; \sqrt{1 + (t - 32°) \times 0.00208}$$

*t* being the temperature in degrees Fahrenheit.

The velocity of sound through the air is independent of the barometric pressure, and experiments show it to be sensibly un-affected by its hygrometrical state of moisture and dryness; by the nature of the sound itself, whether produced by a blow, gun-shot, the voice, or a musical instrument; by the original direction of the sound, whether, for instance, the muzzle of a gun is turned one way or the other; or by the nature and position of the ground over which the sound is conveyed.

It is affected by the wind; but, in ordinary cases, likely to be selected for experiment, its influence would be almost inappre-ciable.

*Velocity and Force of the Wind.*

| Velocity in— | | Pressure on 1 square foot. | Common designations of the force of the winds. |
|---|---|---|---|
| 1 hour. | 1 second. | | |
| *Miles.* | *Feet.* | *Pounds.* | |
| 1 | 1.47 | 0.005 | Hardly perceptible. |
| 2 | 2.93 | 0.020 | Just perceptible. |
| 3 | 4.40 | 0.044 | |
| 4 | 5.87 | 0.079 | Gentle, pleasant wind. |
| 5 | 7.33 | 0.123 | |
| 10 | 14.67 | 0.492 | Pleasant, brisk breeze. |
| 15 | 22.00 | 1.107 | |
| 20 | 29.34 | 1.968 | Very brisk. |
| 25 | 36.67 | 3.075 | |
| 30 | 44.01 | 4.429 | High wind. |
| 35 | 51.34 | 6.027 | |
| 40 | 58.68 | 7.873 | Very high. |
| 45 | 66.01 | 9.963 | |
| 50 | 73.35 | 12.300 | A storm or tempest. |
| 60 | 88.02 | 17.715 | A great storm. |
| 80 | 117.36 | 31.490 | A hurricane. |
| 100 | 146.70 | 49.200 | A hurricane that tears up trees, carries buildings before it, &c. |

### XXXIV.—*For Reconnaissances.*

#### "THREE-POINT PROBLEM."

At a point, P, from whence are to be seen three points, A, C, B, forming a triangle, the elements (*i. e.*, the angles and sides) of which are known, measure the angles A P C and C P B; then, required to determine the direction and distance of the point P from each object.

Make—

$$A \, C = a;$$
$$B \, C = b;$$
$$B C A = C;$$
$$A P C = P; \text{ and}$$
$$C P B = P';$$

also, make—

$$R = 360° - P - P' - C;$$
$$x = C \, A \, P;$$
$$y = P \, B \, C.$$

Then will—

$$\cot x = \cot R \left( \frac{a \sin P'}{b \sin P \cos R} + 1 \right)$$

$$y = R - x$$

The use of these formulæ need not be embarrassing if care is taken in properly applying the signs of cos R and cot R. When R is less than 90° both cos R and cot R are plus; between 90° and 180° both are minus; between 180° and 270° cos R is minus and cot R plus; between 270° and 360° cos R is plus and cot R minus.

This problem is indeterminate when P falls upon the circumference of the circle passing through A B C. A case of this nature is of rare occurrence, however, in practice.

For the more general form of this problem, where the angles are measured from the point sought to any number of given points, to fix its position, see Coast Survey Report of 1864.

**XXXV.**—*For Computing the Principal Geodetic Quantities Depending on the Spheroidal Figure of the Earth at any Given Latitude.*

$$\text{Eccentricity of the Earth} = e = \left( \frac{a^2 - b^2}{a^2} \right)^{\frac{1}{2}}$$

$$= \left( 1 - \frac{b^2}{a^2} \right)^{\frac{1}{2}} = 2\,E - E^2$$

$$\text{Ellipticity} = E = \frac{a - b}{a} = 1 - \frac{b}{a}$$

or, very nearly—

$$e^2 = 2E \; ; \qquad E = \frac{e^2}{2}$$

Normal ending at minor axis (or radius of curvature of a section perpendicular to the meridian) $= N$

$$= \frac{a}{(1 - e^2 \sin^2 L.)^{\frac{1}{2}}}$$

Normal ending at major axis ..... $= N' = N\,(1 - e^2)$

$$= \frac{a\,(1 - e^2)}{(1 - e^2 \sin^2 L.)^{\frac{1}{2}}}$$

Tangent ending at minor axis... $= t = N \cot L.$

Tangent ending at major axis.... $= T = N \tan L.\,(1 - e^2)$

Radius of the parallel......... $= \rho = N \cos L.$

Radius of curvature of the meridian $= R = \dfrac{N^3}{a^2}\,(1 - e^2)$

$$= \frac{a\,(1 - e^2)}{(1 - e^2 \sin^2 L.)^{\frac{3}{2}}}$$

Radius of curvature of a section making an angle Z with the meridian........... $= R_z$

$$= \frac{N\,R}{N \cos^2 Z + R \sin^2 Z}$$

Radius of the earth........... $= r$

$$= a \left( 1 - \frac{e^2\,(1 - e^2)\sin^2 L}{1 - e^2 \sin^2 L} \right)^{\frac{1}{2}}$$

Equatorial radius............... $= a$

Polar radius................. $= b$

The given latitude.............. $= L$

XXXVI.—*Numerical Values of Some of the Preceding Quantities, from a Discussion by* BESSEL *in the "Astronomische Nachrichten," No.* 438.

$$a = \text{equatorial radius} = 3272077.14 \text{ toises}$$
$$\log = 6.5148235337$$
$$b = \text{polar radius} = 3261139.33 \text{ toises}$$
$$\log = 6.5133693539$$

Ratio of the toise to the metre, law of France, December 10, 1799:

$$T = 1^{m}.9490363; \qquad \log = 0.2898199300$$

whence in metres—

$$a = 6377397^{m}.15; \qquad \log = 6.8046434637$$
$$b = 6356078^{m}.96; \qquad \log = 6.8031892839$$

Ratio of the axes:

$$a : b :: 299.1528 : 298.1528$$

Mean uncertainty $= \pm 4.667$ units.

Length of the earth's quadrant $= 5131179^{t}.81 = 10000855^{m}.76$

Mean uncertainty $= \pm 498^{m}.23$

$$e = \text{eccentricity} = \left(1 - \frac{b^2}{a^2}\right)^{\frac{1}{2}} = 0.0816967$$
$$\log = 8.9122052271$$
$$E = \text{ellipticity} = \tfrac{1}{2} e^2 \qquad \log = 7.5233789824$$

Length, in toises, of a meridional degree whose middle latitude is $\varphi$:

$$D_m = 57013^{t}.109 - 286^{t}.337 \cos 2\varphi + 0^{t}.611 \cos 4\varphi \Big\} + 0^{t}.001 \cos 6\varphi$$

Length of a degree of the parallel, in toises:

$$D_p = 57156^{t}.285 \cos \varphi - 47^{t}.825 \cos 3\varphi + 0^{t}.060 \cos 5\varphi$$

or, making $\sin \psi = e \sin \varphi$—

$$\log D_p = 4.7567009.0 + \log \cos \varphi - \log \cos \psi$$

### XXXVII.—*Relative Lengths of the Yard and the Metre.*

1. From Clarke's comparisons referred to the present parliamentary standard, (*Comparison of Standards of Length made at the Ordnance Survey Office by Captain A. R. Clarke, R. E., F. R. S., published by authority*, 1866:)

*Values Adopted in the Measurement, now in Progress, of an Arc of Parallel Extending from Ireland to the River Ural in Russia, as "the Exact Relative Lengths of Standards" used as the Units of Measure in the Triangulations of England, France, Belgium, Prussia, and Russia :*

| Standards. | Expressed in terms of the standard yard. | Expressed in inches. | Expressed in lines of the toise. | Expressed in millimetres. |
|---|---|---|---|---|
| The yard ......... | 1. 00000000 | 36. 000000 | 405. 34622 | 914. 39180 |
| The toise ......... | 2. 13151116 | 76. 734402 | 864. 00000 | 1949. 03632 |
| The metre ....... | 1. 09362311 | 39. 370432 | 443. 29600 | 1000. 00000 |

$$\log 39.370432 = 1.595170\dot{1}816$$

2. From Kater's comparisons with the Shuckburg scale, (*Phil. Trans. for* 1818:)

1 metre at 32° F. = 39.370790 inches of the old imperial standard at 62° F.

$$\log = 1.595174\dot{1}293$$

3. From Hassler's comparisons of the Troughton 82-inch scale, with the iron standard *committee metre* of the American Philosophical Society, (*Report of the Secretary of the Treasury on the Comparison of Weights and Measures, Twenty-Second Congress, First Session, June* 20, 1832:)

*Value adopted by the United States Coast Survey.*

1 metre at 32° F. = 39.36850535 inches of the Troughton 82-inch scale at 62° F.

$$\log = 1.5951489586$$

a value materially smaller than the preceding.

There is a doubt whether this discordance is to be attributed to inaccuracy in the length of the Troughton scale or in errors in the co-efficients of expansion used by Mr. Hassler.

## XXXVII.—*Relative Lengths of the Yard and the Metre*—Con'd.

### *Logarithms to Reduce Metres to Yards.*

| Clarke. | Kater. | Coast Survey. |
|---|---|---|
| 0.0388676809 | 0.0388716286 | 0.0388464579 |

$$\log 3 = 0.4771212547$$
$$\log 12 = 1.0791812460$$
$$\log 5280 = 3.7226339225$$

Kater's length of the metre in English inches was adopted in the preparation of the first edition of this Collection as being at the time most generally in use. It has been retained throughout the present volume, although the results of Clarke's comparisons should now be universally adopted.

The *committee metre* of the American Philosophical Society is the unit of length to which all linear measures of the Coast Survey are referred. It was compared August 24, 1867, at Paris, directly with the standard platinum metre of the *Conservatoire des arts et métiers*, and was found (at the temperature of melting ice) $= 1^m.00000336$ of the *platinum metre of the archives*.

The French *standard metre* has its normal length at zero centigrade, or the freezing-point. It was intended to be a natural standard, and to represent the ten-millionth part of the terrestrial arc between the equator and the pole, which was assumed to be 5130740 toises, and the length of the metre 443.29596 lines of the *toise du Pérou;* which quantity was declared by law in 1799 to be the length of the *legal* metre, and is the length of the standard *platinum metre of the archives*.

The *toise du Pérou*, made in 1735, is a bar of iron, and has its standard length at 13° Reaumur, (61°.25 Fahrenheit.) It was used by La Condamine in the measurement of an arc of the meridian in Peru in 1744. As the above determination of the length of the quarter of the meridian is now known to be erroneous, the *legal* metre becomes, in fact, but a legalized part of the *toise du Pérou*, and this last remains the primitive standard. It is the unit of length in which the greater part of the European geodetic measurements are expressed.

The standard *Klafter* of *Vienna* has its normal length at 13° Reaumur, and is $= 840.76134$ lines of the *toise du Pérou*.

The standard *Prussian foot* is a standard also at 13° Reaumur, and was declared by law to be $= 139.13$ lines of the *toise du Pérou*.

XXXVIII.—*Numerical Values of Bessel's Terrestrial Elements in English Yards, adopting Kater's Value of the Metre, viz:* 39.37079 *English Inches ;* $\log$ 1.5951741293.

Log. to reduce toises to yards $= 0.3286915586$

Log. to reduce metres to yards $= 0.0388716286$

$a =$ equatorial radius $= 6974532^y.339$

$\log = 6.8435150923$

$b =$ polar radius $= 6951218^y.059$

$\log = 6.8420609125$

Length, in yards, of a meridional degree whose middle latitude is $\varphi$:

$$D_m = 121525^y.183 - 610^y.336 \cos 2\,\varphi + 1^y.302 \cos 4\,\varphi \atop + 0^y.002 \cos 6\,\varphi$$

Length, in yards, of a degree of the parallel :

$$D_p = 121830^y.366 \cos \varphi - 101^y.941 \cos 3\,\varphi + 0^y.128 \cos 5\,\varphi$$

or, making $\sin \psi = e \sin \varphi$—

$$\log D_p = 5.0853925 + \log \cos \varphi - \log \cos \psi$$

or, using the logarithms of the numerical co-efficients—

$$D_m = 121525^y.183 - (2.7855691) \cos 2\,\varphi + (0.1147) \cos 4\,\varphi \atop + (7.3287) \cos 6\,\varphi$$

$$D_p = (5.0857556) \cos \varphi - (2.00835) \cos 3\,\varphi + (9.1069) \cos 5\,\varphi$$

or—

$$D_p = \frac{(5.0853925) \cos \varphi}{\cos \psi}$$

### Constant Logarithms.

$e^2 = 0.00667435 \dots\dots\dots\dots\dots\dots \log = 7.8244104542$

$\frac{1}{2} e^2 = E =$ ellipticity $= \dfrac{1}{299.66} \dots\dots\dots = 7.5233789824$

$\sin 1'' \dots\dots\dots\dots\dots\dots\dots\dots\dots = 4.6855748668$

$\frac{1}{2} \sin 1'' \dots\dots\dots\dots\dots\dots\dots\dots = 4.3845448711$

$3 \dfrac{e^2}{2} \sin 1'' \dots\dots\dots\dots\dots\dots = 2.6860751039$

$(1 - e^2) = 0.99332565 \dots\dots\dots\dots = 9.9970916404$

$a (1 - e^2) \dots\dots\dots\dots\dots\dots\dots = 6.8406067325$

$a \sin 1'' \dots\dots\dots\dots\dots\dots\dots\dots = 1.5290899591$

$a \sin 1''$, (arithmetical complement) $\dots\dots = 8.4709100409$

XXXIX.—*For Computing the Geodetic Latitudes, Longitudes, and Azimuths of Points of a Triangulation.*

1. For distances not exceeding one hundred miles:

$$- d\, L = K\, B \cos Z + K^2\, C \sin^2 Z + (\delta L)^2 D - K^2\, h\, E \sin^2 Z$$

where—

$$B = \frac{1}{R \text{ arc } 1''}$$

$$C = \frac{\tan L}{2\, N\, R \text{ arc } 1''}$$

$$D = \frac{\tfrac{3}{2}\, e^2 \sin L \cos L \text{ arc } 1''}{(1 - e^2 \sin^2 L)^{\frac{3}{2}}}$$

$$E = \frac{1 + 3 \tan^2 L}{6\, N^2}$$

$h = K\, B \cos Z$, or first term;

$\delta L$ an approximate value for $- d\, L$, computed from first and second term;

$$d\, M = \frac{A'\, K \sin Z}{\cos L'}$$

$A' = \dfrac{1}{N \text{ arc } 1''}$ referred to second point;

$$- d\, Z = \frac{d\, M \sin \tfrac{1}{2}\, (L + L')}{\cos \tfrac{1}{2}\, d\, L} + d\, M^3\, F$$

log F for latitude $25^\circ = 7.8324$; for latitude $45^\circ = 7.8404$.

2. For distances not exceeding twenty miles:

In terms of the sides of the triangles:

$$u'' = \frac{K}{N \sin 1''} = \frac{K\, (1 - e^2 \sin^2 L)^{\frac{1}{2}}}{a \sin 1''}.$$

$$L' = L - (1 + e^2 \cos^2 L)\, u'' \cos Z$$
$$\qquad - (1 + e^2 \cos^2 L)\, (u'' \sin Z)^2 \tan L \times \tfrac{1}{2} \sin 1'' \Big\}$$

$$M' = M + \frac{u'' \sin Z}{\cos L'}$$

$$Z' = 180^\circ + Z - \frac{u'' \sin Z}{\cos L'} \sin \tfrac{1}{2}\, (L + L')$$

### XXXIX.—*For Computing Geodetic Latitudes, &c.*—Continued.

In terms of the co-ordinates of rectangular axes referred to one of the points of the triangulation, the latitude and longitude of which are known; $y$ being the ordinate in the direction of the meridian, and $x$ the ordinate perpendicular to it:

$$L' = L \pm \frac{y}{R \sin 1''} - \tfrac{1}{2} \sin 1'' \left( \frac{x}{N \sin 1''} \right)^2 \times \tan \left( L \pm \frac{y}{R \sin 1''} \right)$$

$$M' = M \pm \left( \frac{x}{N \sin 1''} \right) \times \frac{1}{\cos L'}$$

$$Z' = (180 + Z) \pm \frac{x}{N \sin 1''} \tan L'$$

K = distance in yards between two stations, the latitude and longitude of one of which is known, and $n''$ this same distance converted to seconds of arc;

L = latitude of 1st station;

M = longitude of 1st station, + if west;

Z = azimuth of 2d station at 1st, counted from the south round by the west, from 0° to 360°; the algebraic signs of the sine and cosine of this angle must be carefully attended to;

L', M', Z,' the same things at 2d station, or quantities required;

$a$ = the equatorial radius;

$e$ = the eccentricity;

R = the radius of curvature of the meridian; and

N = the radius of curvature of a section perpendicular to the meridian.

The quantity—

$$\frac{n'' \sin Z}{\cos L'} \sin \tfrac{1}{2} (L + L')$$

or—

$$(M' - M) \sin \tfrac{1}{2} (L + L')$$

by which the azimuth at one end of a line exceeds the azimuth at the other, is called *the convergence of the meridians.*

XL.—*To Compute the Length and Direction of a Line Joining Two Points, the Latitudes and Longitudes of which are Known, or Measurement of a Base by Astronomical Observations.*

$$\frac{\beta}{2} = \frac{e^2 (L - L') \cos^2 \frac{1}{2} (L + L')}{2}$$

$$N = \frac{a}{[1 - e^2 \sin^2 \frac{1}{2} (L + L')]^{\frac{3}{2}}}$$

$$l = L - \frac{\beta}{2}$$

$$x'' = (M' - M) \cos l'$$

$$l' = L' + \frac{\beta}{2}$$

$$y'' = (l - l') - \frac{1}{2} \sin 1'' \, x''^2 \tan l$$

$$\tan Z = \frac{x''}{y''}$$

$$x = x'' N \sin 1''$$

$$u'' = \frac{x''}{\sin Z} = \frac{y''}{\cos Z}$$

$$y = y'' N \sin 1''$$

$$K = u'' N \sin 1''$$

in which—

L, L', M, M', represent the latitudes and longitudes of the two points;

$u''$, the distance between these points in seconds of arc ;

K, the distance between these points in linear units ;

$x''$, the number of seconds in the arc passing through the point of which L' is the latitude, and perpendicular to the meridian of the point of which L is the latitude;

$y''$, the seconds in the portion of this meridian between L and the foot of this perpendicular ;

$x, y$, the same quantities in linear units ;

Z, the azimuth of the second point, L', from the first, L; and

N, the normal at the middle latitude.

XL.—*To Compute the Length and Direction of a Line, &c.*—Con'd.

Particular attention must be paid to the sign $(L - L')$, for upon this depends the sign of $\frac{\beta}{2}$, and also to that of $(l - l')$ in the value of $y''$, so as to know whether the small quantity—

$$(-\tfrac{1}{2} \sin 1'' \; x''^2 \tan l)$$

is to be added to or substracted from $(l - l')$.

The azimuth Z is counted from the south, round by the west, from 0° to 360°.

The azimuth Z' (if required) is to be computed from Z, as in XXXIX, (2.)

This can be presented in a different form, thus :

$$\text{as, } (M' - M) \cos L' = u'' \sin Z$$

and,

$$u'' \cos Z = \frac{(L - L') - \tfrac{1}{4} u''^2 \sin^2 Z \cos^2 L' \tan L \sin 1'' (1 + e^2 \cos^2 L)}{1 + e^2 \cos^2 L}$$

Substituting, in this last, the value of $u'' \sin Z$, and dividing one by the other:

$$\tan Z = \frac{(M' - M) \cos L' (1 + e^2 \cos^2 L)}{(L - L') - \tfrac{1}{2}(M' - M)^2 \cos^2 L' \tan L \sin 1'' (1 + e^2 \cos^2 L)}$$

Then, knowing Z—

$$u'' = \frac{(M' - M) \cos L'}{\sin Z}$$

and—

$$K = u'' \, N \sin 1''$$

N being the normal for the mean latitude.

XLI.—*To Compute the Distance between Two Points, knowing their Latitudes and the Azimuth of one from the other.*

$$\frac{\beta}{2} = \frac{e^2 (L - L') \cos^2 \tfrac{1}{2} (L + L')}{2}$$

$$N = \frac{a}{[1 - e^2 \sin^2 \tfrac{1}{2} (L + L')]^{\frac{1}{2}}}$$

$$l = L - \frac{\beta}{2}; \qquad \tan \varphi = \frac{\tan l}{\cos Z}; \qquad l' = L' + \frac{\beta}{2}$$

$$\sin (\varphi - u'') = \frac{\sin l'}{\sin l} \sin \varphi ; \quad K = u'' \, N \sin 1''$$

See the note to the preceding formulæ. The algebraic sign of the azimuth, Z, will determine the sign of $\varphi$, and consequently whether the quantity $u''$ is to be added to or subtracted from $\varphi$.

XLII.—*To Compute the Distance between Two Points, knowing the Latitude of one, the Azimuth from this to the other, and the Difference of their Longitudes.*

$$\tan \varphi = \sin L \tan Z \qquad \tan L'' = \frac{\tan L \sin (\varphi - m)}{\sin \varphi}$$

$$\beta = e^2 (L - L'') \cos^2 \tfrac{1}{2} (L + L'')$$

$$L' = L'' - \beta \qquad l = L - \frac{\beta}{2} \qquad l' = L' + \frac{\beta}{2}$$

$$u'' = \frac{m \cos l'}{\sin Z} \qquad\qquad K = u'' N \sin 1''$$

$m =$ the difference of longitude.

The azimuth, Z, is, as before, counted from the south round by the west; its algebraic sign will determine the sign of $\varphi$, and consequently whether it is to be increased or diminished by $m$.

*Elements of the Figure of the Earth, Deduced by Captain A. R. Clarke, Royal Engineers, in Computing the Figures of the Meridians and of the Equator for Several Measured Arcs of Meridian, ( Comparison of Standards of Length, &c., London, 1866.)*

| Semi-axes. | | Length. | |
|---|---|---|---|
| | *Feet.* | *Toises.* | *Metres.* |
| Major semi-axis = $a$, of equator, (longitude 15° 34′ east)........................ | 20926350 | 3272537.3 | 6378294.0 |
| Minor semi-axis = $b$, of equator, (longitude 105° 34′ east)........................ | 20919972 | 3271540.1 | 6376350.4 |
| Polar semi-axis = $c$............................ | 20853429 | 3261133.8 | 6356068.1 |

$$\frac{a-c}{c} = \frac{1}{285.97} \qquad \frac{b-c}{c} = \frac{1}{313.38} \qquad \frac{a-b}{c} = \frac{1}{3269.5}$$

The length of the meridian quadrant passing through Paris is.......10001472.5 métres. and the minimum quadrant, in longitude 105° 34′, is.......10000024.5 metres.

*For a Spheroid of Revolution more nearly Corresponding to the Same Geodetic Measurements.*

| Semi-axes. | Length. | | |
|---|---|---|---|
| | *Feet.* | *Toises.* | *Metres.* |
| Equatorial semi-axis = $a$ ........................ | 20926062 | 3272492.3 | 6378206.4 |
| Polar semi-axis = $b$............................ | 20855121 | 3261398.4 | 6356583.8 |

$$\frac{b}{a} = \frac{293.98}{294.98} \qquad \frac{a-b}{a} = \frac{1}{294.98} = \text{ellipticity.}$$

*Survey of* ...........................

| No. of tri-angle. | Position. | Names of stations. | No. of obs. | Observed angles. | Errors and their dis-tribution. | Spherical angles. | Spherical excess. | Final plane angles. |
|---|---|---|---|---|---|---|---|---|
|  |  |  |  | °  ′  ″ | ″ | ″ | ″ | °  ′  ″ |
|  | Sought. | Cedar Point.. | 18 | 66 34 04.80 | − 0.36 | 04.44 | 1.58 | 66 34 02.86 |
| XIII | Right.. | Buck Hill ... | 18 | 64 08 37.78 | − 0.36 | 37.43 | 1.58 | 64 08 35.84 |
| *(Known side.)* |  |  |  |  |  |  |  |  |
|  | Left ... | Fort Flats ... | 18 | 49 17 23.24 | − 0.36 | 22.88 | 1.58 | 47 17 21.30 |
|  |  |  |  |  |  |  |  | 180 0 0.00 |

*Survey of* ...........................

|  | Latitudes. |
|---|---|

$$L' = L - u''.(1 + e^2 \cos^2 L) \cos Z$$
$$- \tfrac{1}{2} \sin 1'' \sin^2 Z\, u''^2 (1 + e^2 \cos^2 L) \tan L$$

**Fort Flats...**

Latitude $L$ ....... $= 45°\ 39'\ 13''.89$

$\log K$ (yards)......... $= 4.7295212$     $\tfrac{1}{2} \sin 1''$ ........ $= 4.38454$

$\log \dfrac{1}{N \sin 1''}$ ......... $= 8.4701676$     $2 \log \sin Z$ ...... $= 9.09522$

$\log u''$ ............... $= 3.1996888$     $2 \log u''$ ........ $= 6.39936$

$\log (1 + e^2 \cos^2 L)$ .... $= 0.0014140$     ............... $= 0.00141$

$\log \cos Z$ ........ $(-) = 9.9711240$     $\log \tan L$ ....... $= 0.00991$

$\log$ 1st term ........ $= 3.1722268$     $\log$ 2d term ..... $= 9.89034$

1st term........ $(+) = +1486''.71$     2d term ........ $= 0''.77$

2d term ........ $(-) = - 0''.77$

$\delta L$ ............... $= 0°\ 24'\ 45''.94$     $L + L'$ ... $= 91°\ 43'\ 13''.72$

$L$ ............... $= 45°\ 39'\ 13''.89$     $\dfrac{L + L'}{2}$ ... $= 45°\ 51'\ 36''.86$

**Cedar Point .** Latitude $L'$ ....... $= 46°\ 03'\ 59''.83$

AND COMPUTATION.

## Calculation of Triangles.

| Computing letter. | Logarithms of their sines. | Calculation of the sides. | Sides in yards. | Designation. |
|---|---|---|---|---|
| S | 9.9626198 | log R L .........= 4.7379524 <br> comp log sin S ..= 0.0373802 <br> log sin R .......= 9.9541886 | = 54695.61 | { Buck Hill — Fort { Flats. |
| R | 9.9541886 | log L S ........= 4.7295212 | = 53644.00 | { Fort Flats—Cedar { Point. |
| | | log R L + } = 4.7753326 <br> comp log sin S } <br> log sin L .......= 9.8796760 | | |
| L | 9.8796760 | log R S ........= 4.6550086 | = 45186 49 | { Buck Hill—Cedar { Point. |

## Geodetic Determination of Positions.          (Secondary.)

| Longitudes. | Azimuths. | Remarks. |
|---|---|---|
| $M' = M + \dfrac{u'' \sin Z}{\cos L'}$ | $Z' = 180° + Z - (\delta M) \sin \dfrac{L + L'}{2}$ | |
| Longitude M = 84° 42′ 22″.19 | Azimuth Z .....= 159° 20′ 13″.62 | |
| log sin Z ....(+)=9.5476117 | 180° | |
| log $u''$ .........= 3.1996888 | 180° + Z ......= 339° 20′ 13″.62 <br> 20° 39′ 46″.38 | |
| | 2.7473005 | |
| log cos L' .......= 9.8412474 | log sin $\dfrac{L + L'}{2}$ ......= 9.8559089 | |
| log $\delta$ M .........= 2.9060529 | ................(+)= 2.9060529 | |
| | log $\delta$ Z .............= 2.7619618 | |
| | − 578″.05 | |
| $\delta$ M ........= 0° 13′ 25″.48 | $\delta$ Z.............= 0° 09′ 38″.05 | |
| M ..........= 84° 42′ 22″.19 | 180° + Z ......= 339° 20′ 13″.62 | |
| Longitude M = 84° 55′ 47″.67 | Azimuth Z′ ....= 339° 10′ 35″.57 | |

### Normal or Radius of Curvature of the Perpendicular to the Meridian.

Ellipticity $= \frac{1}{300}$; equatorial radius $= 6974532$ yards; $\log = 6.8435151$.
Log. to reduce yards to metres $= 9.9611283714$.

| Latitude | $N = \dfrac{a}{(1 - e^2 \sin^2 L)^{\frac{1}{2}}}$ | | | $\log(1 + e^2 \cos^2 L)$ | Difference for 10'. |
|---|---|---|---|---|---|
| | $\log N$ | Com. diff. for 10'. | $\log \dfrac{1}{N \sin 1''}$ | | |
| **°   '** | | | | | |
| 20   0 | 6.8436847 | | 8.4707404 | 0.0025521 | |
| | | 27.3 | | | 55 |
| 15 | 6888 | 27.6 | 7363 | 5439 | 56 |
| 30 | 6929 | 27.8 | 7322 | 5356 | 55 |
| 45 | 6971 | 28.1 | 7280 | 5274 | 56 |
| 21   0 | 7013 | | 7238 | 5191 | |
| | | 28.4 | | | 57 |
| 15 | 7056 | 28.7 | 7196 | 5106 | 57 |
| 30 | 7099 | 29.0 | 7153 | 5021 | 58 |
| 45 | ·7142 | 29.3 | 7109 | 4934 | 58 |
| 22   0 | 7186 | | 7066 | 4847 | |
| | | 29.5 | | | 58 |
| 15 | 7230 | 29.7 | 7021 | 4760 | 59 |
| 30 | 7274 | 30.0 | 6977 | 4671 | 59 |
| 45 | 7319 | 30.2 | 6932 | 4582 | 60 |
| 23   0 | 7365 | | 6887 | 4492 | |
| | | 30.5 | | | 61 |
| 15 | 7410 | 30.7 | 6841 | 4401 | 62 |
| 30 | 7457 | 31.0 | 6795 | 4309 | 62 |
| 45 | 7503 | 31.2 | 6748 | 4217 | 62 |
| 24   0 | 7550 | | 6701 | 4124 | |
| | | 31.5 | | | 63 |
| 15 | 7597 | 31.7 | 6654 | 4030 | 63 |
| 30 | 7645 | 32.0 | 6607 | 3935 | 63 |
| 45 | 7693 | 32.2 | 6559 | 3840 | 64 |
| 25   0 | 7741 | | 6510 | 3744 | |
| | | 32.5 | | | 64 |
| 15 | 7790 | 32.7 | 6462 | 3648 | 65 |
| 30 | 7839 | 32.9 | 6413 | 3550 | 65 |
| 45 | 7888 | 33.2 | 6363 | 3452 | 66 |
| 26   0 | 7938 | | 6313 | 3353 | |
| | | 33.4 | | | 66 |
| 15 | 7988 | 33.6 | ·6263 | 3254 | 67 |
| 30 | 8038 | 33.8 | 6213 | 3154 | 67 |
| 45 | 8089 | 34.0 | 6162 | 3053 | 68 |
| 27   0 | 8140 | | 6111 | 2951 | |
| | | 34.3 | | | 68 |
| 15 | 8192 | 34.5 | 6060 | 2849 | 69 |
| 30 | 8243 | 34.7 | 6008 | 2746 | 69 |
| 45 | 8295 | 34.9 | 5956 | 2643 | 69 |
| 28   0 | 8348 | | 5904 | 2539 | |
| | | 35.1 | | | 70 |
| 15 | 8400 | 35.3 | 5851 | 2434 | 70 |
| 30 | 8453 | 35.5 | 5798 | 2329 | 71 |
| 45 | 6.8438507 | 35.7 | 8.4705745 | 0.0022223 | 71 |

## Normal, &c.—Continued.

| Latitude. ° | ′ | $N = \dfrac{a}{(1 - e^2 \sin^2 L)^{\frac{1}{2}}}$ log N | Com. diff. for 10′. | $\log \dfrac{1}{N \sin 1''}$ | log $(1 + e^2 \cos^2 L)$ | Difference for 10′. |
|---|---|---|---|---|---|---|
| 29 | 0 | 6.8438560 | 35.9 | 8.4705691 | 0,0022117 | 71 |
|  | 15 | 8614 | 36.1 | 5637 | 2010 | 72 |
|  | 30 | 8668 | 36.3 | 5583 | 1902 | 72 |
|  | 45 | 8723 | 36.5 | 5529 | 1794 | 72 |
| 30 | 0 | 8777 | 36.7 | 5474 | 1686 | 73 |
|  | 15 | 8832 | 36.9 | 5419 | 1576 | 73 |
|  | 30 | 8888 | 37.1 | 5364 | 1466 | 73 |
|  | 45 | 8943 | 37.2 | 5308 | 1356 | 74 |
| 31 | 0 | 8999 | 37.4 | 5252 | 1245 | 74 |
|  | 15 | 9055 | 37.6 | 5196 | 1134 | 75 |
|  | 30 | 9111 | 37.7 | 5140 | 1022 | 75 |
|  | 45 | 9168 | 37.9 | 5084 | 0910 | 75 |
| 32 | 0 | 9225 | 38.1 | 5027 | 0797 | 75 |
|  | 15 | 9282 | 38.2 | 4970 | 0684 | 76 |
|  | 30 | 9339 | 38.4 | 4912 | 0570 | 77 |
|  | 45 | 9397 | 38.5 | 4855 | 0455 | 77 |
| 33 | 0 | 9454 | 38.7 | 4797 | 0340 | 77 |
|  | 15 | 9512 | 38.9 | 4737 | 0225 | 77 |
|  | 30 | 9571 | 39.0 | 4681 | .0020109 | 77 |
|  | 45 | 9629 | 39.1 | 4622 | .0019993 | 77 |
| 34 | 0 | 9688 | 39.2 | 4564 | 9877 | 78 |
|  | 15 | 9747 | 39.4 | 4505 | 9760 | 78 |
|  | 30 | 9806 | 39.5 | 4446 | 9643 | 79 |
|  | 45 | 9865 | 39.7 | 4387 | 9525 | 79 |
| 35 | 0 | 9924 | 39.8 | 4327 | 9407 | 80 |
|  | 15 | .8439984 | 40.0 | 4267 | 9288 | 80 |
|  | 30 | .8440044 | 40.0 | 4208 | 9169 | 80 |
|  | 45 | 0104 | 40.1 | 4148 | 9050 | 80 |
| 36 | 0 | 0164 | 40.3 | 4087 | 8931 | 80 |
|  | 15 | 0224 | 40.3 | 4027 | 8811 | 81 |
|  | 30 | 0285 | 40.5 | 3966 | 8690 | 80 |
|  | 45 | 0346 | 40.6 | 3906 | 8570 | 81 |
| 37 | 0 | 0406 | 40.7 | 3845 | 8449 | 81 |
|  | 15 | 0467 | 40.7 | 3784 | 8328 | 81 |
|  | 30 | 0529 | 40.9 | 3723 | 8206 | 81 |
|  | 45 | 0590 | 41.0 | 3661 | 8084 | 81 |
| 38 | 0 | 0651 | 41.1 | 3600 | 7963 | 82 |
|  | 15 | 0713 | 41.1 | 3538 | 7840 | 82 |
|  | 30 | 0775 | 41.2 | 3477 | 7717 | 82 |
|  | 45 | 0837 | 41.4 | 3415 | 7594 | 82 |
| 39 | 0 | 0898 | 41.4 | 3353 | 7471 | 82 |
|  | 15 | 0961 | 41.4 | 3291 | 7348 | 83 |
|  | 30 | 1023 | 41.5 | 3229 | 7224 | 82 |
|  | 45 | 6.8441085 | 41.6 | 8.4703166 | 0,0017101 | 83 |

## Normal, &c.—Continued.

| Latitude | $N = \dfrac{a}{(1 - e^2 \sin^2 L)^{\frac{1}{2}}}$ | | $\log \dfrac{1}{N \sin 1''}$ | $.\ \log (1 + e^2 \cos^2 L)$ | Difference for 10'. |
|---|---|---|---|---|---|
| ° ′ | log $N$ | Com. diff. for 10'. | | | |
| 40  0 | 6.8441147 | | 8.4703104 | 0.0016977 | |
|   15 | 1210 | 41.7 | 3041 | 6853 | 83 |
|   30 | 1273 | 41.8 | 2979 | 6728 | 84 |
|   45 | 1335 | 41.8 | 2916 | 6604 | 83 |
|      |      | 41.9 |      |      | 84 |
| 41  0 | 1398 | | 2853 | 6479 | |
|   15 | 1461 | 41.9 | 2791 | 6354 | 84 |
|   30 | 1524 | 41.9 | 2728 | 6229 | 84 |
|   45 | 1587 | 42.0 | 2665 | 6104 | 84 |
|      |      | 42.1 |      |      | 84 |
| 42  0 | 1650 | | 2602 | 5979 | |
|   15 | 1713 | 42.1 | 2539 | 5853 | 84 |
|   30 | 1776 | 42.1 | 2475 | 5728 | 84 |
|   45 | 1839 | 42.2 | 2412 | 5602 | 84 |
|      |      | 42.1 |      |      | 84 |
| 43  0 | 1903 | | 2349 | 5477 | |
|   15 | 1967 | 42.2 | 2286 | 5351 | 84 |
|   30 | 2029 | 42.2 | 2222 | 5225 | 84 |
|   45 | 2093 | 42.3 | 2159 | 5099 | 84 |
|      |      | 42.3 |      |      | 84 |
| 44  0 | ˙2156 | | 2095 | 4973 | |
|   15 | 2219 | 42.3 | 2032 | 4847 | 84 |
|   30 | 2283 | 42.3 | 1969 | 4721 | 84 |
|   45 | 2346 | 42.3 | 1905 | 4595 | 84 |
|      |      | 42.3 |      |      | 84 |
| 45  0 | 2410 | | 1842 | 4469 | |
|   15 | 2473 | 42.3 | 1778 | 4343 | 84 |
|   30 | 2537 | 42.3 | 1715 | 4217 | 84 |
|   45 | 2600 | 42.3 | 1651 | 4091 | 84 |
|      |      | 42.3 |      |      | 84 |
| 46  0 | 2663 | | 1588 | 3965 | |
|   15 | 2727 | 42.3 | 1525 | 3839 | 84 |
|   30 | 2790 | 42.3 | 1461 | 3713 | 84 |
|   45 | 2854 | 42.3 | 1398 | 3587 | 84 |
|      |      | 42.2 |      |      | 84 |
| 47  0 | 2917 | | 1334 | 3461 | |
|   15 | 2980 | 42.2 | 1271 | 3336 | 84 |
|   30 | 3043 | 42.1 | 1208 | 3210 | 84 |
|   45 | 3107 | 42.1 | 1145 | 3084 | 84 |
|      |      | 42.1 |      |      | 84 |
| 48  0 | 3170 | | 1082 | 2959 | |
|   15 | 3233 | 42.1 | 1018 | 2833 | 84 |
|   30 | 3296 | 42.0 | 0955 | 2708 | 84 |
|   45 | 3359 | 42.0 | 0892 | 2583 | 84 |
|      |      | 41.9 |      |      | 84 |
| 49  0 | 3422 | | 0830 | 2458 | |
|   15 | 3485 | 41.9 | 0767 | 2333 | 84 |
|   30 | 3547 | 41.9 | 0704 | 2209 | 83 |
|   45 | 3610 | 41.8 | 0641 | 2084 | 84 |
|      |      | 41.8 |      |      | 83 |
| 50  0 | 6.8443673 | | 8.4700579 | 0.0011960 | |

### Radius of Curvature of the Meridian.

Ellipticity $= \frac{1}{300}$; equatorial radius $= 6974532$ yards.

| Latitude. | $R = \dfrac{a(1-e^2)}{(1-e^2\sin^2 L)^{\frac{3}{2}}}$ | | |
|---|---|---|---|
| | log R | Com. diff. for 10'. | $\log \dfrac{1}{R \sin 1''}$ |
| ° ' | | | |
| 20 0 | 6. 8411155 | 81. 9 | 8. 4733096 |
| 15 | 1278 | 82. 7 | 2973 |
| 30 | 1402 | 83. 5 | 2849 |
| 45 | 1527 | 84. 3 | 2724 |
| 21 0 | 1654 | 85. 1 | 2598 |
| 15 | 1781 | 86. 0 | 2470 |
| 30 | 1910 | 86. 8 | 2341 |
| 45 | 2040 | 87. 5 | 2211 |
| 22 0 | 2172 | 88. 4 | 2080 |
| 15 | 2304 | 89. 1 | 1947 |
| 30 | 2438 | 90. 0 | 1813 |
| 45 | 2573 | 91. 0 | 1679 |
| 23 0 | 2709 | 91. 5 | 1543 |
| 15 | 2846 | 92. 3 | 1405 |
| 30 | 2984 | 93. 0 | 1267 |
| 45 | 3124 | 93. 6 | 1128 |
| 24 0 | 3264 | 94. 6 | 0987 |
| 15 | 3406 | 95. 2 | 0845 |
| 30 | 3549 | 96. 0 | 0702 |
| 45 | 3693 | 96. 7 | 0559 |
| 25 0 | 3838 | 97. 4 | 0414 |
| 15 | 3984 | 98. 1 | 0268 |
| 30 | 4131 | 98. 8 | . 4730120 |
| 45 | 4279 | 99. 4 | . 4729972 |
| 26 0 | 4428 | 100. 1 | 9823 |
| 15 | 4578 | 100. 9 | 9673 |
| 30 | 4730 | 101. 5 | 9522 |
| 45 | 4882 | 102. 1 | 9370 |
| 27 0 | 5035 | 102. 8 | 9216 |
| 15 | 5189 | 103. 4 | 9062 |
| 30 | 5344 | 104. 1 | 8907 |
| 45 | 5500 | 104. 7 | 8751 |
| 28 0 | 5657 | 105. 3 | 8594 |
| 15 | 5815 | 106. 0 | 8436 |
| 30 | 5974 | 106. 5 | 8277 |
| 45 | 6. 8416134 | 107. 1 | 8. 4728117 |

## Radius of Curvature of the Meridian—Continued.

$$R = \frac{a(1 - e^2)}{(1 - e^2 \sin^2 L)^{\frac{3}{2}}}$$

| Latitude. | log R | Com. diff. for 10'. | $\log \dfrac{1}{R \sin 1''}$ |
|---|---|---|---|
| ° ' | | | |
| 29   0 | 6. 8416295 | 107. 1 | 8. 4727956 |
| 15 | 6456 | 108. 2 | 7795 |
| 30 | 6619 | 108. 6 | 7632 |
| 45 | 6782 | 109. 4 | 7469 |
| 30   0 | 6946 | 110. 0 | 7305 |
| 15 | 7111 | 110. 5 | 7140 |
| 30 | 7277 | 111. 1 | 6974 |
| 45 | 7444 | 111. 6 | 6808 |
| 31   0 | 7611 | 112. 2 | 6640 |
| 15 | 7779 | 112. 7 | 6472 |
| 30 | 7948 | 113. 1 | 6303 |
| 45 | 8118 | 113. 7 | 6133 |
| 32   0 | 8288 | 114. 1 | 5963 |
| 15 | 8460 | 114. 6 | 5792 |
| 30 | 8632 | 115. 1 | 5620 |
| 45 | 8804 | 115. 6 | 5447 |
| 33   0 | 8978 | 116. 0 | 5274 |
| 15 | 9152 | 116. 5 | 5100 |
| 30 | 9326 | 116. 9 | 4925 |
| 45 | 9502 | 117. 3 | 4750 |
| 34   0 | 9678 | 117. 7 | 4574 |
| 15 | . 8419854 | 118. 1 | 4397 |
| 30 | . 8420031 | 118. 5 | 4220 |
| 45 | 0209 | 119. 1 | 4042 |
| 35   0 | 0387 | 119. 3 | 3864 |
| 15 | 0566 | 119. 7 | 3685 |
| 30 | 0746 | 120. 0 | 3506 |
| 45 | 0926 | 120. 4 | 3325 |
| 36   0 | 1107 | 120. 7 | 3145 |
| 15 | 1288 | 121. 1 | 2964 |
| 30 | 1469 | 121. 4 | 2782 |
| 45 | 1651 | 121. 7 | 2600 |
| 37   0 | 1834 | 122. 0 | 2417 |
| 15 | 2017 | 122. 3 | 2234 |
| 30 | 2200 | 122. 7 | 2051 |
| 45 | 2384 | 122. 9 | 1867 |
| 38   0 | 2569 | 123. 1 | 1683 |
| 15 | 2753 | 123. 5 | 1498 |
| 30 | 2939 | 123. 7 | 1313 |
| 45 | 3124 | 124. 0 | 1127 |
| 39   0 | 3310 | 124. 1 | 0941 |
| 15 | 3496 | 124. 4 | 0755 |
| 30 | 3683 | 124. 6 | 0568 |
| 45 | 6. 8423870 | 124. 8 | 8. 4720382 |

*Radius of Curvature of the Meridian*—Continued.

$$R = \frac{a\,(1-e^2)}{(1-e^2\sin^2 L)^{\frac{3}{2}}}$$

| Latitude. | | log R | Com. diff. for 10'. | $\log \dfrac{1}{R \sin 1''}$ |
|---|---|---|---|---|
| ° | ′ | | | |
| 40 | 0 | 6.8424057 | 125.0 | 8.4720194 |
| | 15 | 4244 | 125.1 | .4720007 |
| | 30 | 4432 | 125.3 | .4719819 |
| | 45 | 4620 | 125.5 | 9631 |
| 41 | 0 | 4808 | 125.7 | 9443 |
| | 15 | 4997 | 125.8 | 9254 |
| | 30 | 5186 | 126.0 | 9066 |
| | 45 | 5375 | 126.2 | 8877 |
| 42 | 0 | 5564 | 126.2 | 8687 |
| | 15 | 5753 | 126.3 | 8498 |
| | 30 | 5943 | 126.4 | 8309 |
| | 45 | 6132 | 126.6 | 8119 |
| 43 | 0 | 6322 | 126.6 | 7929 |
| | 15 | 6512 | 126.7 | 7739 |
| | 30 | 6702 | 126.7 | 7549 |
| | 45 | 6892 | 126.8 | 7359 |
| 44 | 0 | 7082 | 126.8 | 7169 |
| | 15 | 7273 | 126.9 | 6979 |
| | 30 | 7463 | 127.0 | 6788 |
| | 45 | 7653 | 126.9 | 6598 |
| 45 | 0 | 7844 | 127.0 | 6408 |
| | 15 | 8034 | 126.9 | 6217 |
| | 30 | 8224 | 126.9 | 6027 |
| | 45 | 8415 | 126.9 | 5837 |
| 46 | 0 | 8605 | 126.8 | 5647 |
| | 15 | 8795 | 126.8 | 5456 |
| | 30 | 8985 | 126.7 | 5266 |
| | 45 | 9175 | 126.7 | 5076 |
| 47 | 0 | 9365 | 126.6 | 4886 |
| | 15 | 9555 | 126.6 | 4696 |
| | 30 | 9745 | 126.4 | 4506 |
| | 45 | .8429934 | 126.3 | 4317 |
| 48 | 0 | .8430124 | 126.2 | 4127 |
| | 15 | 0313 | 126.1 | 3938 |
| | 30 | 0502 | 126.0 | 3749 |
| | 45 | 0691 | 125.8 | 3560 |
| 49 | 0 | 0880 | 125.7 | 3371 |
| | 15 | 1068 | 125.5 | 3183 |
| | 30 | 1257 | 125.3 | 2995 |
| | 45 | 1445 | 125.0 | 2807 |
| 50 | 0 | 6.8431632 | | 8.4712619 |

## XLIII.—*Projection of Maps.*

### POLYCONIC PROJECTION.

In this development of the earth's surface each parallel of latitude is supposed to be represented on a plane by the development of a cone having the parallel for its base and its vertex in the point where a tangent to the parallel intersects the earth's axis. The map thus becomes the development of the surfaces of several successive cones, and the degrees of the parallel preserve their true length.

Normal .......................... $N = \dfrac{a}{(1 - e^2 \sin^2 L.)^{\frac{1}{2}}}$

Radius of the meridian .............. $R_m = N^3 \dfrac{(1 - e^2)}{a^2}$

Radius of the parallel ............... $R_p = N \cos L$

Degree of the meridian ............. $D_m = \dfrac{\pi}{180} R_m$

$$= 3600\, R_m \sin 1''$$

Degree of the parallel .............. $D_p = \dfrac{\pi}{180} R_p$

$$= 3600\, R_p \sin 1''$$

Radius of the developed parallel or side of tangent cone................ $r = N \cot L$

Designating by $n$ any arc of the parallel, or difference of longitude, to be developed, and by $\theta$ the corresponding angle subtended by the developed parallel at the vertex of the cone; then the length of the given arc will be :

$$n\, R_p = n\, N \cos L$$

and also—

$$\theta r = \theta\, N \cot L .$$

whence—

Angle of the developed parallel, $\theta = n \sin L$

and as the developed parallels are circular arcs, the co-ordinates of curvature are :

$d_m$, difference of meridians, $= x = r \sin \theta$

$d_p$, difference of parallels, $= y = r \,\text{ver} \sin \theta = x \tan \frac{1}{2} \theta$

For surfaces of small extent the arc of the parallel may be considered coincident with its chord ; and as the angle between a

### XLIII.—*Projection of Maps*—Continued.

tangent and a chord is half the angle at the center subtended by the chord,

$\delta_m$, difference of meridians, $= x = D_p \cos \frac{1}{2} \theta$

$\delta_p$, difference of parallels, $= y = D_p \sin \frac{1}{2} \theta$

The values of $\delta_m$ and $\delta_p$ and of $D_m$ and $D_p$ will be found in the following tables.

*Example of their Use.*—Let it be required to make a projection containing 40' of longitude between the parallels of 41° 30' and 42° 10', to be subdivided to 5'.

Assume the center of the sheet to be the intersection of the middle parallel with the middle meridian of the proposed map, which point call A; in this case a point in the parallel of 41° 50'.

Through A draw the central meridian and a line at right angles to it.

Beginning at A, lay of above and below, on the central meridian, the values of $D_m$ from 41° 50' to 41° 55'; 41° 55' to 42°; 42° to 42° 5', etc.; and from 41° 50' to 41° 45'; 41° 45' to 41° 40', etc.; these values to be taken from the table of *Meridional Arcs—Values of $D_m$ in Yards*, by interpolation from the values there given for the middle latitudes of 41° and 42°.

Through each of the points ..., $A^{ii}$, $A^i$, A, $A_i$, $A_{ii}$, ..., thus found, lay off perpendiculars to the central meridian.

Now turn to the table of *Co-ordinates, $\delta_m$ and $\delta_p$; in Yards*, and lay off, from each of the points ..., $A^{ii}$, $A^i$, A, $A_i$, $A_{ii}$, ..., to the right and left of the central meridian, the values of $\delta_m$ for successively 5', 10', 15', and 20', corresponding (by interpolation from the columns of 41° 30' and 42°) to each parallel of latitude required; and, from the points thus found, the corresponding values of $\delta_p$ at right angles to the lines already drawn.

Lines passing through the extremities of $\delta_p$ will be the required meridians and parallels.

The projection being made, any point whose latitude and longitude are known will be projected on the map from elements taken from the tables of values of $D_m$ and $D_p$, which are measured from the *meridians* and *parallels*, and not from the axes of co-ordinates used in making the projection.

*Polyconic Projection—Co-ordinates, $\delta_m$, $\delta_\mu$, in Yards.*

| Longitude. | Latitude 22° 0'. | | Latitude 22° 30'. | | Latitude 23° 0'. | |
|---|---|---|---|---|---|---|
| | $\delta_m$ | $\delta_\mu$ | $\delta_m$ | $\delta_\mu$ | $\delta_m$ | $\delta_\mu$ |
| ° ′ | | | | | | |
| 1 | 1882.0 | 0.1 | 1875.3 | 0.1 | 1868.5 | 0.1 |
| 2 | 3763.9 | 0.4 | 3750.6 | 0.4 | 3737.0 | 0.4 |
| 3 | 5645.9 | 0.9 | 5625.9 | 1.0 | 5605.4 | 1.0 |
| 4 | 7527.8 | 1.6 | 7501.2 | 1.7 | 7473.9 | 1.7 |
| 5 | 9409.8 | 2.6 | 9376.4 | 2.6 | 9342.4 | 2.7 |
| 6 | 11291.8 | 3.7 | 11251.7 | 3.7 | 11210.9 | 3.8 |
| 7 | 13173.7 | 5.0 | 13127.0 | 5.1 | 13079.4 | 5.2 |
| 8 | 15055.7 | 6.6 | 15002.3 | 6.7 | 14947.8 | 6.8 |
| 9 | 16937.6 | 8.3 | 16877.6 | 8.5 | 16816.3 | 8.6 |
| 10 | 18819.6 | 10.3 | 18752.9 | 10.5 | 18684.8 | 10.6 |
| 11 | 20701.6 | 12.4 | 20628.2 | 12.6 | 20553.3 | 12.8 |
| 12 | 22583.5 | 14.8 | 22503.5 | 15.0 | 22421.8 | 15.3 |
| 13 | 24465.5 | 17.3 | 24378.8 | 17.6 | 24290.2 | 17.9 |
| 14 | 26347.4 | 20.1 | 26254.1 | 20.5 | 26158.7 | 20.8 |
| 15 | 28229.4 | 23.1 | 28129.3 | 23.5 | 28027.2 | 23.9 |
| 16 | 30111.4 | 26.2 | 30004.6 | 26.7 | 29895.7 | 27.2 |
| 17 | 31993.3 | 29.6 | 31879.9 | 30.2 | 31764.2 | 30.7 |
| 18 | 33875.3 | 33.2 | 33755.2 | 33.3 | 33632.6 | 34.4 |
| 19 | 35757.2 | 37.0 | 35630.5 | 37.7 | 35501.1 | 38.3 |
| 20 | 37639.2 | 41.0 | 37505.8 | 41.7 | 37369.6 | 42.5 |
| 25 | 47049.0 | 64.1 | 46675.8 | 65.2 | 46712.0 | 66.4 |
| 30 | 56458.7 | 92.3 | 56258.7 | 93.9 | 56054.3 | 95.6 |
| 40 | 75278.2 | 164.1 | 75011.5 | 167.0 | 74739.0 | 169.9 |
| 50 | 94097.7 | 256.3 | 93764.2 | 260.9 | 93423.7 | 265.5 |
| 1 00 | 112917.0 | 369.1 | 112516.9 | 375.8 | 112108.2 | 382.3 |
| 1 20 | 150555.4 | 656.2 | 150021.9 | 668.0 | 149476.9 | 679.6 |
| 1 30 | 169374.4 | 830.5 | 168774.2 | 845.4 | 168161.1 | 860.1 |
| 1 40 | 188193.3 | 1025.4 | 187526.3 | 1043.8 | 186845.1 | 1061.8 |
| 2 00 | 225830.5 | 1476.5 | 225030.0 | 1503.0 | 224212.5 | 1529.1 |
| 2 30 | 282284.7 | 2307.1 | 281284.0 | 2348.5 | 280261.9 | 2389.2 |
| 3 00 | 338736.6 | 3322.2 | 337535.6 | 3381.8 | 336309.0 | 3440.4 |
| 3 30 | 395186.0 | 4521.9 | 393784.5 | 4603.0 | 392353.1 | 4682.8 |
| 4 00 | 451632.0 | 5906.2 | 450029.9 | 6012.1 | 448393.7 | 6116.2 |

*Polyconic Projection—Co ordinates, $\delta_m$, $\delta_p$, in Yards.*

| Longitude. | Latitude 23° 30′. | | Latitude 24° 0′. | | Latitude 24° 30′. | |
|---|---|---|---|---|---|---|
| ° ′ | $\delta_m$ | $\delta_p$ | $\delta_m$ | $\delta_p$ | $\delta_m$ | $\delta_p$ |
| 1 | 1861.5 | 0.1 | 1854.4 | 0.1 | 1847.2 | 0.1 |
| 2 | 3723.1 | 0.4 | 3708.9 | 0.4 | 3694.4 | 0.4 |
| 3 | 5584.6 | 1.0 | 5563.3 | 1.0 | 5541.6 | 1.0 |
| 4 | 7446.1 | 1.8 | 7417.7 | 1.8 | 7388.8 | 1.8 |
| 5 | 9307.6 | 2.7 | 9272.2 | 2.7 | 9236.0 | 2.8 |
| 6 | 11169.2 | 3.9 | 11126.6 | 3.9 | 11083.2 | 4.0 |
| 7 | 13030.7 | 5.3 | 12981.0 | 5.4 | 12930.4 | 5.4 |
| 8 | 14892.2 | 6.9 | 14835.5 | 7.0 | 14777.6 | 7.1 |
| 9 | 16753.7 | 8.7 | 16689.9 | 8.9 | 16624.8 | 9.0 |
| 10 | 18615.3 | 10.8 | 18544.3 | 11.0 | 18472.0 | 11.1 |
| 11 | 20476.8 | 13.1 | 20398.8 | 13.3 | 20319.2 | 13.5 |
| 12 | 22338.3 | 15.6 | 22253.2 | 15.8 | 22166.4 | 16.0 |
| 13 | 24199.9 | 18.2 | 24107.6 | 18.5 | 24013.6 | 18.8 |
| 14 | 26061.4 | 21.2 | 25962.1 | 21.5 | 25860.8 | 21.8 |
| 15 | 27922.9 | 24.3 | 27816.5 | 24.7 | 27708.0 | 25.1 |
| 16 | 29784.4 | 27.7 | 29670.9 | 28.1 | 29555.2 | 28.5 |
| 17 | 31646.0 | 31.2 | 31525.4 | 31.7 | 31402.4 | 32.2 |
| 18 | 33507.5 | 35.0 | 33379.8 | 35.5 | 33249.6 | 36.0 |
| 19 | 35369.0 | 39.0 | 35234.2 | 39.6 | 35096.8 | 40.2 |
| 20 | 37230.5 | 43.2 | 37088.7 | 43.9 | 36944.0 | 44.6 |
| 25 | 46538.1 | 67.5 | 46360.8 | 68.6 | 46179.9 | 69.6 |
| 30 | 55845.8 | 97.2 | 55632.9 | 98.7 | 55415.9 | 100.3 |
| 40 | 74460.9 | 172.7 | 74177.1 | 175.5 | 73887.7 | 178.3 |
| 50 | 93076.0 | 269.9 | 92721.3 | 274.3 | 92359.5 | 278.5 |
| 1 00 | 111691.0 | 388.7 | 111265.3 | 394.9 | 110831.2 | 401.1 |
| 1 20 | 148920.6 | 690.9 | 148353.0 | 702.1 | 147774.1 | 713.0 |
| 1 30 | 167535.2 | 874.5 | 166896.6 | 888.6 | 166245.4 | 902.4 |
| 1 40 | 186149.7 | 1079.6 | 185440.1 | 1097.0 | 184716.5 | 1114.1 |
| 2 00 | 223377.9 | 1554.6 | 222526.4 | 1579.7 | 221658.0 | 1604.3 |
| 2 30 | 279218.6 | 2429.1 | 278154.1 | 2468.3 | 277068.4 | 2506.8 |
| 3 00 | 335056.8 | 3497.9 | 333779.1 | 3554.4 | 332476.1 | 3609.8 |
| 3 30 | 390892.0 | 4761.1 | 389401.1 | 4837.9 | 387880.6 | 4913.3 |
| 4 00 | 446723.4 | 6218.5 | 445019.2 | 6318.9 | 443281.1 | 6417.4 |

*Polyconic Projection—Co-ordinates, $\delta_m$, $\delta_p$, in Yards.*

| Longitude | Latitude 25° 0'. | | Latitude 25° 30'. | | Latitude 26° 0'. | |
|---|---|---|---|---|---|---|
| | $\delta_m$ | $\delta_p$ | $\delta_m$ | $\delta_p$ | $\delta_m$ | $\delta_p$ |
| ° ′ | | | | | | |
| 1 | 1839.8 | 0.1 | 1832.3 | 0.1 | 1824.7 | 0.1 |
| 2 | 3679.6 | 0.5 | 3664.6 | 0.5 | 3649.3 | 0.5 |
| 3 | 5519.5 | 1.0 | 5496.9 | 1.0 | 5474.0 | 1.0 |
| 4 | 7359.3 | 1.8 | 7329.2 | 1.8 | 7298.6 | 1.9 |
| 5 | 9199.1 | 2.8 | 9161.5 | 2.9 | 9123.3 | 2.9 |
| 6 | 11038.9 | 4.1 | 10993.8 | 4.1 | 10947.9 | 4.2 |
| 7 | 12878.8 | 5.5 | 12826.1 | 5.6 | 12772.6 | 5.7 |
| 8 | 14718.6 | 7.2 | 14658.5 | 7.3 | 14597.2 | 7.4 |
| 9 | 16558.4 | 9.2 | 16490.8 | 9.3 | 16421.9 | 9.4 |
| 10 | 18398.2 | 11.3 | 18323.1 | 11.5 | 18246.5 | 11.6 |
| 11 | 20238.0 | 13.7 | 20155.4 | 13.9 | 20071.2 | 14.1 |
| 12 | 22077.9 | 16.3 | 21987.7 | 16.5 | 21895.8 | 16.8 |
| 13 | 23917.7 | 19.1 | 23820.0 | 19.4 | 23720.5 | 19.7 |
| 14 | 25757.5 | 22.2 | 25652.3 | 22.5 | 25545.1 | 22.8 |
| 15 | 27597.3 | 25.4 | 27484.6 | 25.8 | 27369.8 | 26.2 |
| 16 | 29437.1 | 29.0 | 29316.9 | 29.4 | 29194.4 | 29.8 |
| 17 | 31277.0 | 32.7 | 31149.2 | 33.2 | 31019.1 | 33.6 |
| 18 | 33116.8 | 36.6 | 32981.5 | 37.2 | 32843.7 | 37.7 |
| 19 | 34956.6 | 40.8 | 34813.8 | 41.4 | 34668.4 | 42.0 |
| 20 | 36796.4 | 45.2 | 36646.1 | 45.9 | 36493.0 | 46.5 |
| 25 | 45995.5 | 70.7 | 45807.6 | 71.7 | 45616.2 | 72.7 |
| 30 | 55194.6 | 101.8 | 54969.1 | 103.2 | 54739.5 | 104.7 |
| 40 | 73592.7 | 180.9 | 73292.0 | 183.6 | 72985.8 | 186.1 |
| 50 | 91990.7 | 282.7 | 91614.9 | 286.8 | 91232.1 | 290.8 |
| 1 00 | 110388.6 | 407.1 | 109937.6 | 413.0 | 109478.3 | 418.8 |
| 1 20 | 147184.0 | 723.8 | 146582.7 | 734.3 | 145970.3 | 744.6 |
| 1 30 | 165581.5 | 916.0 | 164905.0 | 929.3 | 164216.0 | 942.3 |
| 1 40 | 183978.8 | 1130.9 | 183227.1 | 1147.3 | 182461.5 | 1163.4 |
| 2 00 | 220772.7 | 1628.5 | 219870.6 | 1652.1 | 218951.9 | 1675.2 |
| 2 30 | 275961.6 | 2544.5 | 274833.9 | 2581.4 | 273685.3 | 2617.6 |
| 3 00 | 331147.8 | 3664.1 | 329794.3 | 3717.3 | 328415.8 | 3769.3 |
| 3 30 | 386330.6 | 4987.2 | 384751.3 | 5059.6 | 383142.7 | 5130.5 |
| 4 00 | 441509.4 | 6513.9 | 439704.0 | 6608.5 | 437865.3 | 6701.0 |

*Polyconic Projection—Co-ordinates, $\delta_m$, $\delta_p$, in Yards.*

| Longitude. | Latitude 26° 30'. | | Latitude 27° 0'. | | Latitude 27° 30'. | |
|---|---|---|---|---|---|---|
| | $\delta_m$ | $\delta_p$ | $\delta_m$ | $\delta_p$ | $\delta_m$ | $\delta_p$ |
| ° ′ | | | | | | |
| 1 | 1816.9 | 0.1 | 1808.9 | 0.1 | 1800.9 | 0.1 |
| 2 | 3633.7 | 0.5 | 3617.9 | 0.5 | 3601.7 | 0.5 |
| 3 | 5450.6 | 1.1 | 5426.8 | 1.1 | 5402.6 | 1.1 |
| 4 | 7267.4 | 1.9 | 7235.7 | 1.9 | 7203.4 | 1.9 |
| 5 | 9084.3 | 2.9 | 9044.6 | 3.0 | 9004.3 | 3.0 |
| 6 | 10901.2 | 4.2 | 10853.6 | 4.3 | 10805.1 | 4.4 |
| 7 | 12718.0 | 5.8 | 12662.5 | 5.9 | 12606.0 | 5.9 |
| 8 | 14534.9 | 7.5 | 14471.4 | 7.6 | 14406.9 | 7.7 |
| 9 | 16351.7 | 9.5 | 16280.3 | 9.7 | 16207.7 | 9.8 |
| 10 | 18168.6 | 11.8 | 18089.3 | 11.9 | 18008 6 | 12.1 |
| 11 | 19985.5 | 14.3 | 19898.2 | 14.5 | 19809.4 | 14.6 |
| 12 | 21802.3 | 17.0 | 21707.1 | 17.2 | 21610.3 | 17.4 |
| 13 | 23619.2 | 19.9 | 23516.0 | 20.2 | 23411.1 | 20.4 |
| 14 | 25436.0 | 23.1 | 25325.0 | 23.4 | 25212.0 | 23.7 |
| 15 | 27252.9 | 26.5 | 27133.9 | 26.9 | 27012.8 | 27.2 |
| 16 | 29069.8 | 30.2 | 28942.8 | 30.6 | 28813.7 | 30.9 |
| 17 | 30886.6 | 34.1 | 30751.7 | 34.5 | 30614.6 | 34.9 |
| 18 | 32703.5 | 38.2 | 32560.7 | 38.7 | 32415.4 | 39.2 |
| 19 | 34520.3 | 42.6 | 34369.6 | 43.1 | 34216.3 | 43.6 |
| 20 | 36337.2 | 47.2 | 36178.5 | 47.8 | 36017.1 | 48.4 |
| 25 | 45421.4 | 73.7 | 45223.1 | 74.7 | 45021.4 | 75.6 |
| 30 | 54505.6 | 106.1 | 54267.7 | 107.5 | 54025.6 | 108.8 |
| 40 | 72674.1 | 188.6 | 72356.8 | 191.1 | 72034.0 | 193.5 |
| 50 | 90842.4 | 294.8 | 90445.8 | 298.6 | 90042.4 | 302.4 |
| 1 00 | 109010.7 | 424.5 | 108534.8 | 430.0 | 108050.6 | 435.4 |
| 1 20 | 145346.7 | 754.6 | 144712.1 | 764.4 | 144066.5 | 774.0 |
| 1 30 | 163514.5 | 955.1 | 162800.5 | 967.5 | 162074.2 | 979.6 |
| 1 40 | 181682.0 | 1179.1 | 180888.7 | 1194.4 | 180081.7 | 1209.4 |
| 2 00 | 218016.4 | 1697.9 | 217064.4 | 1720.0 | 216095.9 | 1741.6 |
| 2 30 | 272515.9 | 2652.9 | 271325.7 | 2687.5 | 270114.9 | 2721.2 |
| 3 00 | 327012.2 | 3820.2 | 325583.8 | 3870.0 | 324130.7 | 3918.6 |
| 3 30 | 381505.0 | 5199.8 | 379838.2 | 5267.5 | 378142.5 | 5333.6 |
| 4 00 | 435993.2 | 6791.5 | 434088.0 | 6880.0 | 432149.7 | 6966.3 |

GEODESY.

*Polyconic Projection—Co-ordinates,* $\delta_m$, $\delta_p$, *in Yards.*

| Longitude. | Latitude 28° 0'. | | Latitude 28° 30'. | | Latitude 29°. | |
|---|---|---|---|---|---|---|
| | $\delta_m$ | $\delta_p$ | $\delta_m$ | $\delta_p$ | $\delta_m$ | $\delta_p$ |
| ° ′ | | | | | | |
| 1 | 1792.7 | 0.1 | 1784.3 | 0.1 | 1775.8 | 0.1 |
| 2 | 3585.3 | 0.5 | 3568.6 | 0.5 | 3551.7 | 0.5 |
| 3 | 5378.0 | 1.1 | 5352.9 | 1.1 | 5327.5 | 1.1 |
| 4 | 7170.6 | 2.0 | 7137.2 | 2.0 | 7103.3 | 2.0 |
| 5 | 8963.3 | 3.1 | 8921.5 | 3.1 | 8879.1 | 3.1 |
| 6 | 10755.9 | 4.4 | 10705.8 | 4.5 | 10655.0 | 4.5 |
| 7 | 12548.6 | 6.0 | 12490.2 | 6.1 | 12430.8 | 6.1 |
| 8 | 14341.2 | 7.8 | 14274.5 | 7.9 | 14206.6 | 8.0 |
| 9 | 16133.9 | 9.9 | 16058.8 | 10.0 | 15982.5 | 10.1 |
| 10 | 17926.5 | 12.2 | 17843.1 | 12.4 | 17758.3 | 12.5 |
| 11 | 19719.2 | 14.8 | 19627.4 | 15.0 | 19534.1 | 15.2 |
| 12 | 21511.8 | 17.6 | 21411.7 | 17.8 | 21309.9 | 18.0 |
| 13 | 23304.5 | 20.7 | 23196.0 | 20.9 | 23085.8 | 21.2 |
| 14 | 25097.1 | 24.0 | 24980.3 | 24.3 | 24861.6 | 24.5 |
| 15 | 26889.8 | 27.5 | 26764.6 | 27.9 | 26637.4 | 28.2 |
| 16 | 28682.4 | 31.3 | 28548.9 | 31.7 | 28413.2 | 32.1 |
| 17 | 30475.1 | 35.4 | 30333.2 | 35.8 | 30189.1 | 36.2 |
| 18 | 32267.7 | 39.7 | 32117.5 | 40.1 | 31964.9 | 40.6 |
| 19 | 34060.3 | 44.2 | 33901.8 | 44.7 | 33740.7 | 45.2 |
| 20 | 35853.0 | 49.0 | 35686.1 | 49.5 | 35516.6 | 50.1 |
| 25 | 44816.2 | 76.5 | 44607.6 | 78.4 | 44395.7 | 78.3 |
| 30 | 53779.4 | 110.1 | 53529.1 | 111.4 | 53274.8 | 112.7 |
| 40 | 71705.8 | 195.8 | 71372.1 | 198.1 | 71032.9 | 200.3 |
| 50 | 89632.0 | 306.0 | 89214.9 | 309.6 | 88790.9 | 313.0 |
| 1 00 | 107558.2 | 440.7 | 107057.6 | 445.8 | 106548.8 | 450.8 |
| 1 20 | 143410.0 | 783.4 | 142742.5 | 792.5 | 142064.1 | 801.4 |
| 1 30 | 161335.6 | 991.5 | 160584.6 | 1003.0 | 159821.5 | 1014.3 |
| 1 40 | 179260.9 | 1224.1 | 178426.5 | 1238.3 | 177578.6 | 1252.2 |
| 2 00 | 215110.9 | 1762.6 | 214109.6 | 1783.2 | 213092.0 | 1803.1 |
| 2 30 | 268883.6 | 2754.1 | 267631.8 | 2786.2 | 266359.6 | 2817.4 |
| 3 00 | 322652.8 | 3965.9 | 321150.5 | 4012.1 | 319623.7 | 4057.1 |
| 3 30 | 376418.1 | 5398.1 | 374665.0 | 5460.9 | 372833.4 | 5522.1 |
| 4 00 | 430178.5 | 7050.6 | 428174.6 | 7132 7 | 426138.2 | 7212.6 |

*Polyconic Projection—Co-ordinates, $\delta_m$, $\delta_p$, in Yards.*

| Longitude. | Latitude 29° 30'. | | Latitude 30° 0'. | | Latitude 30° 30'. | |
|---|---|---|---|---|---|---|
| | $\delta_m$ | $\delta_p$ | $\delta_m$ | $\delta_p$ | $\delta_m$ | $\delta_p$ |
| ° ′ | | | | | | |
| 1 | 1767.2 | 0.1 | 1758.5 | 0.1 | 1749.6 | 0.1 |
| 2 | 3534.4 | 0.5 | 3516.9 | 0.5 | 3499.2 | 0.5 |
| 3 | 5301.6 | 1.1 | 5275.4 | 1.2 | 5248.8 | 1.2 |
| 4 | 7068.9 | 2.0 | 7033.9 | 2.0 | 6998.3 | 2.1 |
| 5 | 8836.1 | 3.2 | 8792.3 | 3.2 | 8747.9 | 3.2 |
| 6 | 10603.3 | 4.6 | 10550.8 | 4.6 | 10497.5 | 4.6 |
| 7 | 12370.5 | 6.2 | 12309.3 | 6.3 | 12247.1 | 6.3 |
| 8 | 14137.7 | 8.1 | 14067.8 | 8.2 | 13996.7 | 8.2 |
| 9 | 15904.9 | 10.3 | 15826.2 | 10.4 | 15746.3 | 10.5 |
| 10 | 17672.2 | 12.7 | 17584.7 | 12.8 | 17495.9 | 12.9 |
| 11 | 19439.4 | 15.3 | 19343.1 | 15.5 | 19245.4 | 15.6 |
| 12 | 21206.6 | 18.2 | 21101.6 | 18.4 | 20995.0 | 18.6 |
| 13 | 22973.8 | 21.4 | 22860.1 | 21.6 | 22744.6 | 21.8 |
| 14 | 24741.0 | 24.8 | 24618.5 | 25.1 | 23494.2 | 25.3 |
| 15 | 26508.2 | 28.5 | 26377.0 | 28.8 | 25243.8 | 29.1 |
| 16 | 28275.4 | 32.4 | 28135.5 | 32.7 | 26993.3 | 33.1 |
| 17 | 30042.6 | 36.6 | 29893.9 | 37.0 | 28742.9 | 37.4 |
| 18 | 31809.9 | 41.0 | 31652.4 | 41.4 | 30492.5 | 41.8 |
| 19 | 33577.1 | 45.7 | 33410.9 | 46.2 | 32242.1 | 46.6 |
| 20 | 35344.3 | 51.6 | 35169.3 | 51.2 | 33991.7 | 51.7 |
| 25 | 44180.3 | 79.1 | 43961.6 | 79.9 | 43239.6 | 80.7 |
| 30 | 53016.4 | 113.9 | 52753.9 | 115.1 | 52487.4 | 116.3 |
| 40 | 70688.4 | 202.5 | 70338.4 | 204.6 | 69983.1 | 206.6 |
| 50 | 88360.2 | 316.4 | 87922.8 | 319.7 | 87478.7 | 322.9 |
| 1 00 | 106032.0 | 455.6 | 105507.1 | 460.4 | 104974.1 | 464.9 |
| 1 20 | 141375.0 | 810.0 | 140675.1 | 818.4 | 139964.4 | 826.6 |
| 1 30 | 159046.2 | 1025.2 | 158258.7 | 1035.8 | 157459.2 | 1046.1 |
| 1 40 | 176717.1 | 1265.7 | 175842.2 | 1278.8 | 174953.8 | 1291.5 |
| 2 00 | 212058.1 | 1822.6 | 211008.1 | 1841.5 | 209942.1 | 1859.8 |
| 2 30 | 265067.1 | 2847.8 | 263754.5 | 2877.3 | 262421.8 | 2905.9 |
| 3 00 | 318072.5 | 4100.8 | 316497.1 | 4143.3 | 314897.6 | 4184.5 |
| 3 30 | 371073.4 | 5581.7 | 369235.2 | 5639.5 | 367368.9 | 5695.6 |
| 4 00 | 424069.3 | 7290.3 | 421968.0 | 7365.9 | 419834.7 | 7439.1 |

8

*Polyconic Projection—Co-ordinates, $\delta_m$, $\delta_p$, in Yards.*

| Longitude. | Latitude 31° 0'. | | Latitude 31° 30'. | | Latitude 32° 0'. | |
|---|---|---|---|---|---|---|
| | $\delta_m$ | $\delta_p$ | $\delta_m$ | $\delta_p$ | $\delta_m$ | $\delta_p$ |
| ° ' | | | | | | |
| 1 | 1740.6 | 0.1 | 1731.4 | 0.1 | 1722.1 | 0.1 |
| 2 | 3481.1 | 0.5 | 3462.8 | 0.5 | 3444.3 | 0.5 |
| 3 | 5221.7 | 1.2 | 5194.3 | 1.2 | 5166.4 | 1.2 |
| 4 | 6962.3 | 2.1 | 6925.7 | 2.1 | 6888.6 | 2.1 |
| 5 | 8702.9 | 3.3 | 8657.1 | 3.3 | 8610.7 | 3.3 |
| 6 | 10443.4 | 4.7 | 10388.5 | 4.7 | 10332.8 | 4.8 |
| 7 | 12184.0 | 6.4 | 12119.9 | 6.4 | 12055.0 | 6.5 |
| 8 | 13924.6 | 8.3 | 13851.4 | 8.4 | 13777.1 | 8.5 |
| 9 | 15665.1 | 10.6 | 15582.8 | 10.7 | 15499.3 | 10.8 |
| 10 | 17405.7 | 13.0 | 17314.2 | 13.2 | 17221.4 | 13.3 |
| 11 | 19146.3 | 15.8 | 19045.6 | 15.9 | 18943.6 | 16.1 |
| 12 | 20886.8 | 18.8 | 20777.1 | 18.9 | 20665.7 | 19.1 |
| 13 | 22627.4 | 22.0 | 22508.5 | 22.2 | 22387.8 | 22.4 |
| 14 | 24368.0 | 25.6 | 24239.9 | 25.8 | 24110.0 | 26.0 |
| 15 | 26108.5 | 29.3 | 25971.3 | 29.6 | 25832.1 | 29.9 |
| 16 | 27849.1 | 33.4 | 27702.7 | 33.7 | 27554.3 | 34.0 |
| 17 | 29589.7 | 37.7 | 29434.2 | 38.0 | 29276.4 | 38.4 |
| 18 | 31330.3 | 42.2 | 31165.6 | 42.6 | 30998.5 | 43.0 |
| 19 | 33070.8 | 47.1 | 32897.0 | 47.5 | 32720.7 | 47.9 |
| 20 | 34811.4 | 52.2 | 34628.4 | 52.6 | 34442.8 | 53.1 |
| 25 | 43514.2 | 81.6 | 43286.0 | 82.3 | 43053.5 | 83.0 |
| 30 | 52217.0 | 117.3 | 51942.5 | 118.4 | 51664.1 | 119.5 |
| 40 | 69622.5 | 208.6 | 69256.6 | 210.5 | 68885.4 | 212.4 |
| 50 | 87027.9 | 326.0 | 86570.5 | 328.9 | 86106.5 | 331.8 |
| 1 00 | 104433.2 | 469.4 | 103884.3 | 473.7 | 103327.4 | 477.8 |
| 1 20 | 139243.1 | 834.5 | 138511.2 | 842.1 | 137768.8 | 849.5 |
| 1 30 | 156647.8 | 1056.1 | 155824.4 | 1065.8 | 154989.1 | 1075.1 |
| 1 40 | 174052.6 | 1303.8 | 173137.2 | 1315.8 | 172209.1 | 1327.3 |
| 2 00 | 208860.0 | 1877.5 | 207762.0 | 1894.7 | 206648.2 | 1911.3 |
| 2 30 | 261069.1 | 2933.7 | 259696.5 | 2960.5 | 258304.1 | 2986.5 |
| 3 00 | 313274.2 | 4224.5 | 311626.9 | 4263.1 | 309955.8 | 4300.5 |
| 3 30 | 365474.6 | 5750.0 | 363552.4 | 5802.6 | 361602.5 | 5853.5 |
| 4 00 | 417669.4 | 7510.2 | 415472.3 | 7578.9 | 413243.4 | 7645.3 |

Polyconic Projection—Co-ordinates, $\delta_m$, $\delta_p$, in Yards.

| Longitude | Latitude 32° 30'. | | Latitude 33° 0'. | | Latitude 33° 30'. | |
|---|---|---|---|---|---|---|
| | $\delta_m$ | $\delta_p$ | $\delta_m$ | $\delta_p$ | $\delta_m$ | $\delta_p$ |
| 1 | 1712.7 | 0.1 | 1703.2 | 0.1 | 1693.5 | 0.1 |
| 2 | 3425.5 | 0.5 | 3406.4 | 0.5 | 3387.0 | 0.5 |
| 3 | 5138.2 | 1.2 | 5109.6 | 1.2 | 5080.6 | 1.2 |
| 4 | 6850.9 | 2.1 | 6812.8 | 2.2 | 6774.1 | 2.2 |
| 5 | 8563.7 | 3.3 | 8515.9 | 3.4 | 8467.6 | 3.4 |
| 6 | 10276.4 | 4.8 | 10219.1 | 4.9 | 10161.1 | 4.9 |
| 7 | 11989.1 | 6.6 | 11922.3 | 6.6 | 18154.6 | 6.7 |
| 8 | 13701.8 | 8.6 | 13625.5 | 8.6 | 13548.1 | 8.7 |
| 9 | 15414.6 | 10.8 | 15328.7 | 10.9 | 15241.7 | 11.0 |
| 10 | 17127.3 | 13.4 | 17031.9 | 13.5 | 16935.2 | 13.6 |
| 11 | 18840.0 | 16.2 | 18735.1 | 16.3 | 18628.7 | 16.4 |
| 12 | 20552.8 | 19.3 | 20438.3 | 19.4 | 20322.2 | 19.6 |
| 13 | 22265.5 | 22.6 | 22141.4 | 22.8 | 22015.7 | 23.0 |
| 14 | 23978.2 | 26.2 | 23844.6 | 26.4 | 23709.2 | 26.6 |
| 15 | 25690.9 | 30.1 | 25547.8 | 30.4 | 25402.7 | 30.6 |
| 16 | 27403.7 | 34.3 | 27251.0 | 34.5 | 27096.3 | 34.9 |
| 17 | 29116.4 | 38.7 | 28954.2 | 39.0 | 28789.8 | 39.3 |
| 18 | 30829.1 | 43.4 | 30657.4 | 43.7 | 30483.3 | 44.0 |
| 19 | 32541.9 | 48.5 | 32360.6 | 49.0 | 32176.8 | 49.1 |
| 20 | 34254.6 | 53.5 | 34063.8 | 54.0 | 33870.3 | 54.4 |
| 25 | 42818.2 | 83.6 | 42579.6 | 84.3 | 42337.9 | 85.0 |
| 30 | 51381.8 | 120.5 | 51095.5 | 121.4 | 50805.4 | 122.3 |
| 40 | 68508.9 | 214.1 | 68127.2 | 215.9 | 67740.4 | 217.5 |
| 50 | 85635.9 | 334.6 | 85158.8 | 337.3 | 84675.2 | 339.9 |
| 1 00 | 102762.7 | 481.8 | 102190.2 | 485.7 | 101609.9 | 489.4 |
| 1 20 | 137015.8 | 856.6 | 136252.4 | 863.5 | 135478.6 | 870.1 |
| 1 30 | 154142.0 | 1084.1 | 153283.1 | 1092.8 | 152412.6 | 1101.2 |
| 1 40 | 171267.9 | 1338.4 | 170313.6 | 1349.2 | 169346.3 | 1359.5 |
| 2 00 | 205518.7 | 1927.4 | 204373.5 | 1942.8 | 203212.7 | 1957.7 |
| 2 30 | 256892.0 | 3011.5 | 255460.4 | 3035.6 | 254009.2 | 3058.8 |
| 3 00 | 308261.1 | 4336.6 | 306542.9 | 4371.3 | 304801.4 | 4404.7 |
| 3 30 | 359625.1 | 5902.6 | 357620.3 | 5949.9 | 355588.2 | 5995.3 |
| 4 00 | 410983.2 | 7709.5 | 408691.6 | 7771.2 | 406368.8 | 7830.6 |

*Polyconic Projection—Co-ordinates, $\delta_m$, $\delta_p$, in Yards.*

| Longitude. | Latitude 34° 0'. | | Latitude 34° 30'. | | Latitude 35° 0'. | |
|---|---|---|---|---|---|---|
| | $\delta_m$ | $\delta_p$ | $\delta_m$ | $\delta_p$ | $\delta_m$ | $\delta_p$ |
| 1 | 1683.7 | 0.1 | 1673.8 | 0.1 | 1663.7 | 0.1 |
| 2 | 3367.4 | 0.5 | 3347.6 | 0.6 | 3327.5 | 0.6 |
| 3 | 5051.1 | 1.2 | 5021.4 | 1.2 | 4991.2 | 1.2 |
| 4 | 6734.9 | 2.2 | 6695.1 | 2.2 | 6654.9 | 2.2 |
| 5 | 8418.6 | 3.4 | 8368.9 | 3.4 | 8318.7 | 3.5 |
| 6 | 10102.3 | 4.9 | 10042.7 | 5.0 | 9982.4 | 5.0 |
| 7 | 11786.0 | 6.7 | 11716.5 | 6.8 | 11646.1 | 6.8 |
| 8 | 13469.7 | 8.8 | 13390.3 | 8.8 | 13309.8 | 8.9 |
| 9 | 15153.4 | 11.1 | 15064.1 | 11.2 | 14973.6 | 11.2 |
| 10 | 16837.2 | 13.7 | 16737.9 | 13.8 | 16637.3 | 13.9 |
| 11 | 18520.9 | 16.6 | 18411.6 | 16.7 | 18301.0 | 16.8 |
| 12 | 20204.6 | 19.7 | 20085.4 | 19.9 | 19964.8 | 20.0 |
| 13 | 21888.3 | 23.1 | 21759.2 | 23.3 | 21628.5 | 23.5 |
| 14 | 23572.0 | 26.8 | 23433.0 | 27.0 | 23292.2 | 27.2 |
| 15 | 25255.7 | 30.8 | 25106.8 | 31.0 | 24955.9 | 31.2 |
| 16 | 26939.4 | 35.1 | 26780.6 | 35.3 | 26619.7 | 35.5 |
| 17 | 28623.1 | 39.6 | 28454.4 | 39.8 | 28283.4 | 40.1 |
| 18 | 30306.9 | 44.4 | 30128.1 | 44.7 | 29947.1 | 45.0 |
| 19 | 31990.6 | 49.4 | 31801.9 | 49.8 | 31610.9 | 50.1 |
| 20 | 33674.3 | 54.8 | 33475.7 | 55.2 | 33274.6 | 55.5 |
| 25 | 42092.8 | 85.6 | 41844.6 | 86.2 | 41593.2 | 86.7 |
| 30 | 50511.4 | 123.2 | 50213.5 | 125.1 | 49911.8 | 124.9 |
| 40 | 67348.3 | 219.1 | 66951.1 | 220.6 | 66548.9 | 222.1 |
| 50 | 84185.1 | 342.3 | 83688.7 | 344.7 | 83185.8 | 347.0 |
| 1 00 | 101021.8 | 493.0 | 100426.0 | 496.4 | 99822.6 | 499.7 |
| 1 20 | 134694.5 | 876.4 | 133900.1 | 882.5 | 133095.5 | 888.3 |
| 1 30 | 151530.5 | 1109.2 | 150636.8 | 1116.9 | 149731.5 | 1124.2 |
| 1 40 | 168366.1 | 1369.4 | 167373.1 | 1378.9 | 166367.3 | 1387.9 |
| 2 00 | 202036.4 | 1971.9 | 200844.7 | 1985.6 | 199637.7 | 1998.6 |
| 2 30 | 252538.7 | 3081.1 | 251049.0 | 3102.5 | 249540.1 | 3122.8 |
| 3 00 | 303036.6 | 4436.8 | 301248.7 | 4467.5 | 299437.8 | 4496.9 |
| 3 30 | 353529.0 | 6039.0 | 351442.8 | 6080.8 | 349329.8 | 6120.8 |
| 4 00 | 404015.1 | 7887.7 | 401630.5 | 7942.3 | 399215.4 | 7994.5 |

*Polyconic Projection—Co-ordinates, $\delta_m$, $\delta_p$, in Yards.*

| Longitude. | Latitude 35° 30'. | | Latitude 36° 0'. | | Latitude 36° 30'. | |
|---|---|---|---|---|---|---|
| | $\delta_m$ | $\delta_p$ | $\delta_m$ | $\delta_p$ | $\delta_m$ | $\delta_p$ |
| 1 | 1653.5 | 0.1 | 1643.2 | 0.1 | 1632.8 | 0.1 |
| 2 | 3307.1 | 0.6 | 3286.5 | 0.6 | 3265.6 | 0.6 |
| 3 | 4960.6 | 1.3 | 4929.7 | 1.3 | 4898.4 | 1.3 |
| 4 | 6614.2 | 2.2 | 6572.9 | 2.2 | 6531.2 | 2.3 |
| 5 | 8267.7 | 3.5 | 8216.2 | 3.5 | 8164.0 | 3.5 |
| 6 | 9921.3 | 5.0 | 9859.4 | 5.1 | 9796.8 | 5.1 |
| 7 | 11574.8 | 6.8 | 11502.7 | 6.9 | 11429.6 | 6.9 |
| 8 | 13228.4 | 8.9 | 13145.9 | 9.0 | 13062.4 | 9.0 |
| 9 | 14881.9 | 11.3 | 14789.1 | 11.4 | 14695.2 | 11.4 |
| 10 | 16535.5 | 14.0 | 16432.4 | 14.0 | 16328.0 | 14.1 |
| 11 | 18189.0 | 16.9 | 18075.6 | 17.0 | 17960.8 | 17.0 |
| 12 | 19842.5 | 20.1 | 19718.8 | 20.2 | 19593.6 | 20.3 |
| 13 | 21496.1 | 23.6 | 21362.0 | 23.7 | 21226.4 | 23.9 |
| 14 | 23149.6 | 27.4 | 23005.3 | 27.5 | 22859.2 | 27.7 . |
| 15 | 24803.2 | 31.4 | 24648.5 | 31.6 | 24492.0 | 31.8 |
| 16 | 26456.7 | 35.7 | 26291.8 | 36.0 | 26124.8 | 36.2 |
| 17 | 28110.3 | 40.4 | 27935.0 | 40.6 | 27757.6 | 40.8 |
| 18 | 29763.8 | 45.2 | 29578.2 | 45.5 | 29390.4 | 45.8 |
| 19 | 31417.3 | 50.4 | 31221.5 | 50.7 | 31023.2 | 51.0 |
| 20 | 33070.9 | 55.9 | 32864.7 | 56.2 | 32656.0 | 56.5 |
| 25 | 41338.6 | 87.2 | 41080.8 | 87.8 | 40819.9 | 88.3 |
| 30 | 49606.2 | 125.7 | 49296.9 | 126.4 | 48983.9 | 127.1 |
| 40 | 66141.5 | 223.4 | 65729.1 | 224.8 | 65311.7 | 226.0 |
| 50 | 82676.6 | 349.1 | 82161.1 | 351.2 | 81639.3 | 353.1 |
| 1 00 | 99211.5 | 502.7 | 98592.9 | 505.7 | 97966.7 | 508.5 |
| 1 20 | 132280.7 | 893.8 | 131455.9 | 899.1 | 130621.0 | 904.1 |
| 1 30 | 148814.9 | 1131.2 | 147886.9 | 1137.9 | 146947.7 | 1144.2 |
| 1 40 | 165348.8 | 1396.6 | 164317.7 | 1404.8 | 163274.0 | 1412.6 |
| 2 00 | 198415.4 | 2011.1 | 197178.0 | 2022.9 | 195925.6 | 2034.1 |
| 2 30 | 248012.1 | 3142.3 | 246465.3 | 3160.8 | 244899.6 | 3178.3 |
| 3 00 | 297604.0 | 4524.9 | 295747.6 | 4551.6 | 293868.6 | 4576.8 |
| 3 30 | 347190.2 | 6158.9 | 345024.1 | 6195.2 | 342831.7 | 6229.5 |
| 4 00 | 396769.7 | 8044.3 | 394293.8 | 8091.6 | 391787.8 | 8136.5 |

*Polyconic Projection—Co-ordinates, $\delta_m$, $\delta_p$, in Yards.*

| Longitude | Latitude 37° 0'. | | Latitude 37° 30'. | | Latitude 38° 0'. | |
|---|---|---|---|---|---|---|
| | $\delta_m$ | $\delta_p$ | $\delta_m$ | $\delta_p$ | $\delta_m$ | $\delta_p$ |
| ° ′ | | | | | | |
| 1 | 1622.2 | 0.1 | 1611.6 | 0.1 | 1600.7 | 0.1 |
| 2 | 3244.5 | 0.6 | 3223.1 | 0.6 | 3201.5 | 0.6 |
| 3 | 4866.7 | 1.3 | 4834.7 | 1.3 | 4802.2 | 1.3 |
| 4 | 6489.0 | 2.3 | 6446.2 | 2.3 | 6403.0 | 2.3 |
| 5 | 8111.2 | 3.5 | 8057.8 | 3.6 | 8003.7 | 3.6 |
| 6 | 9733.4 | 5.1 | 9669.3 | 5.1 | 9604.5 | 5.2 |
| 7 | 11355.7 | 7.0 | 11280.9 | 7.0 | 11205.2 | 7.0 |
| 8 | 12977.9 | 9.1 | 12892.4 | 9.1 | 12806.0 | 9.2 |
| 9 | 14600.2 | 11.5 | 14504.0 | 11.6 | 14406.7 | 11.6 |
| 10 | 16222.4 | 14.2 | 16115.6 | 14.3 | 16007.5 | 14.3 |
| 11 | 17844.6 | 17.2 | 17727.1 | 17.3 | 17608.2 | 17.3 |
| 12 | 19466.9 | 20.4 | 19338.6 | 20.5 | 19209.0 | 20.6 |
| 13 | 21089.1 | 24.0 | 20950.2 | 24.1 | 20809.7 | 24.2 |
| 14 | 22711.4 | 27.8 | 22561.8 | 28.0 | 22410.5 | 28.1 |
| 15 | 24333.6 | 31.9 | 24173.3 | 32.1 | 24011.2 | 32.3 |
| 16 | 25955.8 | 36.4 | 25784.9 | 36.5 | 25612.0 | 36.7 |
| 17 | 27578.1 | 41.0 | 27396.4 | 41.2 | 27212.7 | 41.4 |
| 18 | 29200.3 | 46.0 | 29008.0 | 46.2 | 28813.4 | 46.4 |
| 19 | 30822.6 | 51.3 | 30619.5 | 51.5 | 30414.2 | 51.7 |
| 20 | 32444.8 | 56.8 | 32231.1 | 57.1 | 32015.0 | 57.3 |
| 25 | 40555.9 | 88.7 | 40288.8 | 89.1 | 40018.6 | 89.6 |
| 30 | 48667.1 | 127.8 | 48346.5 | 128.4 | 48022.3 | 129.0 |
| 40 | 64889.2 | 227.2 | 64461.9 | 228.3 | 64029.6 | 229.3 |
| 50 | 81111.3 | 355.0 | 80577.1 | 356.7 | 80036.7 | 358.3 |
| 1 00 | 97333.1 | 511.2 | 96692.0 | 513.7 | 96043.6 | 516.0 |
| 1 20 | 129776.1 | 908.8 | 128921.3 | 913.2 | 128056.7 | 917.4 |
| 1 30 | 145997.2 | 1150.2 | 145035.5 | 1155.8 | 144062.8 | 1161.0 |
| 1 40 | 162217.9 | 1419.9 | 161149.4 | 1426.9 | 160068.5 | 1433.4 |
| 2 00 | 194658.2 | 2044.7 | 193375.9 | 2054.7 | 192078.9 | 2064.1 |
| 2 30 | 243315.2 | 3194.9 | 241712.2 | 3210.5 | 240090.8 | 3225.1 |
| 3 00 | 291967.1 | 4600.7 | 290043.4 | 4623.1 | 288097.5 | 4644.1 |
| 3 30 | 340613.1 | 6262.0 | 338368.4 | 6292.6 | 336098.0 | 6321.2 |
| 4 00 | 389251.9 | 8178.9 | 386686.3 | 8218.8 | 384091.2 | 8256.3 |

*Polyconic Projection—Co-ordinates, $\delta_m$, $\delta_p$, in Yards.*

| Longitude | Latitude 38° 30′. | | Latitude 39° 0′. | | Latitude 39° 30′. | |
|---|---|---|---|---|---|---|
| | $\delta_m$ | $\delta_p$ | $\delta_m$ | $\delta_p$ | $\delta_m$ | $\delta_p$ |
| ° ′ | | | | | | |
| 1 | 1589.8 | 0.1 | 1578.8 | 0.1 | 1567.6 | 0.1 |
| 2 | 3179.6 | 0.6 | 3157.5 | 0.6 | 3135.2 | 0.6 |
| 3 | 4769.5 | 1.3 | 4736.3 | 1.3 | 4702.8 | 1.3 |
| 4 | 6359.3 | 2.3 | 6315.1 | 2.3 | 6270.4 | 2.3 |
| 5 | 7949.1 | 3.6 | 7893.8 | 3.6 | 7838.0 | 3.6 |
| 6 | 9538.9 | 5.2 | 9472.6 | 5.2 | 9405.6 | 5.2 |
| 7 | 11128.7 | 7.1 | 11051.4 | 7.1 | 10973.2 | 7.1 |
| 8 | 12718.6 | 9.2 | 12630.1 | 9.2 | 12540.8 | 9.3 |
| 9 | 14308.4 | 11.7 | 14208.9 | 11.7 | 14108.4 | 11.7 |
| 10 | 15898.2 | 14.4 | 15787.7 | 14.5 | 15676.0 | 14.5 |
| 11 | 17488.0 | 17.4 | 17366.4 | 17.5 | 17243.6 | 17.5 |
| 12 | 19077.8 | 20.7 | 18945.2 | 20.8 | 18811.1 | 20.9 |
| 13 | 20667.6 | 24.3 | 20524.0 | 24.4 | 20378.7 | 24.5 |
| 14 | 22257.4 | 28.2 | 22102.7 | 28.3 | 21946.3 | 28.4 |
| 15 | 23847.3 | 32.4 | 23681.5 | 32.5 | 23513.9 | 32.6 |
| 16 | 25437.1 | 36.8 | 25260.3 | 37.0 | 25081.5 | 37.1 |
| 17 | 27026.9 | 41.6 | 26839.0 | 41.8 | 26649.1 | 41.9 |
| 18 | 28616.7 | 46.6 | 28417.8 | 46.8 | 28216.7 | 47.0 |
| 19 | 30206.5 | 52.0 | 29996.6 | 52.2 | 29784.3 | 52.3 |
| 20 | 31796.3 | 57.6 | 31575.3 | 57.8 | 31351.9 | 58.0 |
| 25 | 39745.4 | 90.0 | 39469.1 | 90.3 | 39189.8 | 90.6 |
| 30 | 47694.4 | 129.5 | 47362.9 | 130.1 | 47027.7 | 130.5 |
| 40 | 63592.4 | 230.3 | 63150.3 | 231.2 | 62703.4 | 232.0 |
| 50 | 79490.2 | 359.9 | 78937.6 | 361.3 | 78379.0 | 362.6 |
| 1 00 | 95387.8 | 518.2 | 94724.7 | 520.2 | 94054.4 | 522.1 |
| 1 20 | 127182.3 | 921.2 | 126298.1 | 924.8 | 125404.3 | 928.2 |
| 1 30 | 143079.0 | 1165.9 | 142084.4 | 1170.5 | 141078.8 | 1174.7 |
| 1 40 | 158975.5 | 1439.4 | 157870.3 | 1445.1 | 156753.0 | 1450.2 |
| 2 00 | 190767.1 | 2072.8 | 189440.8 | 2080.9 | 188100.1 | 2088.3 |
| 2 30 | 238451.0 | 3238.7 | 236793.0 | 3251.4 | 235116.9 | 3263.0 |
| 3 00 | 286129.6 | 4663.8 | 284139.8 | 4682.0 | 282128.3 | 4698.8 |
| 3 30 | 333801.8 | 6347.9 | 331480.2 | 6372.7 | 329133.2 | 6395.6 |
| 4 00 | 381466.7 | 8291.2 | 378813.1 | 8323.6 | 376130.5 | 8353.4 |

*Polyconic Projection—Co-ordinates, $\delta_m$, $\delta_p$, in Yards.*

| Longitude | Latitude 40° 0'. | | Latitude 40° 30'. | | Latitude 41° 0'. | |
|---|---|---|---|---|---|---|
| | $\delta_m$ | $\delta_p$ | $\delta_m$ | $\delta_p$ | $\delta_m$ | $\delta_p$ |
| ° ′ | | | | | | |
| 1 | 1556.3 | 0.1 | 1544.9 | 0.1 | 1533.4 | 0.1 |
| 2 | 3112.6 | 0.6 | 3089.8 | 0.6 | 3066.7 | 0.6 |
| 3 | 4668.9 | 1.3 | 4634.7 | 1.3 | 4600.1 | 1.3 |
| 4 | 6225.2 | 2.3 | 6179.6 | 2.3 | 6133.5 | 2.3 |
| 5 | 7781.5 | 3.6 | 7724.5 | 3.6 | 7666.8 | 3.7 |
| 6 | 9337.8 | 5.2 | 9269.4 | 5.3 | 9200.2 | 5.3 |
| 7 | 10894.1 | 7.1 | 10814.3 | 7.2 | 10733.6 | 7.2 |
| 8 | 12450.4 | 9.3 | 12359.2 | 9.3 | 12266.9 | 9.4 |
| 9 | 14006.7 | 11.8 | 13904.0 | 11.8 | 13800.3 | 11.9 |
| 10 | 15563.0 | 14.6 | 15448.9 | 14.6 | 15333.7 | 14.6 |
| 11 | 17119.3 | 17.6 | 16993.8 | 17.7 | 16867.0 | 17.7 |
| 12 | 18675.6 | 21.0 | 18538.7 | 21.0 | 18400.4 | 21.1 |
| 13 | 20231.9 | 24.6 | 20083.6 | 24.7 | 19933.7 | 24.7 |
| 14 | 21788.2 | 28.5 | 21628.5 | 28.6 | 21467.1 | 28.7 |
| 15 | 23344.5 | 32.7 | 23173.4 | 32.8 | 23000.4 | 32.9 |
| 16 | 24900.8 | 37.2 | 24718.3 | 37.4 | 24533.8 | 37.5 |
| 17 | 26457.1 | 42.0 | 26263.2 | 42.2 | 26067.2 | 42.3 |
| 18 | 28013.4 | 47.1 | 27808.1 | 47.3 | 27600.5 | 47.4 |
| 19 | 29569.8 | 52.5 | 29352.9 | 52.7 | 29133.9 | 52.8 |
| 20 | 31126.1 | 58.2 | 30897.8 | 58.4 | 30667.3 | 58.5 |
| 25 | 38907.5 | 90.9 | 38622.2 | 91.2 | 38334.0 | 91.4 |
| 30 | 46689.0 | 130.9 | 46346.7 | 131.3 | 46000.8 | 131.7 |
| 40 | 62251.8 | 232.8 | 61795.3 | 233.5 | 61334.2 | 234.1 |
| 50 | 77814.4 | 363.7 | 77243.9 | 364.8 | 76667.4 | 365.8 |
| 1 00 | 93376.9 | 523.8 | 92692.2 | 525.3 | 92000.4 | 526.7 |
| 1 20 | 124500.9 | 931.2 | 123588.0 | 933.9 | 122665.7 | 936.4 |
| 1 30 | 140062.5 | 1178.5 | 139035.5 | 1182.0 | 137997.8 | 1185.1 |
| 1 40 | 155623.7 | 1455.0 | 154482.6 | 1459.3 | 153329.6 | 1463.1 |
| 2 00 | 186744.9 | 2095.2 | 185375.4 | 2101.4 | 183991.8 | 2106.9 |
| 2 30 | 233422.9 | 3273.7 | 231710.9 | 3283.4 | 229998.3 | 3292.0 |
| 3 00 | 280095.3 | 4714.1 | 278040.9 | 4728.1 | 275965.1 | 4740.6 |
| 3 30 | 326761.1 | 6416.5 | 324364.1 | 6435.4 | 321942.2 | 6452.4 |
| 4 00 | 373419.3 | 8380.7 | 370679.4 | 8405.5 | 367911.3 | 8427.7 |

*Polyconic Projection—Co-ordinates,* $\delta_m$, $\delta_p$, *in Yards.*

| Longitude. | Latitude 41° 30'. | | Latitude 42° 0'. | | Latitude 42° 30'. | |
|---|---|---|---|---|---|---|
| | $\delta_m$ | $\delta_p$ | $\delta_m$ | $\delta_p$ | $\delta_m$ | $\delta_p$ |
| 1 | 1521.7 | 0.1 | 1510.0 | 0.1 | 1498.1 | 0.1 |
| 2 | 3043.4 | 0.6 | 3019.9 | 0.6 | 2996.2 | 0.6 |
| 3 | 4565.2 | 1.3 | 4529.9 | 1.3 | 4494.2 | 1.3 |
| 4 | 6086.9 | 2.3 | 6039.8 | 2.4 | 5992.3 | 2.4 |
| 5 | 7608.6 | 3.7 | 7549.8 | 3.7 | 7490.4 | 3.7 |
| 6 | 9130.3 | 5.3 | 9059.7 | 5.3 | 8988.5 | 5.3 |
| 7 | 10652.0 | 7.2 | 10569.7 | 7.2 | 10486.5 | 7.2 |
| 8 | 12173.8 | 9.4 | 12079.6 | 9.4 | 11984.6 | 9.4 |
| 9 | 13695.5 | 11.9 | 13589.6 | 11.9 | 13482.7 | 11.9 |
| 10 | 15217.2 | 14.7 | 15099.6 | 14.7 | 14980.8 | 14.7 |
| 11 | 16738.9 | 17.7 | 16609.5 | 17.8 | 16478.8 | 17.8 |
| 12 | 18260.6 | 21.1 | 18119.5 | 21.2 | 17976.9 | 21.2 |
| 13 | 19782.3 | 24.8 | 19629.4 | 24.8 | 19475.0 | 24.9 |
| 14 | 21304.0 | 28.7 | 21139.4 | 28.8 | 20973.1 | 28.8 |
| 15 | 22825.8 | 33.0 | 22649.3 | 33.1 | 22471.1 | 33.1 |
| 16 | 24347.5 | 37.5 | 24159.3 | 37.6 | 23969.2 | 37.6 |
| 17 | 25869.2 | 42.4 | 25669.2 | 42.5 | 25467.3 | 42.5 |
| 18 | 27390.9 | 47.5 | 27179.2 | 47.6 | 26965.4 | 47.7 |
| 19 | 28912.6 | 52.9 | 28689.1 | 53.0 | 28463.4 | 53.1 |
| 20 | 30434.3 | 58.6 | 30199.1 | 58.8 | 29961.5 | 58.9 |
| 25 | 38042.9 | 91.6 | 37748.8 | 91.8 | 37451.3 | 92.0 |
| 30 | 45651.4 | 132.0 | 45298.5 | 132.3 | 44942.2 | 132.5 |
| 40 | 60868.3 | 234.6 | 60397.8 | 235.1 | 59922.7 | 235.5 |
| 50 | 76085.1 | 366.6 | 75496.9 | 367.4 | 74903.0 | 368.0 |
| 1 00 | 91301.6 | 527.9 | 90595.9 | 529.0 | 89883.2 | 529.9 |
| 1 20 | 121733.9 | 938.6 | 120792.9 | 940.5 | 119842.6 | 942.1 |
| 1 30 | 136949.6 | 1187.9 | 135890.9 | 1190.3 | 134821.8 | 1192.3 |
| 1 40 | 152164.9 | 1466.5 | 150988.5 | 1469.5 | 149800.6 | 1472.0 |
| 2 00 | 182594.1 | 2111.8 | 181182.5 | 2116.1 | 179756.9 | 2119.7 |
| 2 30 | 228234.1 | 3299.7 | 226469.4 | 3306.4 | 224687.4 | 3312.0 |
| 3 00 | 273868.3 | 4751.6 | 271750.5 | 4761.2 | 269612.0 | 4769.3 |
| 3 30 | 319495.6 | 6467.5 | 317024.7 | 6480.5 | 314529.5 | 6491.6 |
| 4 00 | 365115.0 | 8447.3 | 362290.8 | 8464.4 | 359438.9 | 8478.8 |

*Polyconic Projection—Co-ordinates, $\delta_m$, $\delta_p$, in Yards.*

| Longitude. | Latitude 43° 0'. | | Latitude 43° 30'. | | Latitude 44° 0'. | |
|---|---|---|---|---|---|---|
| | $\delta_m$ | $\delta_p$ | $\delta_m$ | $\delta_p$ | $\delta_m$ | $\delta_p$ |
| ° ' | | | | | | |
| 1 | 1486.1 | 0.1 | 1474.0 | 0.1 | 1461.8 | 0.1 |
| 2 | 2972.2 | 0.6 | 2948.0 | 0.6 | 2923.5 | 0.6 |
| 3 | 4458.3 | 1.3 | 4421.9 | 1.3 | 4385.3 | 1.3 |
| 4 | 5944.3 | 2.4 | 5895.9 | 2.4 | 5847.0 | 2.4 |
| 5 | 7430.4 | 3.7 | 7369.9 | 3.7 | 7308.8 | 3.7 |
| 6 | 8916.5 | 5.3 | 8843.9 | 5.3 | 8770.5 | 5.3 |
| 7 | 10402.6 | 7.2 | 10317.8 | 7.2 | 10232.3 | 7.2 |
| 8 | 11888.7 | 9.4 | 11791.8 | 9.5 | 11694.1 | 9.5 |
| 9 | 13374.8 | 11.9 | 13265.8 | 12.0 | 13155.8 | 12.0 |
| 10 | 14860.8 | 14.7 | 14739.8 | 14.8 | 14617.6 | 14.8 |
| 11 | 16346.9 | 17.8 | 16213.7 | 17.8 | 16079.3 | 17.9 |
| 12 | 17833.0 | 21.2 | 17687.7 | 21.2 | 17541.1 | 21.3 |
| 13 | 19319.1 | 24.9 | 19161.7 | 24.9 | 19002.8 | 25.0 |
| 14 | 20805.2 | 28.9 | 20635.7 | 28.9 | 20464.6 | 28.9 |
| 15 | 22291.2 | 33.2 | 22109.6 | 33.2 | 21926.3 | 33.2 |
| 16 | 23777.3 | 37.7 | 23583.6 | 37.8 | 23388.1 | 37.8 |
| 17 | 25263.4 | 42.6 | 25057.6 | 42.6 | 24849.9 | 42.7 |
| 18 | 26749.5 | 47.8 | 26531.6 | 47.8 | 26311.6 | 47.9 |
| 19 | 28235.6 | 53.2 | 28005.5 | 53.3 | 27773.4 | 53.3 |
| 20 | 29721.7 | 59.0 | 29479.5 | 59.0 | 29235.1 | 59.1 |
| 25 | 37102.0 | 92.1 | 36849.3 | 92.2 | 36543.8 | 92.3 |
| 30 | 44582.4 | 132.7 | 44219.2 | 132.8 | 43852.6 | 132.9 |
| 40 | 59443.0 | 235.8 | 58958.7 | 236.1 | 58469.9 | 236.3 |
| 50 | 74303.4 | 368.5 | 73698.0 | 368.9 | 73087.0 | 369.2 |
| 1 00 | 89163.6 | 530.7 | 88437.1 | 531.2 | 87703.9 | 531.7 |
| 1 20 | 118883.1 | 943.4 | 117914.5 | 944.4 | 116936.9 | 945.2 |
| 1 30 | 133742.4 | 1194.0 | 132652.7 | 1195.3 | 131552.9 | 1196.3 |
| 1 40 | 148601.2 | 1474.1 | 147390.5 | 1475.7 | 146168.4 | 1476.9 |
| 2 00 | 178317.6 | 2122.7 | 176864.7 | 2125.0 | 175398.2 | 2126.7 |
| 2 30 | 222888.2 | 3316.7 | 221071.9 | 3320.3 | 219238.7 | 3323.0 |
| 3 00 | 267452.8 | 4776.0 | 265273.1 | 4781.3 | 263073.1 | 4785.1 |
| 3 30 | 312010.3 | 6500.7 | 309467.1 | 6507.9 | 306900.3 | 6513.0 |
| 4 00 | 356559.5 | 8490.7 | 353652.8 | 8500.1 | 350719.0 | 8506.8 |

*Polyconic Projection—Co ordinates, $\delta_m$, $\delta_p$, in Yards.*

| Longitude | Latitude 44° 30'. | | Latitude 45° 0'. | | Latitude 45° 30'. | |
|---|---|---|---|---|---|---|
| ° ' | $\delta_m$ | $\delta_p$ | $\delta_m$ | $\delta_p$ | $\delta_m$ | $\delta_p$ |
| 1 | 1449.4 | 0.1 | 1437.0 | 0.1 | 1424.4 | 0.1 |
| 2 | 2898.9 | 0.6 | 2874.0 | 0.6 | 2848.9 | 0.6 |
| 3 | 4348.3 | 1.3 | 4311.0 | 1.3 | 4273.3 | 1.3 |
| 4 | 5797.7 | 2.4 | 5747.9 | 2.4 | 5697.7 | 2.4 |
| 5 | 7247.1 | 3.7 | 7184.9 | 3.7 | 7122.2 | 3.7 |
| 6 | 8696.6 | 5.3 | 8621.9 | 5.3 | 8546.6 | 5.3 |
| 7 | 10146.0 | 7.2 | 10058.9 | 7.2 | 9971.0 | 7.2 |
| 8 | 11595.4 | 9.5 | 11495.9 | 9.5 | 11395.4 | 9.5 |
| 9 | 13044.8 | 12.0 | 12932.9 | 12.0 | 12819.9 | 12.0 |
| 10 | 14494.3 | 14.8 | 14369.8 | 14.8 | 14244.3 | 14.8 |
| 11 | 15943.7 | 17.9 | 15806.8 | 17.9 | 15668.7 | 17.9 |
| 12 | 17393.1 | 21.3 | 17243.8 | 21.3 | 17093.1 | 21.3 |
| 13 | 18842.5 | 25.0 | 18680.8 | 25.0 | 18517.6 | 25.0 |
| 14 | 20292.0 | 29.0 | 20117.7 | 29.0 | 19942.0 | 29.0 |
| 15 | 21741.4 | 33.2 | 21554.7 | 33.3 | 21366.4 | 33.2 |
| 16 | 23190.8 | 37.8 | 22991.7 | 37.8 | 22790.8 | 37.8 |
| 17 | 24640.2 | 42.7 | 24428.7 | 42.7 | 24215.3 | 42.7 |
| 18 | 26089.7 | 47.9 | 25865.7 | 47.9 | 25639.7 | 47.9 |
| 19 | 27539.1 | 53.3 | 27302.7 | 53.4 | 27064.1 | 53.3 |
| 20 | 28988.5 | 59.1 | 28739.6 | 59.1 | 28488.6 | 59.1 |
| 25 | 36235.6 | 92.3 | 35924.5 | 92.4 | 35610.6 | 92.3 |
| 30 | 43482.6 | 133.0 | 43109.3 | 133.0 | 42732.7 | 133.0 |
| 40 | 57976.6 | 236.4 | 57478.9 | 236.5 | 56976.8 | 236.4 |
| 50 | 72470.4 | 369.4 | 71848.3 | 369.5 | 71220.6 | 369.4 |
| 1 00 | 86964.0 | 531.9 | 86217.4 | 532.0 | 85464.2 | 532.0 |
| 1 20 | 115950.3 | 945.7 | 114954.9 | 945.8 | 113950.6 | 945.7 |
| 1 30 | 130442.9 | 1196.8 | 129323.0 | 1197.1 | 128193.2 | 1196.9 |
| 1 40 | 144935.2 | 1477.6 | 143690.8 | 1477.9 | 142435.5 | 1477.7 |
| 2 00 | 173918.3 | 2127.7 | 172425.0 | 2128.1 | 170918.5 | 2127.8 |
| 2 30 | 217388.7 | 3324.6 | 215522.0 | 3325.2 | 213638.8 | 3324.8 |
| 3 00 | 260853.0 | 4787.4 | 258612.0 | 4788.3 | 256352.9 | 4787.7 |
| 3 30 | 304310.0 | 6516.2 | 301696.3 | 6517.3 | 299059.6 | 6516.5 |
| 4 00 | 347758.4 | 8510.9 | 344771.2 | 8512.5 | 341757.6 | 8511.4 |

*Polyconic Projection—Co-ordinates,* $\delta_m$, $\delta_p$, *in Yards.*

| Longitude. | Latitude 46° 0'. | | Latitude 46° 30'. | | Latitude 47° 0'. | |
|---|---|---|---|---|---|---|
| | $\delta_m$ | $\delta_p$ | $\delta_m$ | $\delta_p$ | $\delta_m$ | $\delta_p$ |
| ° ' | | | | | | |
| 1 | 1411.8 | 0.1 | 1399.0 | 0.1 | 1386.1 | 0.1 |
| 2 | 2823.5 | 0.6 | 2798.0 | 0.6 | 2772.2 | 0.6 |
| 3 | 4235.3 | 1.3 | 4197.0 | 1.3 | 4158.4 | 1.3 |
| 4 | 5647.1 | 2.4 | 5596.0 | 2.4 | 5544.5 | 2.4 |
| 5 | 7058.8 | 3.7 | 6995.0 | 3.7 | 6930.6 | 3.7 |
| 6 | 8470.6 | 5.3 | 8394.0 | 5.3 | 8316.7 | 5.3 |
| 7 | 9882.4 | 7.2 | 9793.0 | 7.2 | 9702.8 | 7.2 |
| 8 | 11294.1 | 9.5 | 11192.0 | 9.4 | 11089.0 | 9.4 |
| 9 | 12705.9 | 12.0 | 12591.0 | 12.0 | 12475.1 | 11.9 |
| 10 | 14117.7 | 14.8 | 13990.0 | 14.8 | 13861.2 | 14.7 |
| 11 | 15529.4 | 17.9 | 15389.0 | 17.9 | 15247.3 | 17.8 |
| 12 | 16941.2 | 21.3 | 16788.0 | 21.3 | 16633.4 | 21.2 |
| 13 | 18353.0 | 25.0 | 18186.9 | 24.9 | 18019.5 | 24.9 |
| 14 | 19764.7 | 28.9 | 19585.9 | 28.9 | 19405.7 | 28.9 |
| 15 | 21176.5 | 33.2 | 20984.9 | 33.2 | 20791.8 | 33.2 |
| 16 | 22588.3 | 37.8 | 22383.9 | 37.8 | 22177.9 | 37.7 |
| 17 | 24000.0 | 42.7 | 23782.9 | 42.7 | 23564.0 | 42.6 |
| 18 | 25411.8 | 47.9 | 25181.9 | 47.8 | 24950.1 | 47.8 |
| 19 | 26823.6 | 53.3 | 26580.9 | 53.3 | 26336.2 | 53.2 |
| 20 | 28235.3 | 59.1 | 27979.9 | 59.0 | 27722.4 | 59.0 |
| 25 | 35294.1 | 92.3 | 24974.8 | 92.2 | 34652.9 | 92.2 |
| 30 | 42352.9 | 132.9 | 41969.8 | 132.8 | 41583.4 | 132.7 |
| 40 | 56470.3 | 236.3 | 55959.5 | 236.2 | 55444.3 | 235.9 |
| 50 | 70587.5 | 369.3 | 69949.0 | 369.0 | 69305.1 | 368.6 |
| 1 00 | 84704.5 | 531.7 | 83938.2 | 531.3 | 83165.6 | 530.8 |
| 1 20 | 112937.6 | 945.3 | 111915.9 | 944.6 | 110885.7 | 943.6 |
| 1 30 | 127053.6 | 1196.4 | 125904.2 | 1195.5 | 124747.2 | 1194.3 |
| 1 40 | 141169.2 | 1477.0 | 139892.1 | 1476.0 | 138604.3 | 1474.4 |
| 2 00 | 169399.0 | 2126.9 | 167866.4 | 2125.4 | 166321.0 | 2123.2 |
| 2 30 | 211739.3 | 3323.3 | 209823.5 | 3320.9 | 207891.7 | 3317.5 |
| 3 00 | 254073.4 | 4785.6 | 251774.4 | 4782.1 | 249456.0 | 4777.2 |
| 3 30 | 296400.0 | 6513.8 | 293717.6 | 6509.0 | 291012.8 | 6502.7 |
| 4 00 | 338717.8 | 8507.8 | 335652.1 | 8501.5 | 332560.6 | 8492.7 |

*Polyconic Projection—Co-ordinates, $\delta_m$, $\delta_p$, in Yards.*

| Longitude. | Latitude 47° 30'. $\delta_m$ | $\delta_p$ | Latitude 48° 0'. $\delta_m$ | $\delta_p$ | Latitude 48° 30'. $\delta_m$ | $\delta_p$ |
|---|---|---|---|---|---|---|
| ° ′ | | | | | | |
| 1 | 1373.1 | 0.1 | 1360.0 | 0.1 | 1346.9 | 0.1 |
| 2 | 2746.3 | 0.6 | 2720.1 | 0.6 | 2693.7 | 0.6 |
| 3 | 4119.4 | 1.3 | 4080.1 | 1.3 | 4040.6 | 1.3 |
| 4 | 5492.5 | 2.4 | 5440.2 | 2.4 | 5387.4 | 2.4 |
| 5 | 6865.7 | 3.7 | 6800.2 | 3.7 | 6734.3 | 3.7 |
| 6 | 8238.8 | 5.3 | 8160.3 | 5.3 | 8081.1 | 5.3 |
| 7 | 9612.0 | 7.2 | 9520.3 | 7.2 | 9428.0 | 7.2 |
| 8 | 10985.1 | 9.4 | 10880.4 | 9.4 | 10774.8 | 9.4 |
| 9 | 12358 2 | 11.9 | 12240.4 | 11.9 | 12121.7 | 11.9 |
| 10 | 13731.4 | 14.7 | 13600.5 | 14.7 | 13468.5 | 14.7 |
| 11 | 15104.5 | 17.8 | 14960.5 | 17.8 | 14815.4 | 17.8 |
| 12 | 16477.6 | 21.2 | 16320.5 | 21.2 | 16162.2 | 21.1 |
| 13 | 17850.8 | 24.9 | 17680.6 | 24.8 | 17509.1 | 24.8 |
| 14 | ·19223.9 | 28.9 | 19040.6 | 28.8 | 18855.9 | 28.8 |
| 15 | 20597.0 | 33.1 | 20400.7 | 33.1 | 20202.8 | 33.0 |
| 16 | 21970.1 | 37.7 | 21760.7 | 37.6 | 21549.6 | 37.6 |
| 17 | 23343.3 | 42.6 | 23120.7 | 42.5 | 22896.5 | 42.4 |
| 18 | 24716.4 | 47.7 | 24480.8 | 47.6 | 24243.3 | 47.5 |
| 19 | 26089.5 | 53.1 | 25840.8 | 53.1 | 25590.2 | 53.0 |
| 20 | 27462.7 | 58.9 | 27200.9 | 58.8 | 26937.0 | 58.7 |
| 25 | 34328.3 | 92.0 | 34001.0 | 91.9 | 33671.2 | 91.7 |
| 30 | 41193.9 | 132.5 | 40801.2 | 132.3 | ·40405.4 | 132.0 |
| 40 | 54925.0 | 235.6 | 54401.4 | 235.2 | 53873.6 | 234.7 |
| 50 | 68655.8 | 368.1 | 68001.4 | 367.5 | 67341.7 | 366.8 |
| 1 00 | 82386.5 | 530.1 | 81601.1 | 529.2 | 80809.5 | 528.2 |
| 1 20 | 109846.9 | 942.4 | 108799.7 | 940.8 | 107744.2 | 939.0 |
| 1 30 | 123576.6 | 1192.7 | 122398.5 | 1190.7 | 121211.0 | 1188.4 |
| 1 40 | 137305.8 | 1472.4 | 135996.8 | 1470.0 | 134677.3 | 1467.1 |
| 2 00 | 164762.8 | 2120.3 | 163191.9 | 2116.8 | 161608.6 | 2112.7 |
| 2 30 | 205943.9 | 3313.0 | 203980.3 | 3307.5 | 202001.0 | 3301.1 |
| 3 00 | 247118.6 | 4770.7 | 244762.2 | 4762.9 | 242387.0 | 4753.5 |
| 3 30 | 288285.6 | 6493.5 | 285536.3 | 6482.8 | 282765.2 | 6470.1 |
| 4 00 | 329443.7 | 8481.3 | 326301.5 | 8467.3 | 323134.3 | 8450.7 |

*Polyconic Projection—Co-ordinates, $\delta_m$, $\delta_p$, in Yards.*

| Longitude. | Latitude 49° 0'. | | Latitude 49° 30'. | | Latitude 50° 0'. | |
|---|---|---|---|---|---|---|
| | $\delta_m$ | $\delta_p$ | $\delta_m$ | $\delta_p$ | $\delta_m$ | $\delta_p$ |
| ° ' | | | | | | |
| 1 | 1333.6 | 0.1 | 1320.2 | 0.1 | 1306.7 | 0.1 |
| 2 | 2667.1 | 0.6 | 2640.3 | 0.6 | 2213.3 | 0.6 |
| 3 | 4000.7 | 1.3 | 2960.5 | 1.3 | 3920.0 | 1.3 |
| 4 | 5334.2 | 2.3 | 5280.6 | 2.3 | 5226.6 | 2.3 |
| 5 | 6667.8 | 3.7 | 6600.8 | 3.7 | 6533.3 | 3.6 |
| 6 | 8001.3 | 5.3 | 7920.9 | 5.3 | 7839.9 | 5.2 |
| 7 | 9334.9 | 7.2 | 9241.1 | 7.2 | 9146.6 | 7.1 |
| 8 | 10668.4 | 9.4 | 10561.2 | 9.3 | 10453.2 | 9.3 |
| 9 | 12002.0 | 11.9 | 11881.4 | 11.8 | 11759.9 | 11.8 |
| 10 | 13335.5 | 14.6 | 13201.6 | 14.6 | 13066.5 | 14.6 |
| 11 | 14669.1 | 17.7 | 14521.7 | 17.7 | 14373.2 | 17.6 |
| 12 | 16002.7 | 21.1 | 15841.8 | 21.0 | 15679.8 | 21.0 |
| 13 | 17336.2 | 24.7 | 17162.0 | 24.7 | 16986.5 | 24.6 |
| 14 | 18669.7 | 28.7 | 18482.1 | 28.6 | 18293.1 | 28.5 |
| 15 | 20003.3 | 32.9 | 19802.3 | 32.8 | 19599.8 | 32.8 |
| 16 | 21336.8 | 37.5 | 21122.4 | 37.4 | 20906.4 | 37.3 |
| 17 | 22670.4 | 42.3 | 22442.6 | 42.2 | 22213.1 | 42.1 |
| 18 | 24004.0 | 47.4 | 23752.8 | 47.3 | 23519.7 | 47.2 |
| 19 | 25337.5 | 52.8 | 25082.9 | 52.7 | 24826.4 | 52.6 |
| 20 | 26671.1 | 58.6 | 26403.1 | 58.4 | 26133.0 | 58.2 |
| 25 | 33338.8 | 91.5 | 33003.8 | 91.2 | 32666.2 | 91.0 |
| 30 | 40006.5 | 131.7 | 39604.5 | 131.4 | 39199.4 | 131.0 |
| 40 | 53341.7 | 234.2 | 52805.7 | 233.6 | 52265.7 | 232.9 |
| 50 | 66676.8 | 365.9 | 66006.8 | 365.0 | 65331.7 | 364.0 |
| 1 00 | 80011.6 | 527.0 | 79207.6 | 525.6 | 78397.5 | 524.1 |
| 1 20 | 106680.4 | 936.8 | 105608.3 | 934.4 | 104528.2 | 931.7 |
| 1 30 | 120074.2 | 1185.7 | 118808.2 | 1182.6 | 117593.0 | 1179.2 |
| 1 40 | 133347.5 | 1463.8 | 132007.5 | 1460.0 | 130657.4 | 1455.8 |
| 2 00 | 160012.8 | 2107.9 | 158404.8 | 2102.5 | 156784.7 | 2096.4 |
| 2 30 | 200006.3 | 3293.6 | 197996.2 | 3285.1 | 195970.8 | 3275.6 |
| 3 00 | 239993.2 | 4742.8 | 237581.0 | 4730.5 | 235150.6 | 4716.9 |
| 3 30 | 279972.3 | 6455.4 | 277158.0 | 6438.8 | 274322.4 | 6420.2 |
| 4 00 | 319942.3 | 8431.6 | 316725.8 | 8409.9 | 313485.0 | 8335.6 |

*Arcs of Parallel—Values of D$_p$ in Yards.*

| | L. 20° 30'. | L. 21° 0'. | L. 21° 30'. | L. 22° 0'. | L. 22° 30'. | L. 23° 0'. |
|---|---|---|---|---|---|---|
| ' '' | | | | | | |
| 7 | 221.8 | 221.1 | 220.3 | 219.6 | 218.8 | 218.0 |
| 8 | 253.5 | 252.6 | 251.8 | 250.9 | 250.0 | 249.1 |
| 9 | 285.2 | 284.2 | 283.3 | 282.3 | 281.3 | 280.3 |
| 10 | 316.9 | 315.8 | 314.7 | 313.7 | 312.5 | 311.4 |
| 20 | 633.7 | 631.6 | 629.5 | 627.3 | 625.1 | 622.8 |
| 30 | 950.6 | 947.4 | 944.2 | 941.0 | 937.6 | 934.2 |
| 40 | 1267.4 | 1263.2 | 1259.0 | 1254.6 | 1250.2 | 1245.7 |
| 50 | 1584.3 | 1579.1 | 1573.7 | 1568.3 | 1562.7 | 1557.1 |
| 60 | 1901.1 | 1894.9 | 1888.5 | 1881.9 | 1875.3 | 1868.5 |
| 7 00 | 13307.8 | 13264.1 | 13219.4 | 13173.7 | 13127.0 | 13079.4 |
| 8 00 | 15208.9 | 15159.0 | 15107.9 | 15055.7 | 15002.3 | 14947.9 |
| 9 00 | 17110.0 | 17053.8 | 16996.4 | 16937.6 | 16877.6 | 16816.3 |
| 10 00 | 19011.1 | 18948.7 | 18884.9 | 18819.6 | 18752.9 | 18684.8 |
| 20 00 | 38022.1 | 37897.4 | 37769.7 | 37639.2 | 37505.8 | 37369.6 |
| 30 00 | 57033.2 | 56846.1 | 56654.6 | 56458.8 | 56258.8 | 56054.4 |
| 40 00 | 76044.3 | 75794.8 | 75539.5 | 75278.4 | 75011.7 | 74739.3 |
| 50 00 | 95055.4 | 94743.4 | 94424.3 | 94098.0 | 93764.6 | 93424.1 |
| 60 00 | 114066.4 | 113692.1 | 113309.2 | 112917.7 | 112517.6 | 112108.9 |

*Meridional Arcs—Values of D$_m$ in Yards.*

| | L. 21° 0'. | L. 22° 0'. | L. 23° 0'. |
|---|---|---|---|
| ' '' | | | |
| 7 | 235.4 | 235.4 | 235.5 |
| 8 | 269.0 | 269.1 | 269.1 |
| 9 | 302.7 | 302.7 | 302.8 |
| 10 | 336.3 | 336.4 | 336.4 |
| 20 | 672.6 | 672.7 | 672.8 |
| 30 | 1008.9 | 1009.1 | 1009.2 |
| 40 | 1345.2 | 1345.4 | 1345.6 |
| 50 | 1681.6 | 1681.8 | 1682.0 |
| 60 | 2017.9 | 2018.1 | 2018.4 |
| 7 00 | 14125.0 | 14126.7 | 14128.5 |
| 8 00 | 16142.9 | 16144.8 | 16146.8 |
| 9 00 | 18160.8 | 18162.9 | 18165.2 |
| 10 00 | 20178.6 | 20181.0 | 20183.5 |
| 20 00 | 40357.2 | 40362.1 | 40367.1 |
| 30 00 | 60535.9 | 60543.1 | 60550.6 |
| 40 00 | 80714.5 | 80724.1 | 80734.1 |
| 50 00 | 100893.1 | 100905.2 | 100917.6 |
| 60 00 | 121071.7 | 121086.2 | 121101.2 |

* Intermediate minutes and seconds will be found by moving the decimal point.

### Arcs of Parallel—Values of $D_p$ in Yards.

| ′ ″ | L. 23° 30′. | L. 24° 0′. | L. 24° 30′. | L. 25° 0′. | L. 25° 30′. | L. 26° 0′. |
|---|---|---|---|---|---|---|
| 7 | 217.2 | 216.4 | 215.4 | 214.6 | 213.8 | 212.9 |
| 8 | 248.2 | 247.3 | 246.3 | 245.3 | 244.3 | 243.3 |
| 9 | 279.2 | 278.2 | 277.1 | 276.0 | 274.8 | 273.7 |
| 10 | 310.3 | 309.1 | 307.9 | 306.6 | 305.4 | 304.1 |
| 20 | 620.5 | 618.1 | 615.7 | 613.3 | 610.8 | 608.2 |
| 30 | 930.8 | 927.6 | 923.6 | 919.9 | 916.2 | 912.3 |
| 40 | 1241.0 | 1236.3 | 1231.5 | 1226.5 | 1221.5 | 1216.4 |
| 50 | 1551.3 | 1545.4 | 1539.3 | 1533.2 | 1526.9 | 1520.5 |
| 60 | 1861.5 | 1854.4 | 1847.2 | 1839.8 | 1832.3 | 1824.7 |
| 7 00 | 13030.7 | 12981.0 | 12930.4 | 12878.8 | 12826.1 | 12772.6 |
| 8 00 | 14892.2 | 14835.5 | 14777.6 | 14718.6 | 14658.5 | 14597.2 |
| 9 00 | 16753.8 | 16689.9 | 16624.8 | 16558.4 | 16490.8 | 16421.9 |
| 10 00 | 18615.3 | 18544.3 | 18472.0 | 18398.2 | 18323.1 | 18246.5 |
| 20 00 | 37230.6 | 37088.7 | 36944.0 | 36796.5 | 36646.1 | 36493.0 |
| 30 00 | 55845.8 | 55633.0 | 55416.0 | 55194.7 | 54969.2 | 54739.6 |
| 40 00 | 74461.1 | 74177.4 | 73887.9 | 73592.9 | 73292.3 | 72986.1 |
| 50 00 | 93076.4 | 92721.7 | 92359.9 | 91991.1 | 91615.3 | 91232.6 |
| 60 00 | 111691.7 | 111266.0 | 110831.9 | 110389.4 | 109938.4 | 109479.1 |

### Meridional Arcs—Values of $D_m$ in Yards.

| ′ ″ | L. 24° 0′. | L. 25° 0′. | L. 26° 0′. |
|---|---|---|---|
| 7 | 235.5 | 235.5 | 235.6 |
| 8 | 269.1 | 269.2 | 269.2 |
| 9 | 302.8 | 302.8 | 302.9 |
| 10 | 336.4 | 336.5 | 336.5 |
| 20 | 672.9 | 673.0 | 673.1 |
| 30 | 1009.3 | 1009.4 | 1009.6 |
| 40 | 1345.7 | 1345.9 | 1346.1 |
| 50 | 1682.2 | 1682.4 | 1682.6 |
| 60 | 2018.6 | 2018.9 | 2019.2 |
| 7 00 | 14130.3 | 14132.1 | 14134.1 |
| 8 00 | 16148.9 | 16151.0 | 16153.2 |
| 9 00 | 18167.5 | 18169.9 | 18172.4 |
| 10 00 | 20186.1 | 20188.8 | 20191.5 |
| 20 00 | 40372.2 | 40377.5 | 40383.0 |
| 30 00 | 60558.3 | 60566.3 | 60574.6 |
| 40 00 | 80744.4 | 80755.1 | 80766.1 |
| 50 00 | 100930.5 | 100943.9 | 100957.6 |
| 60 00 | 121116.6 | 121132.6 | 121149.1 |

Intermediate minutes and seconds will be found by moving the decimal-point.

### Arcs of Parallel—Values of $D_p$ in Yards.

|        | L. 26° 30′. | L. 27° 0′. | L. 27° 30′. | L. 28° 0′. | L. 28° 30′. | L. 29° 0 . |
|--------|-------------|------------|-------------|------------|-------------|------------|
| ′  ″   |             |            |             |            |             |            |
| 7      | 212.0       | 211.0      | 210.1       | 209.1      | 208.2       | 207.2      |
| 8      | 242.2       | 241.2      | 240.1       | 239.0      | 237.9       | 236.8      |
| 9      | 272.5       | 271.3      | 270.1       | 268.9      | 267.6       | 266.4      |
| 10     | 302.8       | 301.5      | 300.1       | 298.8      | 297.4       | 296.0      |
| 20     | 605.6       | 603.0      | 600.3       | 597.6      | 594.8       | 591.9      |
| 30     | 908.4       | 904.5      | 900.4       | 896.3      | 892.2       | 887.9      |
| 40     | 1211.2      | 1206.0     | 1200.6      | 1195.1     | 1189.5      | 1183.9     |
| 50     | 1514.0      | 1507.4     | 1500.7      | 1493.9     | 1486.9      | 1479.9     |
| 60     | 1816.9      | 1808.9     | 1800.9      | 1792.7     | 1784.3      | 1775.8     |
| 7  00  | 12718.0     | 12662.5    | 12606.0     | 12548.6    | 12490.1     | 12430.8    |
| 8  00  | 14534.9     | 14471.4    | 14406.9     | 14341.2    | 14274.5     | 14206.6    |
| 9  00  | 16351.7     | 16280.3    | 16207.7     | 16133.9    | 16058.8     | 15982.5    |
| 10 00  | 18168.6     | 18089.3    | 18008.6     | 17926.5    | 17843.1     | 17758.3    |
| 20 00  | 36337.2     | 36178.5    | 36017.1     | 35853.0    | 35686.2     | 35516.6    |
| 30 00  | 54505.8     | 54267.8    | 54025.7     | 53779.5    | 53529.3     | 53274.9    |
| 40 00  | 72674.4     | 72357.1    | 72034.3     | 71706.1    | 71372.4     | 71033.2    |
| 50 00  | 90843.0     | 90446.3    | 90042.9     | 89632.6    | 89215.4     | 88791.5    |
| 60 00  | 109011.5    | 108535.6   | 108051.4    | 107559.1   | 107058.5    | 106549.8   |

### Meridional Arcs—Values of $D_m$ in Yards.

|        | L. 27° 0′. | L. 28° 0′. | L. 29° 0′. |
|--------|------------|------------|------------|
| ′  ″   |            |            |            |
| 7      | 235.6      | 235.6      | 235.7      |
| 8      | 269.3      | 269.3      | 269.3      |
| 9      | 302.9      | 303.0      | 303.0      |
| 10     | 336.6      | 336.6      | 336.7      |
| 20     | 673.1      | 673.2      | 673.3      |
| 30     | 1009.7     | 1009.9     | 1010.0     |
| 40     | 1346.3     | 1346.5     | 1346.7     |
| 50     | 1682.9     | 1683.1     | 1683.3     |
| 60     | 2019.4     | 2019.7     | 2020.0     |
| 7  00  | 14136.0    | 14138.1    | 14140.1    |
| 8  00  | 16155.5    | 16157.8    | 16160.2    |
| 9  00  | 18174.9    | 18177.5    | 18180.2    |
| 10 00  | 20194.3    | 20197.2    | 20200.2    |
| 20 00  | 40388.7    | 40394.5    | 40400.4    |
| 30 00  | 60583.0    | 60591.7    | 60600.6    |
| 40 00  | 80777.4    | 80788.9    | 80800.8    |
| 50 00  | 100971.7   | 100986.2   | 101001.0   |
| 60 00  | 121166.0   | 121183.4   | 121201.2   |

Intermediate minutes and seconds will be found by moving the decimal-point.

### Arcs of Parallel—Values of $D_p$ in Yards.

|  | L. 29° 30'. | L. 30° 0'. | L. 30° 30'. | L. 31° 0'. | L. 31° 30'. | L. 32° 0'. |
|---|---|---|---|---|---|---|
| ' " | | | | | | |
| 7 | 206.2 | 205.2 | 204.1 | 203.1 | 202.0 | 200.9 |
| 8 | 235.6 | 234.5 | 233.3 | 232.1 | 230.9 | 229.6 |
| 9 | 265.1 | 263.8 | 262.4 | 261.1 | 259.7 | 258.3 |
| 10 | 294.5 | 293.1 | 291.6 | 290.1 | 288.6 | 287.0 |
| 20 | 589.1 | 586.2 | 583.2 | 580.2 | 577.1 | 574.0 |
| 30 | 883.6 | 879.2 | 874.8 | 870.3 | 865.7 | 861.1 |
| 40 | 1178.1 | 1172.3 | 1166.4 | 1160.4 | 1154.3 | 1148.1 |
| 50 | 1472.7 | 1465.4 | 1458.0 | 1450.5 | 1442.9 | 1435.1 |
| 60 | 1767.2 | 1758.5 | 1749.6 | 1740.6 | 1731.4 | 1722.1 |
| 7 00 | 12370.5 | 12309.3 | 12247.1 | 12184.0 | 12120.0 | 12055.0 |
| 8 00 | 14137.7 | 14067.7 | 13996.7 | 13924.6 | 13851.4 | 13777.1 |
| 9 00 | 15904.9 | 15826.2 | 15746.3 | 15665.1 | 15582.8 | 15499.3 |
| 10 00 | 17672.2 | 17584.7 | 17495.9 | 17405.7 | 17314.2 | 17221.4 |
| 20 00 | 35344.3 | 35169.4 | 34991.7 | 34811.4 | 34628.4 | 34442.9 |
| 30 00 | 53016.5 | 52754.0 | 52487.6 | 52217.1 | 51942.7 | 51664.3 |
| 40 00 | 70688.7 | 70338.7 | 69983.4 | 69622.8 | 69256.9 | 68885.7 |
| 50 00 | 88360.8 | 87923.4 | 87479.3 | 87028.5 | 86571.1 | 86107.1 |
| 60 00 | 106033.0 | 105508.1 | 104975.2 | 104434.2 | 103885.3 | 103328.6 |

### Meridional Arcs—Values of $D_m$ in Yards.

|  | L. 30° 0'. | L. 31° 0'. | L. 32° 0'. |
|---|---|---|---|
| ' " | | | |
| 7 | 235.7 | 235.7 | 235.8 |
| 8 | 269.4 | 269.4 | 269.5 |
| 9 | 303.0 | 303.1 | 303.1 |
| 10 | 336.7 | 336.8 | 336.8 |
| 20 | 673.4 | 673.5 | 673.6 |
| 30 | 1010.2 | 1010.3 | 1010.5 |
| 40 | 1346.9 | 1347.1 | 1347.3 |
| 50 | 1683.6 | 1683.9 | 1684.1 |
| 60 | 2020.3 | 2020.6 | 2020.9 |
| 7 00 | 14142.3 | 14144.4 | 14146.6 |
| 8 00 | 16162.6 | 16165.1 | 16167.6 |
| 9 00 | 18182.9 | 18185.7 | 18188.5 |
| 10 00 | 20203.2 | 20206.3 | 20209.5 |
| 20 00 | 40406.5 | 40412.6 | 40418.9 |
| 30 00 | 60609.7 | 60619.0 | 60628.4 |
| 40 00 | 80812.9 | 80825.3 | 80837.9 |
| 50 00 | 101016.1 | 101031.6 | 101047.4 |
| 60 00 | 121219.4 | 121237.9 | 121256.8 |

Intermediate minutes and seconds will be found by moving the decimal-point.

## Arcs of Parallel—Values of $D_p$ in Yards.

| | L. 32° 30'. | L. 33° 0'. | L. 33° 30'. | L. 34° 0'. | L. 34° 30'. | L. 35° 0'. |
|---|---|---|---|---|---|---|
| ' " | | | | | | |
| 7 | 199.8 | 198.7 | 197.6 | 196.4 | 195.3 | 194.1 |
| 8 | 228.4 | 227.1 | 225.8 | 224.5 | 223.2 | 221.8 |
| 9 | 256.9 | 254.5 | 254.0 | 252.6 | 251.1 | 249.6 |
| 10 | 285.5 | 283.9 | 282.3 | 280.6 | 279.0 | 277.3 |
| 20 | 570.9 | 567.7 | 564.5 | 561.2 | 557.9 | 554.6 |
| 30 | 856.4 | 851.6 | 846.8 | 841.9 | 836.9 | 831.9 |
| 40 | 1141.8 | 1135.5 | 1129.0 | 1122.5 | 1115.9 | 1109.2 |
| 50 | 1427.3 | 1419.3 | 1411.3 | 1403.1 | 1394.8 | 1386.4 |
| 60 | 1712.7 | 1703.2 | 1693.5 | 1683.7 | 1673.8 | 1663.7 |
| | | | | | | |
| 7 00 | 11989.1 | 11922.3 | 11854.6 | 11786.0 | 11716.5 | 11646.1 |
| 8 00 | 13701.9 | 13625.5 | 13548.1 | 13469.7 | 13390.3 | 13309.8 |
| 9 00 | 15414.6 | 15328.7 | 15241.7 | 15153.5 | 15064.1 | 14973.6 |
| 10 00 | 17127.3 | 17031.9 | 16935.2 | 16837.2 | 16737.9 | 16637.3 |
| 20 00 | 34254.6 | 34063.8 | 33870.4 | 33674.3 | 33475.8 | 33274.6 |
| 30 00 | 51381.9 | 51095.7 | 50805.5 | 50511.5 | 50213.6 | 49911.9 |
| 40 00 | 68509.3 | 68127.6 | 67740.7 | 67348.7 | 66951.5 | 66549.2 |
| 50 00 | 85636.6 | 85159.5 | 84675.9 | 84185.8 | 83689.4 | 83186.5 |
| 60 00 | 102763.9 | 102191.4 | 101611.1 | 101023.0 | 100427.3 | 99823.8 |

## Meridional Arcs—Values of $D_m$ in Yards.

| | L. 33° 0'. | L. 34° 0'. | L. 35° 0'. |
|---|---|---|---|
| ' " | | | |
| 7 | 235.8 | 235.9 | 235.9 |
| 8 | 269.5 | 269.5 | 269.6 |
| 9 | 303.2 | 303.2 | 303.3 |
| 10 | 336.9 | 336.9 | 337.0 |
| 20 | 673.8 | 673.9 | 674.0 |
| 30 | 1010.6 | 1010.8 | 1011.0 |
| 40 | 1347.5 | 1347.7 | 1347.9 |
| 50 | 1684.4 | 1684.7 | 1684.9 |
| 60 | 2021.3 | 2021.6 | 2021.9 |
| | | | |
| 7 00 | 14148.9 | 14151.2 | 14153.5 |
| 8 00 | 16170.1 | 16172.7 | 16175.4 |
| 9 00 | 18191.4 | 18194.3 | 18197.3 |
| 10 00 | 20212.7 | 20215.9 | 20219.2 |
| 20 00 | 40425.4 | 40431.9 | 40438.5 |
| 30 00 | 60638.0 | 60647.8 | 60657.7 |
| 40 00 | 80850.7 | 80863.7 | 80877.0 |
| 50 00 | 101063.4 | 101079.7 | 101096.2 |
| 60 00 | 121276.1 | 121295.6 | 121315.4 |

Intermediate minutes and seconds will be found by moving the decimal-point.

## Arcs of Parallel—Values of $D_p$ in Yards.

|  ′ ″ | L. 35° 30′. | L. 36° 0′. | L. 36° 30′. | L. 37° 0′. | L. 37° 30′. | L. 38° 0′. |
|---|---|---|---|---|---|---|
| 7 | 192.9 | 191.7 | 190.5 | 189.3 | 188.0 | 186.8 |
| 8 | 220.5 | 219.1 | 217.7 | 216.3 | 214.9 | 213.4 |
| 9 | 248.0 | 246.5 | 244.9 | 243.3 | 241.7 | 240.1 |
| 10 | 275.6 | 273.9 | 272.1 | 270.4 | 268.6 | 266.8 |
| 20 | 551.2 | 547.7 | 544.3 | 540.7 | 537.2 | 533.6 |
| 30 | 826.9 | 821.6 | 816.4 | 811.1 | 805.8 | 800.4 |
| 40 | 1102.4 | 1095.5 | 1088.5 | 1081.5 | 1074.4 | 1067.2 |
| 50 | 1378.0 | 1369.4 | 1360.7 | 1351.9 | 1343.0 | 1334.0 |
| 60 | 1653.5 | 1643.2 | 1632.8 | 1622.2 | 1611.6 | 1600.7 |
| 7 00 | 11574.8 | 11502.7 | 11429.6 | 11355.7 | 11280.9 | 11205.2 |
| 8 00 | 13228.4 | 13145.9 | 13062.4 | 12977.9 | 12892.5 | 12806.0 |
| 9 00 | 14881.9 | 14789.1 | 14695.2 | 14600.2 | 14504.0 | 14406.7 |
| 10 00 | 16535.5 | 16432.4 | 16328.0 | 16222.4 | 16115.6 | 16007.5 |
| 20 00 | 33070.9 | 32864.7 | 32656.0 | 32444.8 | 32231.1 | 32015.0 |
| 30 00 | 49606.4 | 49297.1 | 48984.0 | 48667.2 | 48346.7 | 48022.5 |
| 40 00 | 66141.9 | 65729.5 | 65312.1 | 64889.6 | 64462.3 | 64030.0 |
| 50 00 | 82677.3 | 82161.8 | 81640.1 | 81112.0 | 80577.8 | 80037.5 |
| 60 00 | 99212.8 | 98594.2 | 97968.1 | 97334.4 | 96693.4 | 96045.0 |

## Meridional Arcs—Values of $D_m$ in Yards.

|  ′ ″ | L. 36° 0′. | L. 37° 0′. | L. 38° 0′. |
|---|---|---|---|
| 7 | 235.9 | 236.0 | 236.0 |
| 8 | 269.6 | 269.7 | 269.7 |
| 9 | 303.3 | 303.4 | 303.4 |
| 10 | 337.0 | 337.1 | 337.2 |
| 20 | 674.1 | 674.2 | 674.3 |
| 30 | 1011.1 | 1011.3 | 1011.5 |
| 40 | 1348.2 | 1348.4 | 1348.6 |
| 50 | 1685.2 | 1685.5 | 1685.8 |
| 60 | 2022.3 | 2022.6 | 2022.9 |
| 7 00 | 14155.8 | 14158.2 | 14160.6 |
| 8 00 | 16178.1 | 16180.8 | 16183.5 |
| 9 00 | 18200.3 | 18203.4 | 18206.5 |
| 10 00 | 20222.6 | 20226.0 | 20229.4 |
| 20 00 | 40445.2 | 40451.9 | 40458.8 |
| 30 00 | 60667.5 | 60677.9 | 60688.2 |
| 40 00 | 80890.3 | 80903.9 | 80917.6 |
| 50 00 | 101112.9 | 101129.0 | 101147.0 |
| 60 00 | 121335.5 | 121355.8 | 121376.4 |

Intermediate minutes and seconds will be found by moving the decimal-point.

## Arcs of Parallel—Values of $D_p$ in Yards.

| | L. 38° 30'. | L. 39° 0'. | L. 39° 30'. | L. 40° 0'. | L. 40° 30'. | L. 41° 0'. |
|---|---|---|---|---|---|---|
| ' '' | | | | | | |
| 7 | 185.5 | 184.2 | 182.9 | 181.6 | 180.2 | 178.9 |
| 8 | 212.0 | 210.5 | 209.0 | 207.5 | 206.0 | 204.4 |
| 9 | 238.5 | 236.8 | 235.1 | 233.4 | 231.7 | 230.0 |
| 10 | 265.0 | 263.1 | 261.3 | 259.4 | 257.5 | 255.6 |
| 20 | 529.9 | 526.3 | 522.5 | 518.8 | 515.0 | 511.1 |
| 30 | 794.9 | 789.4 | 783.8 | 778.2 | 772.4 | 766.7 |
| 40 | 1059.9 | 1052.5 | 1045.1 | 1037.5 | 1029.9 | 1022.2 |
| 50 | 1324.9 | 1315.6 | 1306.3 | 1296.9 | 1287.4 | 1277.8 |
| 60 | 1589.8 | 1578.8 | 1567.6 | 1556.3 | 1544.9 | 1533.4 |
| 7 00 | 11128.7 | 11051.4 | 10973.2 | 10894.1 | 10814.3 | 10733.6 |
| 8 00 | 12718.6 | 12630.2 | 12540.8 | 12450.4 | 12359.2 | 12266.9 |
| 9 00 | 14308.4 | 14208.9 | 14108.4 | 14006.8 | 13904.1 | 13800.3 |
| 10 00 | 15898.2 | 15787.7 | 15676.0 | 15563.1 | 15448.9 | 15333.4 |
| 20 00 | 31796.4 | 31575.4 | 31351.9 | 31126.1 | 30897.9 | 30667.3 |
| 30 00 | 47694.6 | 47363.1 | 47027.9 | 46689.2 | 46346.8 | 46001.0 |
| 40 00 | 63592.8 | 63150.8 | 62703.9 | 62252.2 | 61795.8 | 61334.6 |
| 50 00 | 79491.0 | 78938.4 | 78379.9 | 77815.3 | 77244.7 | 76668.3 |
| 60 00 | 95389.2 | 94726.1 | 94055.8 | 93378.3 | 92693.7 | 92001.9 |

## Meridional Arcs—Values of $D_m$ in Yards.

| | L. 39° 0'. | | L. 40° 0'. | | L. 41° 0'. |
|---|---|---|---|---|---|
| ' '' | | | | | |
| 7 | 236.0 | | 236.1 | | 236.1 |
| 8 | 269.8 | | 269.8 | | 269.9 |
| 9 | 303.5 | | 303.5 | | 303.6 |
| 10 | 337.2 | | 337.3 | | 337.3 |
| 20 | 674.4 | | 674.5 | | 674.7 |
| 30 | 1011.6 | | 1011.8 | | 1012.0 |
| 40 | 1348.9 | | 1349.1 | | 1349.3 |
| 50 | 1686.1 | | 1686.4 | | 1686.7 |
| 60 | 2023.3 | | 2023.6 | | 2024.0 |
| 7 00 | 14163.0 | | 14165.4 | | 14167.9 |
| 8 00 | 16186.3 | | 16189.1 | | 16191.9 |
| 9 00 | 18209.6 | | 18212.7 | | 18215.8 |
| 10 00 | 20232.8 | | 20236.3 | | 20239.8 |
| 20 00 | 40465.7 | | 40472.7 | | 40479.7 |
| 30 00 | 60698.5 | | 60709.0 | | 60719.5 |
| 40 00 | 80931.4 | | 80945.3 | | 80959.3 |
| 50 00 | 101164.2 | | 101181.6 | | 101199.2 |
| 60 00 | 121397.1 | | 121418.0 | | 121439.0 |

**Intermediate minutes and seconds will be found by moving the decimal-point.**

### Arcs of Parallel—Values of $D_p$ in Yards.

| | L. 41° 30'. | L. 42° 0'. | L. 42° 30'. | L. 43° 0'. | L. 43° 30'. | L. 44° 0'. |
|---|---|---|---|---|---|---|
| ' '' | | | | | | |
| 7 | 177.5 | 176.2 | 174.8 | 173.4 | 172.0 | 170.5 |
| 8 | 202.9 | 201.3 | 199.7 | 198.1 | 196.5 | 194.9 |
| 9 | 228.3 | 226.5 | 224.7 | 222.9 | 221.1 | 219.3 |
| 10 | 253.6 | 251.7 | 249.7 | 247.7 | 245.7 | 243.6 |
| 20 | 507.2 | 503.3 | 449.4 | 495.4 | 491.3 | 487.3 |
| 30 | 760.9 | 755.0 | 749.0 | 743.0 | 737.0 | 730.9 |
| 40 | 1014.5 | 1006.6 | 998.7 | 990.7 | 982.7 | 974.5 |
| 50 | 1268.1 | 1258.3 | 1248.4 | 1238.4 | 1228.3 | 1218.1 |
| 60 | 1521.7 | 1510.0 | 1498.1 | 1486.1 | 1474.0 | 1461.8 |
| 7 00 | 10652.0 | 10569.7 | 10486.6 | 10402.6 | 10317.9 | 10232.3 |
| 8 00 | 12173.8 | 12079.7 | 11984.6 | 11888.7 | 11791.8 | 11694.1 |
| 9 00 | 13695.5 | 13589.6 | 13482.7 | 13374.8 | 13265.8 | 13155.8 |
| 10 00 | 15217.2 | 15099.6 | 14980.8 | 14860.9 | 14739.8 | 14617.6 |
| 20 00 | 30434.4 | 30199.1 | 29961.6 | 29721.7 | 29479.6 | 29235.2 |
| 30 00 | 45651.6 | 45298.7 | 44942.4 | 44582.6 | 44219.4 | 43852.8 |
| 40 00 | 60868.8 | 60398.3 | 59923.2 | 59443.4 | 58959.2 | 58470.4 |
| 50 00 | 76086.0 | 75497.9 | 74903.9 | 74304.3 | 73698.9 | 73088.0 |
| 60 00 | 91303.2 | 90597.4 | 89884.7 | 89165.1 | 88438.7 | 87705.6 |

### Meridional Arcs—Values of $D_m$ in Yards.

| | L. 42° 0'. | L. 43° 0'. | L. 44° 0'. |
|---|---|---|---|
| ' '' | | | |
| 7 | 236.2 | 236.2 | 236.3 |
| 8 | 269.9 | 270.0 | 270.0 |
| 9 | 303.7 | 303.7 | 303.8 |
| 10 | 337.4 | 337.4 | 337.5 |
| 20 | 674.8 | 674.9 | 675.0 |
| 30 | 1012.2 | 1012.3 | 1012.5 |
| 40 | 1349.6 | 1349.8 | 1350.0 |
| 50 | 1686.9 | 1687.2 | 1687.5 |
| 60 | 2024.3 | 2024.7 | 2025.0 |
| 7 00 | 14170.3 | 14172.8 | 14175.3 |
| 8 00 | 16194.7 | 16197.5 | 16200.3 |
| 9 00 | 18219.0 | 18222.2 | 18225.4 |
| 10 00 | 20243.4 | 20246.9 | 20250.4 |
| 20 00 | 40486.7 | 40493.8 | 40500.9 |
| 30 00 | 60730.1 | 60740.7 | 60751.3 |
| 40 00 | 80973.4 | 80987.5 | 81001.7 |
| 50 00 | 101216.8 | 101234.4 | 101252.2 |
| 60 00 | 121460.1 | 121481.3 | 121502.6 |

Intermediate minutes and seconds will be found by moving the decimal-point.

### Arcs of Parallel—Values of $D_p$ in Yards.

| | L. 44° 30'. | L. 45° 0'. | L. 45° 30'. | L. 46° 0'. | L. 46° 30'. | L. 47° 0'. |
|---|---|---|---|---|---|---|
| ' " | | | | | | |
| 7 | 169.1 | 167.6 | 166.2 | 164.7 | 163.2 | 161.7 |
| 8 | 193.3 | 191.6 | 189.9 | 188.2 | 186.5 | 184.8 |
| 9 | 217.4 | 215.5 | 213.7 | 211.8 | 209.8 | 207.9 |
| 10 | 241.6 | 239.5 | 237.4 | 235.3 | 233.2 | 331.0 |
| 20 | 483.1 | 479.0 | 474.8 | 470.6 | 466.3 | 462.0 |
| 30 | 724.7 | 718.5 | 712.2 | 705.9 | 699.5 | 693.1 |
| 40 | 966.3 | 958.0 | 949.6 | 941.2 | 932.7 | 924.1 |
| 50 | 1207.9 | 1197.5 | 1187.0 | 1176.5 | 1165.8 | 1155.1 |
| 60 | 1449.4 | 1437.0 | 1424.4 | 1411.8 | 1399.0 | 1386.1 |
| 7 00 | 10146.0 | 10058.9 | 9971.0 | 9882.4 | 9793.0 | 9702.8 |
| 8 00 | 11595.4 | 11495.9 | 11395.5 | 11294.2 | 11192.0 | 11089.0 |
| 9 00 | 13044.8 | 12932.9 | 12819.9 | 12705.9 | 12591.0 | 12475.1 |
| 10 00 | 14494.3 | 14369.8 | 14244.3 | 14117.7 | 13990.0 | 13861.2 |
| 20 00 | 28988.6 | 28739.7 | 28488.6 | 28235.4 | 27980.0 | 27722.4 |
| 30 00 | 43482.8 | 43109.5 | 42733.0 | 42353.1 | 41970.0 | 41583.6 |
| 40 00 | 57977.1 | 57479.4 | 56977.3 | 56470.8 | 55960.0 | 55444.8 |
| 50 00 | 72471.4 | 71849.2 | 71221.6 | 70588.5 | 69950.0 | 69306.0 |
| 60 00 | 86965.7 | 86219.1 | 85465.9 | 84706.2 | 83940.0 | 83167.3 |

### Meridional Arcs—Values of $D_m$ in Yards.

| | L. 45° 0'. | L. 46° 0'. | L. 47° 0'. |
|---|---|---|---|
| ' " | | | |
| 7 | 236.3 | 236.3 | 236.4 |
| 8 | 270.1 | 270.1 | 270.1 |
| 9 | 303.8 | 303.9 | 303.9 |
| 10 | 337.6 | 337.6 | 337.7 |
| 20 | 675.1 | 675.3 | 675.4 |
| 30 | 1012.7 | 1012.9 | 1013.1 |
| 40 | 1350.3 | 1350.5 | 1350.7 |
| 50 | 1687.8 | 1688.1 | 1688.4 |
| 60 | 2025.4 | 2025.8 | 2026.1 |
| 7 00 | 14177.8 | 14180.3 | 14182.8 |
| 8 00 | 16203.2 | 16206.0 | 16208.9 |
| 9 00 | 18228.6 | 18231.8 | 18235.0 |
| 10 00 | 20254.0 | 20257.5 | 20261.1 |
| 20 00 | 40508.0 | 40515.1 | 40522.2 |
| 30 00 | 60761.9 | 60772.6 | 60783.2 |
| 40 00 | 81015.9 | 81030.1 | 81044.3 |
| 50 00 | 101269.9 | 101287.7 | 101305.4 |
| 60 00 | 121523.9 | 121545.2 | 121566.5 |

Intermediate minutes and seconds will be found by moving the decimal point.

## Arcs of Parallel—Values of $D_p$ in Yards.

|  | L. 47° 30'. | L. 48° 0'. | L. 48° 30'. | L. 49° 0'. | L. 49° 30'. | L. 50° 0'. |
|---|---|---|---|---|---|---|
| ' '' |  |  |  |  |  |  |
| 7 | 160.2 | 158.7 | 157.1 | 155.6 | 154.0 | 152.4 |
| 8 | 183.1 | 181.3 | 179.6 | 177.8 | 176.0 | 174.2 |
| 9 | 206.0 | 204.0 | 202.0 | 200.0 | 198.0 | 196.0 |
| 10 | 228.9 | 226.7 | 224.5 | 222.3 | 220.0 | 217.8 |
| 20 | 457.7 | 453.3 | 449.0 | 444.5 | 440.1 | 435.6 |
| 30 | 686.6 | 680.0 | 673.4 | 666.8 | 660.1 | 653.3 |
| 40 | 915.4 | 906.7 | 897.9 | 889.0 | 880.1 | 871.1 |
| 50 | 1144.3 | 1133.4 | 1122.4 | 1111.3 | 1100.1 | 1088.9 |
| 60 | 1373.1 | 1360.0 | 1346.9 | 1333.6 | 1320.2 | 1306.7 |
| 7 00 | 9612.0 | 9520.3 | 9428.0 | 9334.9 | 9241.1 | 9146.6 |
| 8 00 | 10985.1 | 10880.4 | 10774.8 | 10668.5 | 10561.2 | 10453.2 |
| 9 00 | 12358.2 | 12240.4 | 12121.7 | 12002.0 | 11881.4 | 11759.9 |
| 10 00 | 13731.4 | 13600.5 | 13468.5 | 13335.6 | 13201.6 | 13066.5 |
| 20 00 | 27462.7 | 27200.9 | 26937.1 | 26671.1 | 26403.1 | 26133.1 |
| 30 00 | 41194.1 | 40801.4 | 40405.6 | 40006.7 | 39604.7 | 39199.6 |
| 40 00 | 54925.5 | 54401.9 | 53874.1 | 53342.3 | 52806.2 | 52266.2 |
| 50 00 | 68656.8 | 68002.4 | 67342.7 | 66677.8 | 66007.8 | 65332.7 |
| 60 00 | 82388.2 | 81602.8 | 80811.2 | 80013.4 | 79209.4 | 78399.3 |

## Meridional Arcs—Values of $D_m$ in Yards.

|  |  | L. 48° 0'. |  | L. 49° 0'. |  | L. 50° 0'. |
|---|---|---|---|---|---|---|
| ' '' |  |  |  |  |  |  |
| 7 |  | 236.4 |  | 236.5 |  | 236.5 |
| 8 |  | 270.2 |  | 270.2 |  | 270.3 |
| 9 |  | 304.0 |  | 304.0 |  | 304.1 |
| 10 |  | 337.7 |  | 337.8 |  | 337.9 |
| 20 |  | 675.5 |  | 675.6 |  | 675.7 |
| 30 |  | 1013.2 |  | 1013.4 |  | 1013.6 |
| 40 |  | 1351.0 |  | 1351.2 |  | 1351.4 |
| 50 |  | 1688.7 |  | 1689.0 |  | 1689.3 |
| 60 |  | 2026.5 |  | 2026.8 |  | 2027.2 |
| 7 00 |  | 14185.2 |  | 14187.7 |  | 14190.2 |
| 8 00 |  | 16211.7 |  | 16214.5 |  | 16217.3 |
| 9 00 |  | 18238.2 |  | 18241.3 |  | 18244.5 |
| 10 00 |  | 20264.6 |  | 20268.1 |  | 20271.7 |
| 20 00 |  | 40529.2 |  | 40536.3 |  | 40543.3 |
| 30 00 |  | 60793.9 |  | 60804.4 |  | 60815.0 |
| 40 00 |  | 81058.5 |  | 81072.6 |  | 81086.6 |
| 50 00 |  | 101323.1 |  | 101340.7 |  | 101358.9 |
| 60 00 |  | 121587.7 |  | 121608.9 |  | 121629 9 |

Intermediate minutes and seconds will be found by moving the decimal-point.

*Lengths in Nautical Miles and Statute Miles of Degrees of Latitude and Longitude in Different Latitudes.*

| | DEGREE OF THE PARALLEL. | | | DEGREE OE THE MERIDIAN. | |
|---|---|---|---|---|---|
| Latitude of Parallel. | Nautical miles. | Statute miles. | Latitude of middle point. | Nautical miles. | Statute miles. |
| ° | | | ° | | |
| 20 | 56.404 | 65.018 | 20 | 59.664 | 68.777 |
| 21 | 56.039 | 64.598 | | | |
| 22 | 55.657 | 64.158 | | | |
| 23 | 55.258 | 63.698 | | | |
| 24 | 54.843 | 63.219 | | | |
| 25 | 54.411 | 62.721 | 25 | 59.706 | 68.825 |
| 26 | 53.962 | 62.204 | | | |
| 27 | 53.497 | 61.668 | | | |
| 28 | 53.016 | 61.113 | | | |
| 29 | 52.518 | 60.540 | | | |
| 30 | 52.005 | 59.948 | 30 | 59.749 | 68.875 |
| 31 | 51.476 | 59.338 | | | |
| 32 | 50.931 | 58.709 | | | |
| 33 | 50.370 | 58.063 | | | |
| 34 | 49.794 | 57.399 | | | |
| 35 | 49.203 | 56.718 | 35 | 59.796 | 68.929 |
| 36 | 48.597 | 56.019 | | | |
| 37 | 47.976 | 55.304 | | | |
| 38 | 47.341 | 54.571 | | | |
| 39 | 46.690 | 53.822 | | | |
| 40 | 46.026 | 53.056 | 40 | 59.847 | 68.987 |
| 41 | 45.348 | 52.274 | | | |
| 42 | 44.654 | 51.476 | | | |
| 43 | 43.949 | 50.662 | | | |
| 44 | 43.230 | 49.833 | | | |
| 45 | 42.497 | 48.988 | 45 | 59.899 | 69.048 |
| 46 | 41.752 | 48.128 | | | |
| 47 | 40.993 | 47.254 | | | |
| 48 | 40.222 | 46.365 | | | |
| 49 | 39.439 | 45.462 | | | |
| 50 | 38.643 | 44.545 | 50 | 59.951 | 69.108 |

A degree of longitude at the equator = 69.163 statute miles.
A second of time at the equator      = 1521.6 feet.

*Co-ordinates of Curvature in Statute Miles for Maps of Large Extent.*

| Longitude. | Latitude 20°. | | Latitude 22°. | | Latitude 24°. | | Latitude 26°. | |
|---|---|---|---|---|---|---|---|---|
| | $d_m$ | $d_p$ | $d_m$ | $d_p$ | $d_m$ | $d_p$ | $d_m$ | $d_p$ |
| 2 | 130.0 | 0.8 | 128.3 | 0.8 | 126.4 | 0.9 | 124.4 | 0.9 |
| 4 | 260.0 | 3.1 | 256.6 | 3.3 | 252.8 | 3.6 | 248.8 | 3.8 |
| 6 | 390.0 | 6.9 | 384.8 | 7.5 | 379.2 | 8.1 | 373.1 | 8.6 |
| 8 | 520.0 | 12.4 | 513.0 | 13.4 | 505.5 | 14.4 | 497.3 | 15.2 |
| 10 | 649.8 | 19.4 | 641.1 | 21.0 | 631.7 | 22.4 | 621.4 | 23.8 |
| 12 | 779.7 | 27.8 | 769.1 | 30.2 | 757.9 | 32.2 | 745.4 | 34.2 |
| 14 | 909.2 | 38.0 | 896.9 | 41.0 | 883.6 | 43.9 | 869.2 | 46.6 |
| 16 | 1039.2 | 49.6 | 1024.5 | 53.6 | 1009.9 | 57.4 | 992.8 | 60.8 |
| 18 | 1168.1 | 62.8 | 1152.2 | 67.9 | 1134.8 | 72.6 | 1116.1 | 77.0 |
| 20 | 1298.0 | 77.6 | 1279.5 | 83.8 | 1261.2 | 89.7 | 1239.2 | 95.0 |
| $r$ | 10892 | | 9813 | | 8905 | | 8130 | |

| Longitude. | Latitude 28°. | | Latitude 30°. | | Latitude 32°. | | Latitude 34°. | |
|---|---|---|---|---|---|---|---|---|
| | $d_m$ | $d_p$ | $d_m$ | $d_p$ | $d_m$ | $d_p$ | $d_m$ | $d_p$ |
| 2 | 122.2 | 1.0 | 119.8 | 1.0 | 117.4 | 1.1 | 114.8 | 1.1 |
| 4 | 244.4 | 4.0 | 239.7 | 4.2 | 234.8 | 4.3 | 229.5 | 4.5 |
| 6 | 366.5 | 9.0 | 359.5 | 9.4 | 352.0 | 9.8 | 344.2 | 10.1 |
| 8 | 488.6 | 16.0 | 479.2 | 16.7 | 469.3 | 17.3 | 458.7 | 17.9 |
| 10 | 610.4 | 25.0 | 598.7 | 26.1 | 586.3 | 27.1 | 573.1 | 28.0 |
| 12 | 732.4 | 36.0 | 718.0 | 37.6 | 703.5 | 39.1 | 687.2 | 40.3 |
| 14 | 853.7 | 49.0 | 837.1 | 51.2 | 819.6 | 53.1 | 801.1 | 54.8 |
| 16 | 975.7 | 64.1 | 956.0 | 66.9 | 936.8 | 69.5 | 914.7 | 71.6 |
| 18 | 1096.0 | 80.9 | 1074.6 | 84.6 | 1051.9 | 87.8 | 1027.9 | 90.5 |
| 20 | 1218.8 | 100.1 | 1192.9 | 104.3 | 1169.2 | 108.6 | 1140.7 | 111.7 |
| $r$ | 7458 | | 6869 | | 6348 | | 5881 | |

*Co-ordinates of Curvature in Statute Miles for Maps of Large Extent.*

| Longitude | Latitude 36°. | | Latitude 38°. | | Latitude 40°. | | Latitude 42°. | |
|---|---|---|---|---|---|---|---|---|
| | $d_m$ | $d_p$ | $d_m$ | $d_p$ | $d_m$ | $d_p$ | $d_m$ | $d_p$ |
| o | | | | | | | | |
| 2 | 112.0 | 1.2 | 109.1 | 1.2 | 106.1 | 1.2 | 102.9 | 1.2 |
| 4 | 224.0 | 4.6 | 218.2 | 4.7 | 212.2 | 4.8 | 205.8 | 4.8 |
| 6 | 335.9 | 10.3 | 327.2 | 10.6 | 318.1 | 10.7 | 308.6 | 10,8 |
| 8 | 447.7 | 18.4 | 436.0 | 18.8 | 423.9 | 18.9 | 411.2 | 19.2 |
| 10 | 559.2 | 28.7 | 544.7 | 29.3 | 529.4 | 29.7 | 513.6 | 30.0 |
| 12 | 670.5 | 41.3 | 653.0 | 42.2 | 634.7 | 42.8 | 615.7 | 43.2 |
| 14 | 781.6 | 56.2 | 761.1 | 57.4 | 739.7 | 58.2 | 717.5 | 58.8 |
| 16 | 892.3 | 73.4 | 868.8 | 74.9 | 844.3 | 76.0 | 818.8 | 76.7 |
| 18 | 1002.6 | 92.8 | 976.2 | 94.7 | 948.5 | 96.1 | 919.8 | 97.0 |
| 20 | 1112.5 | 114.5 | 1083.0 | 116.8 | 1052.3 | 118.5 | 1020.2 | 119.7 |
| r | 5461 | | · 5079 | | 4729 | | 4408 | |

| Longitude | Latitude 44°. | | Latitude 46°. | | Latitude 48°. | | Latitude 50°. | |
|---|---|---|---|---|---|---|---|---|
| | $d_m$ | $d_p$ | $d_m$ | $d_p$ | $d_m$ | $d_p$ | $d_m$ | $d_p$ |
| o | | | | | | | | |
| 2 | 99.7 | 1.2 | 96.2 | 1.2 | 92.7 | 1.2 | 89.1 | 1.2 |
| 4 | 198.9 | 4.8 | 192.4 | 4.8 | 185.4 | 4.8 | 178.1 | 4.8 |
| 6 | 298.7 | 10.9 | 288.5 | 10.9 | 277.9 | 10.8 | 267.0 | 10.7 |
| 8 | 398.0 | 19.3 | 384.4 | 19.3 | 370.3 | 19.2 | 355.7 | 19.0 |
| 10 | 497.1 | 30.2 | 480.0 | 30.2 | 462.3 | 30.0 | 444.1 | 29.7 |
| 12 | 595.9 | 43.4 | 575.4 | 43.4 | 554.1 | 43.2 | 532.3 | 42.8 |
| 14 | 694.3 | 59.1 | 670.3 | 59.1 | 645.6 | 58.8 | 620.0 | 58.2 |
| 16 | 792.3 | 77.1 | 764.9 | 77.1 | 736.5 | 76.7 | 707.3 | 75.9 |
| 18 | 889.9 | 97.5 | 859.0 | 97.5 | 827.0 | 97.0 | 794.1 | 96.0 |
| 20 | 986.9 | 120.2 | 952.5 | 120.2 | 916.9 | 119.6 | 880.3 | 118.4 |
| r | 4110 | | 3833 | | 3575 | | 3332 | |

## XLIV.—*Trigonometrical Leveling.*

Reciprocal zenith-distances measured at two stations at the same time give the best results. When reciprocal, but not simultaneous, the observations should be made on different days, in order to obtain as far as possible an average value of the refraction as well as the mean value of the difference between the respective angles; the same with zenith-distances measured at one station only.

The refraction being greater and more variable at sunrise and sunset, and comparatively stationary between the hours of 10 a. m. and 4 p. m., the best time for observation is between those hours.

The condition of the atmosphere and the relative refraction may be so different at stations more than twenty miles apart, that, as a general rule, the difference of level, determined even by reciprocal observations, cannot be relied upon for much accuracy at greater distances unless a very large number of measurements have been made under the most favorable circumstances. The higher the elevations the more reliable the results.

1.—*To obtain the Difference of Level of Two Points from Reciprocal Zenith-Distances, simultaneous or not.*

Let—

$Z$, $Z'$ = the measured zenith-distances of the telescopes at the two stations, of which $Z$ is the smaller;

$K$ = distance in yards between the two stations;

$R_z$ = radius of curvature of the arc joining the two stations;

$C$ = angle at the earth's center subtended by the arc; and

$dh$ = difference of level of the two stations;

then—

$$C = \frac{K}{R_z \sin 1''}; \qquad dh = \frac{K \sin \frac{1}{2}(Z' - Z)}{\cos \frac{1}{2}(Z' - Z + C)}$$

## XLIV.—*Trigonometrical Leveling*—Continued.

If the telescope is not observed upon, but some other object near, the measured zenith-distance can be reduced to the telescope by the formula :

$$\text{Reduction to telescope, in seconds,} = \pm \frac{r}{K \sin 1''}$$

in which $r$ represents the distance of the object (in yards and decimals) above or below the telescope.

Logarithmic values of $R_z$, in yards, which depend on the inclination to the meridian of the arc joining the two stations and their mean latitude, are given in the following table :

| Angle of inclination. | Lat. 25°. | Lat. 30°. | Lat. 35°. | Lat. 40°. | Lat. 45°. | Lat. 50°. |
|---|---|---|---|---|---|---|
| | log $R_z$ | log ▓ | log $R_z$ | log $R_z$ | log $R_z$ | log $R_z$ |
| 0 | 6.841384 | 6.841695 | 6.842039 | 6.842406 | 6.842785 | 6.843164 |
| 10 | 1456 | 1760 | 2098 | 2458 | 2829 | 3200 |
| 20 | 1663 | 1950 | 2267 | 2606 | 2955 | 3304 |
| 30 | 1981 | 2240 | 2527 | 2833 | 3148 | 3464 |
| 40 | 2370 | 2596 | 2845 | 3111 | 3386 | 3631 |
| 50 | 2785 | 2975 | 3184 | 3408 | 3639 | 3870 |
| 60 | 3176 | 3331 | 3504 | 3687 | 3877 | 4066 |
| 70 | 3494 | 3622 | 3764 | 3915 | 4069 | 4227 |
| 80 | 3702 | 3812 | 3934 | 4063 | 4197 | 4331 |
| 90 | 6.843774 | 6.843878 | 6.843993 | 6.844115 | 6.844241 | 6.844368 |

$$\log \sin 1'' = 4.685575$$

2.—*By the Zenith-Distance measured at one Station.*

Let—

$Z$ = the measured zenith-distance of the signal or object ;

$K$ = the distance between the two stations in yards ;

$m$ = the co-efficient of refraction = 0.071 ; and

$dh$ = difference in height between the two stations ;

then—

$$C = \frac{K}{R_z \sin 1''} ; \qquad dh = \frac{K \cos (Z + m\,C - \tfrac{1}{2}\,C)}{\sin (Z + m\,C - C)}$$

## XLIV.—*Trigonometrical Leveling*—Continued.

3.—*By the Observed Zenith-Distance of the Sea-Horizon.*

Let—

Z = the measured zenith-distance;

$R_s$ = the radius of curvature of the arc; and

$m$ = the co-efficient of refraction = 0.078;

then—

$$h = \frac{R_s}{2\,(1 - m)^2}\,\tan^2 (Z - 90°)$$

4.—*By Observed Angles of Elevation or Depression.*

Let—

A = the observed angle expressed in seconds; and

K = the distance in yards between the two stations.

$$dh = 0.00000485 \; K\,A \pm 0.000000667 \; K^2$$
$$\left[\log 4.68574\right] \qquad \left[\log 2.82413\right]$$

This gives the difference in heights between stations not more than ten or fifteen miles apart, with a probable error less than the uncertainty in the co-efficient of refraction.

### *Co-efficient of Refraction.*

The co-efficient of refraction, or proportion of the intercepted arc, is determined from the observed zenith-distances of two stations, the relative altitudes of which have been determined by the spirit-level; or from reciprocal zenith-distances, simultaneous or not, under the assumption that the mean of a number of observations taken under favorable conditions will eliminate the difference of refraction which is found to exist, even at the same moment at two stations a few miles apart.

The longer the distance the greater is the error caused by any uncertainty in the co-efficient, or in the actual refraction; consequently there is a limit to the distance for which any assumed mean value of the refraction can be depended on for accurate results.

### XLIV.—*Trigonometrical Leveling*—Continued.

The average value of the co-efficient from the Coast Survey observations in the New England States—

Between primary stations $= 0.071$

Of small elevations. . . . . . $= 0.075$

Of the sea-horizon. . . . . . $= 0.078$

In the trigonometrical survey of Massachusetts Mr. Borden used 0.0784 as a mean co-efficient for the sea-coast, and 0.0697 for the interior of the State.

1.—*To determine the Co-efficient of Refraction from Reciprocal Zenith-Distances.*

Let—

$C$ = angle at earth's center subtended by arc ;

$F$ = angle of refraction ; and

$m$ = co-efficient of refraction ;

then—

$$C = \frac{K}{R\,\sin 1''}; \quad F = \frac{C}{2} - \tfrac{1}{2}(Z' + Z - 180°); \quad m = \frac{F}{C}$$

2.—*To determine the Co-efficient from the Zenith-Distance observed at one Station, when the Altitude of the Two Stations above Tide, or their Difference in Height, have been determined by the Leveling-Instrument.*

Compute the true zenith-distances, $Z_0'$ and $Z_0$, of the two given points, and the difference between the true and the observed zenith-distances will be the angle of refraction F.

$$\tfrac{1}{2}(Z_0' + Z_0) = 90° + \frac{C}{2}$$

$$\tfrac{1}{2}(Z_0' - Z_0) = \tan^{-1}\left\{ \frac{h' - h}{K}\left( 1 - \frac{h' + h}{2\,R} - \frac{K^2}{12\,R^2} \right) \right\}$$

$$Z_0 - Z = F; \qquad m = \frac{F}{C}$$

$h$ and $h'$ having been determined by the leveling-instrument.

*Difference between the Apparent and the True Level.*

$$\text{Correction for curvature} \ldots \ldots \ldots = \frac{D^2}{2\,R}$$

$$\text{Correction for refraction} \ldots \ldots \ldots = \frac{D^2}{R} \cdot m$$

$$\text{Correction for curvature and refraction} = (1 - 2\,m)\frac{D^2}{2\,R}$$

where—

    $D$ = the distance;

    $R$ = mean radius of the earth; and

    $m$ = co-efficient of refraction;

or—

    $m$ being $= 0.075$, and log $R$ in feet $= 7.31991307$,

Correction for curvature, in feet, $= \log D^2 - \text{const log } [7.6209430]$

Correction for refraction, in feet, $= \log D^2 + \text{const log } [1.5551483]$

*Corrections for Curvature and Refraction, showing the Differences of the Apparent and True Level, in Feet and Decimals of a Foot, for Distances in Miles.*

| Distance, miles. | Difference in feet for— | | | Distance, miles. | Difference in feet for— | | |
|---|---|---|---|---|---|---|---|
| | Curvature. | Refraction. | Curvature and refraction. | | Curvature. | Refraction. | Curvature and refraction. |
| 1 | 0.7 | 0.1 | 0.6 | 13 | 112.8 | 16.9 | 95.9 |
| 2 | 2.7 | 0.4 | 2.3 | 14 | 130.8 | 19.6 | 111.2 |
| 3 | 6.0 | 0.9 | 5.1 | 15 | 150.2 | 22.5 | 127.7 |
| 4 | 10.6 | 1.6 | 9.0 | 16 | 170.8 | 25.6 | 145.2 |
| 5 | 16.7 | 2.5 | 14.2 | 17 | 192.9 | 28.9 | 164.1 |
| 6 | 24.0 | 3.6 | 20.4 | 18 | 216.2 | 32.4 | 183.8 |
| 7 | 32.7 | 4.9 | 27.8 | 19 | 240.9 | 36.1 | 204.8 |
| 8 | 42.7 | 6.4 | 36.3 | 20 | 266.9 | 40.0 | 226.9 |
| 9 | 54.1 | 8.1 | 44.0 | 21 | 294.3 | 44.1 | 250.2 |
| 10 | 66.7 | 10.0 | 56.7 | 22 | 323.0 | 48.4 | 274.6 |
| 11 | 80.7 | 12.1 | 68.6 | 23 | 353.0 | 52.9 | 300.1 |
| 12 | 96.1 | 14.4 | 81.7 | 24 | 384.4 | 57.7 | 326.7 |

*Reduction, in Feet and Decimals, upon 100 Feet, for the following Vertical Angles.*

| Angle. | Reduc. | Angle. | Reduc. | Angle. | Reduc. | Angle. | Reduc. |
|---|---|---|---|---|---|---|---|
| ° ′ | | ° ′ | | ° ′ | | ° ′ | |
| 3 0 | .137 | 7 30 | .856 | 12 0 | 2.185 | 16 30 | 4.118 |
| 3 15 | .161 | 7 45 | .913 | 12 15 | 2.277 | 16 45 | 4.243 |
| 3 30 | .187 | 8 0 | .973 | 12 30 | 2.370 | 17 0 | 4.370 |
| 3 45 | .214 | 8 15 | 1.035 | 12 45 | 2.466 | 17 15 | 4.498 |
| 4 0 | .244 | 8 30 | 1.098 | 13 0 | 2.553 | 17 30 | 4.628 |
| 4 15 | .275 | 8 45 | 1.164 | 13 15 | 2.662 | 17 45 | 4.760 |
| 4 30 | .308 | 9 0 | 1.231 | 13 30 | 2.763 | 18 0 | 4.894 |
| 4 45 | .343 | 9 15 | 1.300 | 13 45 | 2.866 | 18 15 | 5.030 |
| 5 0 | .381 | 9 30 | 1.371 | 14 0 | 2.970 | 18 30 | 5.168 |
| 5 15 | .420 | 9 45 | 1.444 | 14 15 | 2.077 | 18 45 | 5.307 |
| 5 30 | .460 | 10 0 | 1.519 | 14 30 | 3.185 | 19 0 | 5.448 |
| 5 45 | .503 | 10 15 | 1.596 | 14 45 | 3.295 | 19 15 | 5.591 |
| 6 0 | .548 | 10 30 | 1.675 | 15 0 | 3.407 | 19 30 | 5.736 |
| 6 15 | .594 | 10 45 | 1.755 | 15 15 | 3.521 | 19 45 | 5.882 |
| 6 30 | .643 | 11 0 | 1.837 | 15 30 | 3.637 | 20 0 | 6.031 |
| 6 45 | .693 | 11 15 | 1.921 | 15 45 | 3.754 | | |
| 7 0 | .745 | 11 30 | 2.008 | 16 0 | 3.874 | | |
| 7 15 | .800 | 11 45 | 2.095 | 16 15 | 3.995 | | |

*Ratio of Slopes for the following Vertical Angles.*

| Angle. | To one perpendicular. | Angle. | To one perpendicular. | Angle. | To one perpendicular. | Angle. | To one perpendicular. |
|---|---|---|---|---|---|---|---|
| ° ′ | | ° ′ | | ° ′ | | ° ′ | |
| 0 15 | 229 | 3 35 | 16 | 8 8 | 7 | 18 26 | 3 |
| 0 30 | 115 | 3 49 | 15 | 8 45 | 6½ | 19 59 | 2¾ |
| 0 45 | 76 | 4 6 | 14 | 9 27 | 6 | 21 48 | 2½ |
| 1 0 | 57 | 4 24 | 13 | 9 52 | 5¾ | 23 58 | 2¼ |
| 1 15 | 46 | 4 45 | 12 | 10 18 | 5½ | 26 34 | 2 |
| 1 30 | 39 | 5 0 | 11½ | 10 47 | 5¼ | 29 44 | 1¾ |
| 1 45 | 33 | 5 12 | 11 | 11 19 | 5 | 33 42 | 1½ |
| 2 0 | 28 | 5 27 | 10½ | 11 53 | 4¾ | 38 40 | 1¼ |
| 2 15 | 25 | 5 42 | 10 | 12 32 | 4½ | 45 0 | 1 |
| 2 30 | 23 | 6 0 | 9½ | 13 15 | 4¼ | 53 8 | ¾ |
| 2 45 | 21 | 6 21 | 9 | 14 2 | 4 | 63 28 | ½ |
| 3 0 | 19 | 6 43 | 8½ | 14 55 | 3¾ | 75 58 | ¼ |
| 3 15 | 18 | 7 7 | 8 | 15 56 | 3½ | 78 41 | ⅕ |
| 3 28 | 17 | 7 36 | 7½ | 17 6 | 3¼ | | |

## XLV.—*Barometrical Measurement of Heights.*

### TO OBTAIN THE DIFFERENCE IN THE HEIGHT OF TWO PLACES BY MEANS OF THE BAROMETER.

The following tables have been condensed from those in the appendix of Lieutenant-Colonel Williamson's Treatise on the Use of the Barometer, etc., Professional Papers, Corps of Engineers, No. 15, and are those of Plantamour (Guyot's tables D, 72–79) re-arranged and adapted to English measures.

They are based upon Bessel's formula, which differs from that of La Place principally in containing a factor depending upon the humidity of the air. The modifications introduced by Plantamour consist in some slight changes in the values of the barometric constants in accordance with the supposed more accurate results obtained from the experiments of Regnault.

La Place's formula reduced to English measures, as given by Guyot, (page D, 35,) is:

$$Z = \log \frac{h}{H} \times 60158.6 \text{ English feet} \begin{cases} \left(1 + \frac{t + t' - 64}{900}\right) & (1) \\ (1 + 0.00260 \cos 2 \text{ L}) \\ \left(1 + \frac{Z + 52252}{20886860} + \frac{h}{10443430}\right) \end{cases}$$

Williamson adopts the same convenient form in his reduction of Plantamour's formula to English measures; thus:

$$Z = \log \frac{h}{H} \times 60384.3 \text{ English feet} \begin{cases} \left(1 + \frac{t + t' - 64}{982.2647}\right) & (2) \\ (1 + 0.0026257 \cos 2 \text{ L}) \\ \left(1 + \frac{Z + 52252}{20886860} + \frac{s}{10443430}\right) \\ (1 + m) \end{cases}$$

where $h$ and $H$ are the heights of the barometer reduced to 32° Fahrenheit; $t$, $t'$, the temperatures of the air at the two stations; and $m$, a factor depending upon the humidity of the stratum of air between them; L the latitude of the place.

XLV.—*Barometrical Measurement, &c.*—Continued.

APPLICATION.—1. Reduce the readings of the barometer to 32° Fahrenheit by Table I.

2. Representing by A the constant 60384.3, Table II gives the value of A log $h$ or A log H, and consequently their difference, A $\frac{\log h}{\log H}$. The numbers have been diminished by a constant quantity, which does not affect their differences.

3. Table III, column B, gives values of the factor $\frac{t+t'-64}{982.2647}$ of the temperature-term, to be used only in connection with the tables that give the corrections for humidity.

Column C gives values of the factor $\frac{t+t'-64}{900}$, and is used where no observations of atmospheric humidity are made.

4. Table IV gives A log $\frac{h}{H}$ × 0.0026257 cos 2 L, the correction due to gravity at the sea-level in the mean latitude between the two stations. It is positive from 45° to the equator, and negative from 45° to the poles.

5. Table V shows the correction A log $\frac{h}{H}$ × $\frac{Z+52252}{20886860}$, to be added to the approximate difference of altitude, on account of the decrease of gravity on a vertical acting on the density of the mercurial column.

6. Table VI furnishes the small correction A log $\frac{h}{H}$ × $\frac{s}{10443430}$ for the decrease of gravity on a vertical acting on the density of the air; $s$ representing the approximate difference of altitude. It is always additive

7. Table VII gives the relative atmospheric humidity in fractions of unity. This table is different from any given in Guyot's collection; for, though based upon Regnault's table of maximum force of vapor, and so far the same as Guyot's, it has been calculated with factors determined by Glaisher.

8. Tables VIII and IX are intended to give the correction A log $\frac{h}{H}$ × $m$, due to the relative humidity of the stratum of air between the two stations.

## XLV.—*Barometrical Measurement, &c.*—Continued.

These hypsometrical tables represent the full formula of Planta-mour. If, as is often the case, the observations for the relative humidity are not given with those of the barometer and dry thermometer, then the tables III, (column B,) VIII, and IX should not be used, but the temperature-correction must be calculated from the formula $A \log \frac{h}{H} \times \frac{t + t' - 64}{900}$ with the aid of column C of Table III.

With the temperature-term so calculated, the results will differ from those by Guyot's table, on account of the different value given to the barometric constant of the pressure term.

*Abnormal and Horary Oscillations of the Weight of the Atmosphere.*—The first is a gradual change generally extending over a period of two to seven days, causing the barometer to rise or fall gradually during that time, although sometimes more or less sudden, and occupying perhaps but a few hours; the second, a regular horary oscillation occurring at about the same hours every day, and having a magnitude entirely independent of this gradual change.

*The abnormal change* usually extends over large tracts of country, and in settled weather the barometer rises and falls so gradually that the forces that produce the motion can be separated with more or less accuracy from the horary changes by assuming the portion of this great wave corresponding to 24 hours to be a straight line; generally inclined, however, since the observations at any time of a barometric day differs from that of the next one at the same hour.

To eliminate this movement, subtract the barometric reading (reduced to 32°) at the beginning of one day from that at the same hour on the next succeeding one, and divide the difference by 24. The result is the correction for one hour, to be applied with its proper sign to the hour succeeding the initial hour. The correction for two hours is twice the correction for one hour, etc.

EXAMPLE.—Barometer at 7 a. m., August 7, = 29. 743
Barometer at 7 a. m., August 8, = 29. 487

Difference for 24 hours..... = +0. 256
For 1 hour..... = +0. 0106
and correction at 8 a. m., = + 0.011 ; at 9, = + 0.021 ; at 10, = +0.032, etc.

This correction Williamson names "reduction to level."

XLV.—*Barometrical Measurement, &c.*—Continued.

In the horary oscillation there are two maximum and two minimum points during the 24 hours.   Near the sea-level the barometer attains its maximum about 9 or 10 a. m.   In the afternoon there is a minimum about 3, 4, or 5 p. m.   It then rises until from 10 to midnight, when it falls again until about 4 a. m., and again rises to attain its morning maximum, the day-fluctuations being the larger.   The oscillation is greatest nearer the equator and diminishes toward the poles.   Its amount within the limits of the United States varies from 40 to 120 thousandths of an inch of the barometric column.   It is not equal at all places of the same latitude.

In a series of barometric observations at any place, the mean barometric reading is better obtained from daily horary curves, by plotting the separate readings of each day reduced to 32°, and corrected for the abnormal change by reduction to level. These would present an approximate horary curve for every day of the series, from which erroneous or erratic observations could be detected and rejected if necessary.

### EXAMPLE OF THE USE OF THESE TABLES.

#### Geneva and the Grand St. Bernard.

$h = 28.600$ in.    $t = 48°.2$ F.    Relative humidity, $a = 0.77$
$H = 22.191$ in.    $t' = 28°.6$ F.    Relative humidity, $a' = 0.80$

Lat $= 46°$    $t + t' = 76°.8$ F.    $a + a' = 1.57$

Table II, with argument $h$, gives .................. 27557.3
Table II, with argument $H$, gives .................. 20903.7

Difference = first approximate difference of altitude... 6653.6
Table III, col. B, with argument 76°.8, gives $+ 0.0130$
$$0.0130 \times 6653.6 = +86.5$$

Second approximate difference of altitude.......... $= 6740.1$
Table IV, arguments 46° lat. and 6700 feet, gives.... $-$ 0.6
Table V, argument 6740, gives .................... 19.0
Table VI, argument 6700 feet and 28.6 inches...... 0.8

Third approximate difference of altitude........... $= 6759.3$

XLV.—*Barometrical Measurement, &c.*—Continued.

Table VIII, arguments 22.2 in. and 28.6 in., gives 79
Table IX, arguments 79 and 76°.8 ........... 11.9

$$11.9 \times 1.57 = \text{vapor correction} = \quad 18.7$$

Difference of altitude ............................ 6778.0

The altitude by level is stated to be 6791.5 feet.

The same readings of the barometer and the same air-temperature being used, but calculating the value of the temperature-term from column C, table III,

$$\text{temperature-term} = 6653.6 \times 0.0142 = 94.4$$

The value of the temperature-term, as calculated in the above example, increased by the vapor-correction, is 105.2, a larger result than by the method of La Place, because the sum of the observed relative humidities of the stations was greater than that assumed by him.

In a dry climate the reverse would have been the case.

*Calculation of the same Observations by Guyot's Tables.*

First table of Guyot gives, for $h$ ................. 27454.4
First table of Guyot gives, for $H$ ................. 20825.6

First approximate difference of altitude ........... = 6628.8

$$6628.8 \times \frac{t + t' - 64}{900} \quad = \quad 94.3$$

Second approximate difference of altitude .......... = 6733.1
Table III of Guyot gives ..................... — 0.6
Table IV of Guyot gives ..................... 19.0
Table V of Guyot gives ..................... 0.8

Difference of altitude ............................ 6752.3

This result is 25.6 less than by the following tables, and 39.2 less than by the spirit-level.

*Aneroid Barometer.*

The best aneroids are, as nearly as possible, compensated by the maker for differences of temperature, so that the index shall remain at the same reading on the dial when it is heated and cooled, and are intended to be adjusted to read uniformly inches of mercury at a temperature of 32° Fahrenheit at the level of the sea in 45°.latitude. But before using any aneroid for

XLV.—*Barometrical Measurement, &c.*—Continued.

accurate observations it should be tested under an air-pump, together with a mercurial column, at a known temperature, and its scale-errors carefully noted. In many of them there is an additional scale of altitudes in feet outside of the scale, corresponding to the inches of mercury, generally in the best English instruments divided and marked according to a table prepared for the purpose by Professor Airy. There are some, however, very erroneously marked.

As the aneroid is not affected in its reading by the variation in the force of gravity, it needs no correction for the latitude, nor for the decrease of gravity in altitude acting on the mercurial column. The correction for the decrease of gravity in altitude acting on the density of the air, and the correction for humidity, remain; but the first being small in amount, it can be omitted, and the second combined with the correction for temperature, as is done in the formula of La Place.

The formula for the aneroid would then be:

$$Z = \log \frac{h}{H} \times 60384.3 \, \text{Eng. feet} \left( 1 + \frac{t + t'' - 64}{900} \right)$$

and the tabular quantities may be taken from these tables.

Professor Airy's table, made for the purpose of graduating aneroids to a scale of feet, gives the height of the corrected mercurial column in inches for each fifty feet of altitude at 50° Fahrenheit. The formula is:

$$Z = \log \frac{h}{H} \times 62759 \, \text{Eng. feet} \left( 1 + \frac{t + t'' - 100}{1000} \right)$$

As it is sometimes convenient to have an approximate formula that can be used without any tables whatever, the following may be found useful:

$$Z = 55032 \, \frac{H - h}{H + h} \, \text{Eng. feet, at 55° Fahrenheit,}$$

with a correction of $\pm \dfrac{1}{435}$ for each degree of mean temperature above 55°; or, nearly—

$$Z = 55000 \, \frac{H - h}{H + h} \pm \frac{1}{500}$$

a formula easily remembered, but only useful for altitudes not exceeding 3000 feet.

## TABLE I.—*Reduction of the English Barometer to the Freezing-Point.*

| Degrees of Fahrenheit. | English inches. | | | | | | | Degrees of Fahrenheit. |
|---|---|---|---|---|---|---|---|---|
|  | 17.5 | 18 | 18.5 | 19 | 19.5 | 20 | 20.5 |  |
| 0 | +.045 | +.046 | +.047 | +.049 | +.050 | +.051 | +.053 | 0 |
| 1 | .043 | .045 | .046 | .047 | .048 | .049 | .051 | 1 |
| 2 | .042 | .043 | .044 | .045 | .046 | .048 | .049 | 2 |
| 3 | .040 | .041 | .042 | .044 | .045 | .046 | .047 | 2 |
| 4 | .039 | .040 | .041 | .042 | .043 | .044 | .045 | 4 |
| 5 | .037 | .038 | .039 | .040 | .041 | .042 | .043 | 5 |
| 6 | .035 | .036 | .037 | .038 | .039 | .040 | .041 | 6 |
| 7 | .034 | .035 | .036 | .037 | .038 | .039 | .040 | 7 |
| 8 | .032 | .033 | .034 | .035 | .036 | .037 | .038 | 8 |
| 9 | .031 | .032 | .032 | .033 | .034 | .035 | .036 | 9 |
| 10 | .029 | .030 | .031 | .032 | .032 | .033 | .034 | 10 |
| 11 | .028 | .028 | .029 | .030 | .031 | .031 | .032 | 11 |
| 12 | .026 | .027 | .027 | .028 | .029 | .030 | .030 | 12 |
| 13 | .024 | .025 | .026 | .026 | .027 | .028 | .029 | 13 |
| 14 | .023 | .023 | .024 | .025 | .025 | .026 | .027 | 14 |
| 15 | .021 | .022 | .022 | .023 | .024 | .024 | .025 | 15 |
| 16 | .020 | .020 | .021 | .021 | .022 | .022 | .023 | 16 |
| 17 | .018 | .019 | .019 | .020 | .020 | .021 | .021 | 17 |
| 18 | .017 | .017 | .017 | .018 | .018 | .019 | .019 | 18 |
| 19 | .015 | .015 | .016 | .016 | .017 | .017 | .018 | 19 |
| 20 | .013 | .014 | .014 | .015 | .015 | .015 | .016 | 20 |
| 21 | .012 | .012 | .012 | .013 | .013 | .013 | .014 | 21 |
| 22 | .010 | .011 | .011 | .011 | .011 | .012 | .012 | 22 |
| 23 | .009 | .009 | .009 | .009 | .010 | .010 | .010 | 23 |
| 24 | .007 | .007 | .007 | .008 | .008 | .008 | .008 | 24 |
| 25 | .006 | .006 | .006 | .006 | .006 | .006 | .006 | 25 |
| 26 | .004 | .004 | .004 | .004 | .004 | .005 | .005 | 26 |
| 27 | .002 | .002 | .003 | .003 | .003 | .003 | .003 | 27 |
| 28 | +.001 | +.001 | +.001 | +.001 | +.001 | +.001 | +.001 | 28 |
| 29 | −.001 | −.001 | −.001 | −.001 | −.001 | −.001 | −.001 | 29 |
| 30 | .002 | .002 | .002 | .003 | .003 | .003 | .003 | 30 |
| 31 | .004 | .004 | .004 | .004 | .004 | .004 | .005 | 31 |
| 32 | .005 | .006 | .006 | .006 | .006 | .006 | .006 | 32 |
| 33 | .007 | .007 | .007 | .008 | .008 | .008 | .008 | 33 |
| 34 | .009 | .009 | .009 | .009 | .010 | .010 | .010 | 34 |
| 35 | .010 | .010 | .011 | .011 | .011 | .012 | .012 | 35 |
| 36 | .012 | .012 | .012 | .013 | .013 | .013 | .014 | 36 |
| 37 | .013 | .014 | .014 | .014 | .015 | .015 | .016 | 37 |
| 38 | .015 | .015 | .016 | .016 | .017 | .017 | .017 | 38 |
| 39 | .016 | .017 | .017 | .018 | .018 | .019 | .019 | 39 |
| 40 | .018 | .019 | .019 | .020 | .020 | .021 | .021 | 40 |
| 41 | .020 | .020 | .021 | .021 | .022 | .022 | .023 | 41 |
| 42 | .021 | .022 | .022 | .023 | .024 | .024 | .025 | 42 |
| 43 | .023 | .023 | .024 | .025 | .025 | .026 | .027 | 43 |
| 44 | .024 | .025 | .026 | .026 | .027 | .028 | .028 | 44 |
| 45 | .026 | .027 | .027 | .028 | .029 | .030 | .030 | 45 |
| 46 | .027 | .028 | .029 | .030 | .031 | .031 | .032 | 46 |
| 47 | .029 | .030 | .031 | .031 | .032 | .033 | .034 | 47 |
| 48 | .031 | .031 | .032 | .033 | .034 | .035 | .036 | 48 |
| 49 | .032 | .033 | .034 | .035 | .036 | .037 | .038 | 49 |
| 50 | −.034 | −.035 | −.036 | −.037 | −.038 | −.038 | −.039 | 50 |

TABLE I.—*Reduction of the English Barometer to the Freezing-Point*—Continued.

| Degrees of Fahrenheit | English inches. | | | | | | | Degrees of Fahrenheit |
|---|---|---|---|---|---|---|---|---|
| | 17.5 | 18 | 18.5 | 19 | 19.5 | 20 | 20.5 | |
| 51 | −.035 | −.036 | −.037 | −.038 | −.039 | −.040 | −.041 | 51 |
| 52 | .037 | .038 | .039 | .040 | .041 | .042 | .043 | 52 |
| 53 | .038 | .039 | .041 | .042 | .043 | .044 | .045 | 53 |
| 54 | .040 | .041 | .042 | .043 | .044 | .046 | .047 | 54 |
| 55 | .041 | .043 | .044 | .045 | .046 | .047 | .049 | 55 |
| 56 | .043 | .044 | .045 | .047 | .048 | .049 | .050 | 56 |
| 57 | .045 | .046 | .047 | .048 | .050 | .051 | .052 | 57 |
| 58 | .046 | .047 | .049 | .050 | .051 | .053 | .054 | 58 |
| 59 | .048 | .049 | .050 | .052 | .053 | .055 | .056 | 59 |
| 60 | .049 | .051 | .052 | .054 | .055 | .056 | .058 | 60 |
| 61 | .051 | .052 | .054 | .055 | .057 | .058 | .060 | 61 |
| 62 | .052 | .054 | .055 | .057 | .058 | .060 | .061 | 62 |
| 63 | .054 | .055 | .057 | .059 | .060 | .062 | .063 | 63 |
| 64 | .056 | .057 | .059 | .060 | .062 | .063 | .065 | 64 |
| 65 | .057 | .059 | .060 | .062 | .064 | .065 | .067 | 65 |
| 66 | .059 | .060 | .062 | .064 | .065 | .067 | .069 | 66 |
| 67 | .060 | .062 | .064 | .065 | .067 | .069 | .071 | 67 |
| 68 | .062 | .064 | .065 | .067 | .069 | .071 | .072 | 68 |
| 69 | .063 | .065 | .067 | .069 | .071 | .072 | .074 | 69 |
| 70 | .065 | .067 | .069 | .070 | .072 | .074 | .076 | 70 |
| 71 | .066 | .068 | .070 | .072 | .074 | .076 | .078 | 71 |
| 72 | .068 | .070 | .072 | .074 | .076 | .078 | .080 | 72 |
| 73 | .070 | .072 | .074 | .075 | .077 | .079 | .081 | 73 |
| 74 | .071 | .073 | .075 | .077 | .079 | .081 | .083 | 74 |
| 75 | .073 | .075 | .077 | .079 | .081 | .083 | .085 | 75 |
| 76 | .074 | .076 | .078 | .081 | .083 | .085 | .087 | 76 |
| 77 | .076 | .078 | .080 | .082 | .084 | .087 | .089 | 77 |
| 78 | .077 | .080 | .082 | .084 | .086 | .088 | .091 | 78 |
| 79 | .079 | .081 | .083 | .086 | .088 | .090 | .092 | 79 |
| 80 | .080 | .083 | .085 | .087 | .090 | .092 | .094 | 80 |
| 81 | .082 | .084 | .087 | .089 | .091 | .094 | .096 | 81 |
| 82 | .084 | .086 | .088 | .091 | .093 | .095 | .098 | 82 |
| 83 | .085 | .088 | .090 | .092 | .095 | .097 | .100 | 83 |
| 84 | .087 | .089 | .092 | .094 | .097 | .099 | .101 | 84 |
| 85 | .088 | .091 | .093 | .096 | .098 | .101 | .103 | 85 |
| 86 | .090 | .092 | .095 | .097 | .100 | .103 | .105 | 86 |
| 87 | .091 | .094 | .096 | .099 | .102 | .104 | .107 | 87 |
| 88 | .093 | .095 | .098 | .101 | .103 | .106 | .109 | 88 |
| 89 | .094 | .097 | .100 | .102 | .105 | .108 | .111 | 89 |
| 90 | .096 | .099 | .101 | .104 | .107 | .110 | .112 | 90 |
| 91 | .097 | .100 | .103 | .106 | .109 | .111 | .114 | 91 |
| 92 | .099 | .102 | .105 | .108 | .110 | .113 | .116 | 92 |
| 93 | .101 | .103 | .106 | .109 | .112 | .115 | .118 | 93 |
| 94 | −.102 | −.105 | −.108 | .111 | .114 | .117 | .120 | 94 |
| 95 | | | | .113 | .116 | .118 | .121 | 95 |
| 96 | | | | .114 | .117 | .120 | .123 | 96 |
| 97 | | | | .116 | .119 | .122 | .125 | 97 |
| 98 | | | | .118 | .121 | .124 | .127 | 98 |
| 99 | | | | −.119 | −.122 | −.126 | −.129 | 99 |

TABLE I.—*Reduction of the English Barometer to the Freezing-Point*—Continued.

| Degrees of Fahrenheit | English inches. | | | | | | | Degrees of Fahrenheit |
|---|---|---|---|---|---|---|---|---|
| | 21 | 21.5 | 22 | 22.5 | 23 | 23.5 | 24 | |
| 0 | +.054 | +.055 | +.055 | +.058 | +.059 | +.060 | +.062 | 0 |
| 1 | .052 | .053 | .053 | .056 | .057 | .058 | .059 | 1 |
| 2 | .050 | .051 | .051 | .054 | .055 | .056 | .057 | 2 |
| 3 | .048 | .049 | .049 | .052 | .053 | .054 | .055 | 3 |
| 4 | .046 | .047 | .047 | .050 | .051 | .052 | .053 | 4 |
| 5 | .044 | .045 | .045 | .048 | .049 | .050 | .051 | 5 |
| 6 | .042 | .044 | .044 | .046 | .047 | .048 | .049 | 6 |
| 7 | .041 | .042 | .042 | .044 | .044 | .045 | .046 | 7 |
| 8 | .039 | .040 | .040 | .041 | .042 | .043 | .044 | 8 |
| 9 | .037 | .038 | .038 | .039 | .040 | .041 | .042 | 9 |
| 10 | .035 | .036 | .036 | .037 | .038 | .039 | .040 | 10 |
| 11 | .033 | .034 | .034 | .035 | .036 | .037 | .038 | 11 |
| 12 | .031 | .032 | .032 | .033 | .034 | .035 | .036 | 12 |
| 13 | .029 | .030 | .030 | .031 | .032 | .033 | .033 | 13 |
| 14 | .027 | .028 | .028 | .029 | .030 | .031 | .031 | 14 |
| 15 | .025 | .026 | .026 | .027 | .028 | .029 | .029 | 15 |
| 16 | .024 | .024 | .024 | .025 | .026 | .026 | .027 | 16 |
| 17 | .022 | .022 | .022 | .023 | .024 | .024 | .025 | 17 |
| 18 | .020 | .020 | .020 | .021 | .022 | .022 | .023 | 18 |
| 19 | .018 | .018 | .018 | .019 | .020 | .020 | .020 | 19 |
| 20 | .016 | .016 | .016 | .017 | .018 | .018 | .018 | 20 |
| 21 | .014 | .014 | .014 | .015 | .016 | .016 | .016 | 21 |
| 22 | .012 | .013 | .013 | .013 | .013 | .014 | .014 | 22 |
| 23 | .010 | .011 | .011 | .011 | .011 | .012 | .012 | 23 |
| 24 | .008 | .009 | .009 | .009 | .009 | .010 | .010 | 24 |
| 25 | .007 | .007 | .007 | .007 | .007 | .007 | .008 | 25 |
| 26 | .005 | .005 | .005 | .005 | .005 | .005 | .005 | 26 |
| 27 | .003 | .003 | .003 | .003 | .003 | .003 | .003 | 27 |
| 28 | +.001 | +.001 | +.001 | +.001 | +.001 | +.001 | +.001 | 28 |
| 29 | −.001 | −.001 | −.001 | −.001 | −.001 | −.001 | −.001 | 29 |
| 30 | .003 | .003 | .003 | .003 | .003 | .003 | .003 | 30 |
| 31 | .005 | .005 | .005 | .005 | .005 | .005 | .005 | 31 |
| 32 | .007 | .007 | .007 | .007 | .007 | .007 | .008 | 32 |
| 33 | .008 | .009 | .009 | .009 | .009 | .009 | .010 | 33 |
| 34 | .010 | .011 | .011 | .011 | .011 | .012 | .012 | 34 |
| 35 | .012 | .013 | .013 | .013 | .013 | .014 | .014 | 35 |
| 36 | .014 | .014 | .015 | .015 | .015 | .016 | .016 | 36 |
| 37 | .016 | .016 | .017 | .017 | .018 | .018 | .018 | 37 |
| 38 | .018 | .018 | .019 | .019 | .020 | .020 | .020 | 38 |
| 39 | .020 | .020 | .021 | .021 | .022 | .022 | .023 | 39 |
| 40 | .022 | .022 | .023 | .023 | .024 | .024 | .025 | 40 |
| 41 | .023 | .024 | .025 | .025 | .026 | .026 | .027 | 41 |
| 42 | .025 | .026 | .027 | .027 | .028 | .028 | .029 | 42 |
| 43 | .027 | .028 | .029 | .029 | .030 | .031 | .031 | 43 |
| 44 | .029 | .030 | .031 | .031 | .032 | .033 | .033 | 44 |
| 45 | .031 | .032 | .032 | .033 | .034 | .035 | .035 | 45 |
| 46 | .033 | .034 | .034 | .035 | .036 | .037 | .038 | 46 |
| 47 | .035 | .036 | .036 | .037 | .038 | .039 | .040 | 47 |
| 48 | .037 | .038 | .038 | .039 | .040 | .041 | .042 | 48 |
| 49 | .039 | .039 | .040 | .041 | .042 | .043 | .044 | 49 |
| 50 | −.040 | −.041 | −.042 | −.043 | −.044 | −.045 | −.046 | 50 |

TABLE I.—*Reduction of the English Barometer to the Freezing-Point*—Continued.

| Degrees of Fahrenheit. | English inches. | | | | | | | Degrees of Fahrenheit. |
|---|---|---|---|---|---|---|---|---|
| | 21 | 21.5 | 22 | 22.5 | 23 | 23.5 | 24 | |
| 51 | −.042 | −.043 | −.044 | −.045 | −.046 | −.047 | −.048 | 51 |
| 52 | .044 | .045 | .046 | .047 | .048 | .049 | .050 | 52 |
| 53 | .046 | .047 | .048 | .049 | .050 | .051 | .053 | 53 |
| 54 | .048 | .049 | .050 | .051 | .052 | .054 | .055 | 54 |
| 55 | .050 | .051 | .052 | .053 | .055 | .056 | .057 | 55 |
| 56 | .052 | .053 | .054 | .055 | .057 | .058 | .059 | 56 |
| 57 | .054 | .055 | .056 | .057 | .059 | .060 | .061 | 57 |
| 58 | .055 | .057 | .058 | .059 | .061 | .062 | .063 | 58 |
| 59 | .057 | .059 | .060 | .061 | .063 | .064 | .065 | 59 |
| 60 | .059 | .061 | .062 | .063 | .065 | .066 | .068 | 60 |
| 61 | .061 | .062 | .064 | .065 | .067 | .068 | .070 | 61 |
| 62 | .063 | .064 | .066 | .067 | .069 | .070 | .072 | 62 |
| 63 | .065 | .066 | .068 | .069 | .071 | .072 | .074 | 63 |
| 64 | .067 | .068 | .070 | .071 | .073 | .075 | .076 | 64 |
| 65 | .068 | .070 | .072 | .073 | .075 | .077 | .078 | 65 |
| 66 | .070 | .072 | .074 | .075 | .077 | .079 | .080 | 66 |
| 67 | .072 | .074 | .076 | .077 | .079 | .081 | .083 | 67 |
| 68 | .074 | .076 | .078 | .079 | .081 | .083 | .085 | 68 |
| 69 | .076 | .078 | .080 | .081 | .083 | .085 | .087 | 69 |
| 70 | .078 | .080 | .082 | .083 | .085 | .087 | .089 | 70 |
| 71 | .080 | .082 | .083 | .085 | .087 | .089 | .091 | 71 |
| 72 | .082 | .084 | .085 | .087 | .089 | .091 | .093 | 72 |
| 73 | .083 | .085 | .087 | .089 | .091 | .093 | .095 | 73 |
| 74 | .085 | .087 | .089 | .091 | .093 | .095 | .097 | 74 |
| 75 | .087 | .089 | .091 | .093 | .095 | .098 | .100 | 75 |
| 76 | .089 | .091 | .093 | .095 | .098 | .100 | .102 | 76 |
| 77 | .091 | .093 | .095 | .097 | .100 | .102 | .104 | 77 |
| 78 | .093 | .095 | .097 | .099 | .102 | .104 | .106 | 78 |
| 79 | .095 | .097 | .099 | .101 | .104 | .106 | .108 | 79 |
| 80 | .096 | .099 | .101 | .103 | .106 | .108 | .110 | 80 |
| 81 | .098 | .101 | .103 | .105 | .108 | .110 | .112 | 81 |
| 82 | .100 | .103 | .105 | .107 | .110 | .112 | .115 | 82 |
| 83 | .102 | .105 | .107 | .109 | .112 | .114 | .117 | 83 |
| 84 | .104 | .106 | .109 | .111 | .114 | .116 | .119 | 84 |
| 85 | .106 | .108 | .111 | .113 | .116 | .118 | .121 | 85 |
| 86 | .108 | .110 | .113 | .115 | .118 | .120 | .123 | 86 |
| 87 | .110 | .112 | .115 | .117 | .120 | .123 | .125 | 87 |
| 88 | .111 | .114 | .117 | .119 | .122 | .125 | .127 | 88 |
| 89 | .113 | .116 | .119 | .121 | .124 | .127 | .129 | 89 |
| 90 | .115 | .118 | .121 | .123 | .126 | .129 | .132 | 90 |
| 91 | .117 | .120 | .123 | .125 | .128 | .131 | .134 | 91 |
| 92 | .119 | .122 | .124 | .127 | .130 | .133 | .136 | 92 |
| 93 | .121 | .124 | .126 | .129 | .132 | .135 | .138 | 93 |
| 94 | .123 | .125 | .128 | .131 | .134 | .137 | .140 | 94 |
| 95 | .124 | .127 | .130 | .133 | .136 | .139 | .142 | 95 |
| 96 | .126 | .129 | .132 | .135 | .138 | .141 | .144 | 96 |
| 97 | .128 | .131 | .134 | .137 | .140 | .143 | .146 | 97 |
| 98 | .130 | .133 | .136 | .139 | .142 | .145 | .149 | 98 |
| 99 | −.132 | −.135 | −.138 | −.141 | −.144 | −.148 | −.151 | 99 |

## TABLE I.—*Reduction of the English Barometer to the Freezing-Point*—Continued.

| Degrees of Fahrenheit. | English inches. | | | | | | | Degrees of Fahrenheit. |
|---|---|---|---|---|---|---|---|---|
| | 24.5 | 25 | 25.5 | 26 | 26.5 | 27 | 27.5 | |
| 0 | +.063 | +.064 | +.065 | +.067 | +.068 | +.069 | +.071 | 0 |
| 1 | .061 | .062 | .063 | .064 | .066 | .067 | .068 | 1 |
| 2 | .058 | .060 | .061 | .062 | .063 | .064 | .066 | 2 |
| 3 | .056 | .057 | .058 | .060 | .061 | .062 | .063 | 3 |
| 4 | .054 | .055 | .056 | .057 | .058 | .060 | .061 | 4 |
| 5 | .052 | .053 | .054 | .055 | .056 | .057 | .058 | 5 |
| 6 | .050 | .051 | .052 | .053 | .054 | .055 | .056 | 6 |
| 7 | .047 | .048 | .049 | .050 | .051 | .052 | .053 | 7 |
| 8 | .045 | .046 | .047 | .048 | .049 | .050 | .051 | 8 |
| 9 | .043 | .044 | .045 | .046 | .046 | .047 | .048 | 9 |
| 10 | .041 | .042 | .042 | .043 | .044 | .045 | .046 | 10 |
| 11 | .039 | .039 | .040 | .041 | .042 | .042 | .043 | 11 |
| 12 | .036 | .037 | .038 | .039 | .039 | .040 | .041 | 12 |
| 13 | .034 | .035 | .036 | .036 | .037 | .038 | .038 | 13 |
| 14 | .032 | .033 | .033 | .034 | .035 | .035 | .036 | 14 |
| 15 | .030 | .030 | .031 | .032 | .032 | .033 | .033 | 15 |
| 16 | .028 | .028 | .029 | .029 | .030 | .030 | .031 | 16 |
| 17 | .025 | .026 | .026 | .027 | .027 | .028 | .028 | 17 |
| 18 | .023 | .024 | .024 | .025 | .025 | .025 | .026 | 18 |
| 19 | .021 | .021 | .022 | .022 | .023 | .023 | .023 | 19 |
| 20 | .019 | .019 | .019 | .020 | .020 | .021 | .021 | 20 |
| 21 | .017 | .017 | .017 | .018 | .018 | .018 | .019 | 21 |
| 22 | .014 | .015 | .015 | .015 | .015 | .016 | .016 | 22 |
| 23 | .012 | .012 | .013 | .013 | .013 | .013 | .014 | 23 |
| 24 | .010 | .010 | .010 | .011 | .011 | .011 | .011 | 24 |
| 25 | .008 | .008 | .008 | .008 | .008 | .009 | .009 | 25 |
| 26 | .006 | .006 | .006 | .006 | .006 | .006 | .006 | 26 |
| 27 | .003 | .003 | .003 | .004 | .004 | .004 | .004 | 27 |
| 28 | +.001 | +.001 | +.001 | +.001 | +.001 | +.001 | +.001 | 28 |
| 29 | −.001 | −.001 | −.001 | −.001 | −.001 | −.001 | −.001 | 29 |
| 30 | .003 | .003 | .003 | .003 | .004 | .004 | .004 | 30 |
| 31 | .005 | .006 | .006 | .006 | .006 | .006 | .006 | 31 |
| 32 | .008 | .008 | .008 | .008 | .008 | .008 | .009 | 32 |
| 33 | .010 | .010 | .010 | .010 | .011 | .011 | .011 | 33 |
| 34 | .012 | .012 | .013 | .013 | .013 | .013 | .014 | 34 |
| 35 | .014 | .015 | .015 | .015 | .015 | .016 | .016 | 35 |
| 36 | .016 | .017 | .017 | .017 | .018 | .018 | .018 | 36 |
| 37 | .019 | .019 | .019 | .020 | .020 | .021 | .021 | 37 |
| 38 | .021 | .021 | .022 | .022 | .023 | .023 | .023 | 38 |
| 39 | .023 | .024 | .024 | .024 | .025 | .025 | .026 | 39 |
| 40 | .025 | .026 | .026 | .027 | .027 | .028 | .028 | 40 |
| 41 | .027 | .028 | .029 | .029 | .030 | .030 | .031 | 41 |
| 42 | .030 | .030 | .031 | .031 | .032 | .033 | .033 | 42 |
| 43 | .032 | .032 | .033 | .034 | .034 | .035 | .036 | 43 |
| 44 | .034 | .035 | .035 | .036 | .037 | .037 | .038 | 44 |
| 45 | .036 | .037 | .038 | .038 | .039 | .040 | .041 | 45 |
| 46 | .038 | .039 | .040 | .041 | .042 | .042 | .043 | 46 |
| 47 | .041 | .041 | .042 | .043 | .044 | .045 | .046 | 47 |
| 48 | .043 | .044 | .044 | .045 | .046 | .047 | .048 | 48 |
| 49 | .045 | .046 | .047 | .048 | .049 | .050 | .050 | 49 |
| 50 | −.047 | −.048 | −.049 | −.050 | −.051 | −.052 | −.053 | 50 |

TABLE I.—*Reduction of the English Barometer to the Freezing-Point*—Continued.

| Degrees of Fahrenheit | English inches. | | | | | | | Degrees of Fahrenheit |
|---|---|---|---|---|---|---|---|---|
| | 24.5 | 25 | 25.5 | 26 | 26.5 | 27 | 27.5 | |
| 51 | −.049 | −.050 | −.051 | −.052 | −.053 | −.054 | −.055 | 51 |
| 52 | .052 | .053 | .054 | .055 | .056 | .057 | .058 | 52 |
| 53 | .054 | .055 | .056 | .057 | .058 | .059 | .060 | 53 |
| 54 | .056 | .057 | .058 | .059 | .060 | .062 | .063 | 54 |
| 55 | .058 | .059 | .060 | .062 | .063 | .064 | .065 | 55 |
| 56 | .060 | .061 | .063 | .064 | .065 | .066 | .068 | 56 |
| 57 | .062 | .064 | .065 | .066 | .068 | .069 | .070 | 57 |
| 58 | .065 | .066 | .067 | .069 | .070 | .071 | .073 | 58 |
| 59 | .067 | .068 | .070 | .071 | .072 | .074 | .075 | 59 |
| 60 | .069 | .070 | .072 | .073 | .075 | .076 | .077 | 60 |
| 61 | .071 | .073 | .074 | .076 | .077 | .078 | .080 | 61 |
| 62 | .073 | .075 | .076 | .078 | .079 | .081 | .082 | 62 |
| 63 | .076 | .077 | .079 | .080 | .082 | .083 | .085 | 63 |
| 64 | .078 | .079 | .081 | .082 | .084 | .086 | .087 | 64 |
| 65 | .080 | .082 | .083 | .085 | .086 | .088 | .090 | 65 |
| 66 | .082 | .084 | .085 | .087 | .089 | .090 | .092 | 66 |
| 67 | .084 | .086 | .088 | .089 | .091 | .093 | .095 | 67 |
| 68 | .086 | .088 | .090 | .092 | .093 | .095 | .097 | 68 |
| 69 | .089 | .090 | .092 | .094 | .096 | .098 | .099 | 69 |
| 70 | .091 | .093 | .095 | .096 | .098 | .100 | .102 | 70 |
| 71 | .093 | .095 | .097 | .099 | .101 | .102 | .104 | 71 |
| 72 | .095 | .097 | .099 | .101 | .103 | .105 | .107 | 72 |
| 73 | .097 | .099 | .101 | .103 | .105 | .107 | .109 | 73 |
| 74 | .100 | .102 | .104 | .106 | .108 | .110 | .112 | 74 |
| 75 | .102 | .104 | .106 | .108 | .110 | .112 | .114 | 75 |
| 76 | .104 | .106 | .108 | .110 | .112 | .114 | .117 | 76 |
| 77 | .106 | .108 | .110 | .113 | .115 | .117 | .119 | 77 |
| 78 | .108 | .110 | .113 | .115 | .117 | .119 | .121 | 78 |
| 79 | .110 | .113 | .115 | .117 | .119 | .122 | .124 | 79 |
| 80 | .113 | .115 | .117 | .119 | .122 | .124 | .126 | 80 |
| 81 | .115 | .117 | .119 | .122 | .124 | .126 | .129 | 81 |
| 82 | .117 | .119 | .122 | .124 | .126 | .129 | .131 | 82 |
| 83 | .119 | .122 | .124 | .126 | .129 | .131 | .134 | 83 |
| 84 | .121 | .124 | .126 | .129 | .131 | .134 | .136 | 84 |
| 85 | .123 | .126 | .128 | .131 | .134 | .136 | .139 | 85 |
| 86 | .126 | .128 | .131 | .133 | .136 | .138 | .141 | 86 |
| 87 | .128 | .130 | .133 | .136 | .138 | .141 | .143 | 87 |
| 88 | .130 | .133 | .135 | .138 | .141 | .143 | .146 | 88 |
| 89 | .132 | .135 | .138 | .140 | .143 | .146 | .148 | 89 |
| 90 | .134 | .137 | .140 | .143 | .145 | .148 | .151 | 90 |
| 91 | .136 | .139 | .142 | .145 | .148 | .150 | .153 | 91 |
| 92 | .139 | .141 | .144 | .147 | .150 | .153 | .156 | 92 |
| 93 | .141 | .144 | .147 | .149 | .152 | .155 | .158 | 93 |
| 94 | .143 | .146 | .149 | .152 | .155 | .158 | .160 | 94 |
| 95 | .145 | .148 | .151 | .154 | .157 | .160 | .163 | 95 |
| 96 | .147 | .150 | .153 | .156 | .159 | .162 | .165 | 96 |
| 97 | .149 | .153 | .156 | .159 | .162 | .165 | .168 | 97 |
| 98 | .152 | .155 | .158 | .161 | .164 | .167 | .170 | 98 |
| 99 | −.154 | −.157 | −.160 | −.163 | −.166 | −.170 | −.173 | 99 |

## TABLE I.—*Reduction of the English Barometer to the Freezing-Point*—Continued.

| Degrees of Fahrenheit. | English inches. | | | | | | | Degrees of Fahrenheit. |
| --- | --- | --- | --- | --- | --- | --- | --- | --- |
| | 28 | 28.5 | 29 | 29.5 | 30 | 30.5 | 31 | |
| 0 | +.072 | +.073 | +.074 | +.076 | +.077 | +.078 | +.080 | 0 |
| 1 | .069 | .071 | .072 | .073 | .074 | .075 | .077 | 1 |
| 2 | .067 | .068 | .069 | .070 | .072 | .073 | .074 | 2 |
| 3 | .064 | .065 | .067 | .068 | .069 | .070 | .071 | 3 |
| 4 | .062 | .063 | .064 | .065 | .066 | .067 | .068 | 4 |
| 5 | .059 | .060 | .061 | .062 | .063 | .064 | .066 | 5 |
| 6 | .057 | .058 | .059 | .060 | .061 | .062 | .063 | 6 |
| 7 | .054 | .055 | .056 | .057 | .058 | .059 | .060 | 7 |
| 8 | .052 | .053 | .053 | .054 | .055 | .056 | .057 | 8 |
| 9 | .049 | .050 | .051 | .052 | .053 | .053 | .054 | 9 |
| 10 | .047 | .047 | .048 | .049 | .050 | .051 | .052 | 10 |
| 11 | .044 | .045 | .046 | .046 | .047 | .048 | .049 | 11 |
| 12 | .042 | .042 | .043 | .044 | .044 | .045 | .046 | 12 |
| 13 | .039 | .040 | .040 | .041 | .042 | .042 | .043 | 13 |
| 14 | .036 | .037 | .038 | .038 | .039 | .040 | .040 | 14 |
| 15 | .034 | .035 | .035 | .036 | .036 | .037 | .038 | 15 |
| 16 | .031 | .032 | .033 | .033 | .034 | .034 | .035 | 16 |
| 17 | .029 | .029 | .030 | .030 | .031 | .032 | .032 | 17 |
| 18 | .026 | .027 | .027 | .028 | .028 | .029 | .029 | 18 |
| 29 | .024 | .024 | .025 | .025 | .026 | .026 | .027 | 19 |
| 20 | .021 | .022 | .022 | .023 | .023 | .023 | .024 | 20 |
| 21 | .019 | .019 | .020 | .020 | .020 | .021 | .021 | 21 |
| 22 | .016 | .017 | .017 | .017 | .018 | .018 | .018 | 22 |
| 23 | .014 | .014 | .014 | .015 | .015 | .015 | .015 | 23 |
| 24 | .011 | .012 | .012 | .012 | .012 | .012 | .013 | 24 |
| 25 | .009 | .009 | .009 | .009 | .009 | .010 | .010 | 25 |
| 26 | .006 | .006 | .007 | .007 | .007 | .007 | .007 | 26 |
| 27 | .004 | .004 | .004 | .004 | .004 | .004 | .004 | 27 |
| 28 | +.001 | +.001 | +.001 | +.001 | +.001 | +.001 | +.001 | 28 |
| 29 | −.001 | −.001 | −.001 | −.001 | −.001 | −.001 | −.001 | 29 |
| 30 | .004 | .004 | .004 | .004 | .004 | .004 | .004 | 30 |
| 31 | .006 | .006 | .006 | .007 | .007 | .007 | .007 | 31 |
| 32 | .009 | .009 | .009 | .009 | .009 | .010 | .010 | 32 |
| 33 | .011 | .011 | .012 | .012 | .012 | .012 | .012 | 33 |
| 34 | .014 | .014 | .014 | .015 | .015 | .015 | .015 | 34 |
| 35 | .016 | .017 | .017 | .017 | .017 | .018 | .018 | 35 |
| 36 | .019 | .019 | .019 | .020 | .020 | .020 | .021 | 36 |
| 37 | .021 | .022 | .022 | .022 | .023 | .023 | .024 | 37 |
| 38 | .024 | .024 | .025 | .025 | .026 | .026 | .026 | 38 |
| 39 | .026 | .027 | .027 | .028 | .028 | .029 | .029 | 39 |
| 40 | .029 | .029 | .030 | .030 | .031 | .031 | .032 | 40 |
| 41 | .031 | .032 | .032 | .033 | .034 | .034 | .035 | 41 |
| 42 | .034 | .034 | .035 | .036 | .036 | .037 | .037 | 42 |
| 43 | .036 | .037 | .038 | .038 | .039 | .040 | .040 | 43 |
| 44 | .039 | .040 | .040 | .041 | .042 | .042 | .043 | 44 |
| 45 | .041 | .042 | .043 | .044 | .044 | .045 | .046 | 45 |
| 46 | .044 | .045 | .045 | .046 | .047 | .048 | .049 | 46 |
| 47 | .046 | .047 | .048 | .049 | .050 | .050 | .051 | 47 |
| 48 | .049 | .050 | .051 | .051 | .052 | .053 | .054 | 48 |
| 49 | .051 | .052 | .053 | .054 | .055 | .056 | .057 | 49 |
| 50 | −.054 | −.055 | −.056 | −.057 | −.058 | −.059 | −.060 | 50 |

TABLE I.—*Reduction of the English Barometer to the Freezing-Point*—Continued.

| Degrees of Fahrenheit. | English inches. | | | | | | | Degrees of Fahrenheit. |
|---|---|---|---|---|---|---|---|---|
| | 28 | 28.5 | 29 | 29.5 | 30 | 30.5 | 31 | |
| 51 | −.056 | −.057 | −.058 | −.059 | −.060 | −.061 | −.062 | 51 |
| 52 | .059 | .060 | .061 | .062 | .063 | .064 | .065 | 52 |
| 53 | .061 | .062 | .064 | .065 | .066 | .067 | .068 | 53 |
| 54 | .064 | .065 | .066 | .067 | .068 | .070 | .071 | 54 |
| 55 | .066 | .068 | .069 | .070 | .071 | .072 | .073 | 55 |
| 56 | .069 | .070 | .071 | .073 | .074 | .075 | .076 | 56 |
| 57 | .071 | .073 | .074 | .075 | .076 | .078 | .079 | 57 |
| 58 | .074 | .075 | .076 | .078 | .079 | .080 | .082 | 58 |
| 59 | .076 | .078 | .079 | .080 | .082 | .083 | .085 | 59 |
| 60 | .079 | .080 | .082 | .083 | .084 | .086 | .087 | 60 |
| 61 | .081 | .083 | .084 | .086 | .087 | .089 | .090 | 61 |
| 62 | .084 | .085 | .087 | .088 | .090 | .091 | .093 | 62 |
| 63 | .086 | .088 | .089 | .091 | .092 | .094 | .096 | 63 |
| 64 | .089 | .090 | .092 | .094 | .095 | .097 | .098 | 64 |
| 65 | .091 | .093 | .095 | .096 | .098 | .099 | .101 | 65 |
| 66 | .094 | .095 | .097 | .099 | .101 | .102 | .104 | 66 |
| 67 | .096 | .098 | .100 | .101 | .103 | .105 | .107 | 67 |
| 68 | .099 | .101 | .102 | .104 | .106 | .108 | .109 | 68 |
| 69 | .101 | .103 | .105 | .107 | .109 | .110 | .112 | 69 |
| 70 | .104 | .106 | .107 | .109 | .111 | .113 | .115 | 70 |
| 71 | .106 | .108 | .110 | .112 | .114 | .116 | .118 | 71 |
| 72 | .109 | .111 | .113 | .115 | .117 | .118 | .120 | 72 |
| 73 | .111 | .113 | .115 | .117 | .119 | .121 | .123 | 73 |
| 74 | .114 | .116 | .118 | .120 | .122 | .124 | .126 | 74 |
| 75 | .116 | .118 | .120 | .122 | .125 | .127 | .129 | 75 |
| 76 | .119 | .121 | .123 | .125 | .127 | .129 | .131 | 76 |
| 77 | .121 | .123 | .126 | .128 | .130 | .132 | .134 | 77 |
| 78 | .124 | .126 | .128 | .130 | .133 | .135 | .137 | 78 |
| 79 | .126 | .128 | .131 | .133 | .135 | .137 | .140 | 79 |
| 80 | .129 | .131 | .133 | .136 | .138 | .140 | .143 | 80 |
| 81 | .131 | .133 | .136 | .138 | .141 | .143 | .145 | 81 |
| 82 | .134 | .136 | .138 | .141 | .143 | .146 | .148 | 82 |
| 83 | .136 | .139 | .141 | .143 | .146 | .148 | .151 | 83 |
| 84 | .139 | .141 | .144 | .146 | .148 | .151 | .154 | 84 |
| 85 | .141 | .144 | .146 | .149 | .151 | .154 | .156 | 85 |
| 86 | .144 | .146 | .149 | .151 | .154 | .156 | .159 | 86 |
| 87 | .146 | .149 | .151 | .154 | .156 | .159 | .162 | 87 |
| 88 | .149 | .151 | .154 | .156 | .159 | .162 | .165 | 88 |
| 89 | .151 | .154 | .156 | .159 | .162 | .164 | .167 | 89 |
| 90 | .153 | .156 | .159 | .162 | .164 | .167 | .170 | 90 |
| 91 | .156 | .159 | .162 | .164 | .167 | .170 | .173 | 91 |
| 92 | .158 | .161 | .164 | .167 | .170 | .173 | .175 | 92 |
| 93 | .161 | .164 | .167 | .170 | .172 | .175 | .178 | 93 |
| 94 | .163 | .166 | .169 | .172 | .175 | .178 | .180 | 94 |
| 95 | .166 | .169 | .172 | .175 | .178 | .181 | .183 | 95 |
| 96 | .168 | .171 | .174 | .177 | .180 | .183 | .186 | 96 |
| 97 | .171 | .174 | .177 | .180 | .183 | .186 | .189 | 97 |
| 98 | .173 | .176 | .180 | .183 | .186 | .189 | .191 | 98 |
| 99 | −.176 | −.179 | −.182 | −.185 | −.188 | −.191 | −.194 | −99 |

## TABLE II.

$$D_1 = 60384.3 \times \log H \text{ or } h.$$

Hundredths of an inch.

| Barometer in English Inches. | 0.00 | 0.01 | 0.02 | 0.03 | 0.04 | 0.05 | 0.06 | 0.07 | 0.08 | 0.09 |
|---|---|---|---|---|---|---|---|---|---|---|
| | Feet. | Feet. | Feet. | Feet. | Feet. | Feet. | Feet. | Feet. | Feet. | Feet. |
| 17.0 | 13915.5 | 13930.9 | 13946.3 | 13961.7 | 13977.1 | 13992.5 | 14007.9 | 14023.3 | 14038.6 | 14054.0 |
| .1 | 14069.3 | 14084.6 | 14100.0 | 14115.3 | 14130.6 | 14145.9 | 14161.2 | 14176.4 | 14191.7 | 14207.0 |
| .2 | 14222.2 | 14237.5 | 14252.7 | 14267.9 | 14283.1 | 14298.3 | 14313.5 | 14328.7 | 14343.9 | 14359.1 |
| .3 | 14374.2 | 14389.4 | 14404.5 | 14419.7 | 14434.8 | 14449.9 | 14465.0 | 14480.1 | 14495.2 | 14510.3 |
| .4 | 14525.4 | 14540.5 | 14555.5 | 14570.6 | 14585.6 | 14600.6 | 14615.7 | 14630.7 | 14645.7 | 14660.7 |
| .5 | 14675.7 | 14690.7 | 14705.6 | 14720.6 | 14735.6 | 14750.5 | 14765.4 | 14780.4 | 14795.3 | 14810.2 |
| .6 | 14825.1 | 14840.0 | 14854.9 | 14869.8 | 14884.6 | 14899.5 | 14914.4 | 14929.2 | 14944.0 | 14958.9 |
| .7 | 14973.7 | 14988.5 | 15003.3 | 15018.1 | 15032.9 | 15047.7 | 15062.4 | 15077.2 | 15092.0 | 15106.7 |
| .8 | 15121.4 | 15136.2 | 15150.9 | 15165.6 | 15180.3 | 15195.0 | 15209.7 | 15224.4 | 15239.0 | 15253.7 |
| .9 | 15268.4 | 15283.0 | 15297.6 | 15312.3 | 15326.9 | 15341.5 | 15355.1 | 15370.7 | 15385.3 | 15399.9 |
| 18.0 | 15414.5 | 15429.0 | 15443.6 | 15458.1 | 15472.7 | 15487.2 | 15501.7 | 15516.2 | 15530.7 | 15545.2 |
| .1 | 15559.7 | 15574.2 | 15588.7 | 15603.2 | 15617.6 | 15632.1 | 15646.5 | 15661.0 | 15675.4 | 15689.8 |
| .2 | 15704.2 | 15718.6 | 15733.0 | 15747.4 | 15761.8 | 15776.2 | 15790.5 | 15804.9 | 15819.2 | 15833.6 |
| .3 | 15847.9 | 15862.3 | 15876.6 | 15890.9 | 15905.2 | 15919.5 | 15933.8 | 15948.0 | 15962.3 | 15976.6 |
| .4 | 15990.8 | 16005.1 | 16019.3 | 16033.6 | 16047.8 | 16062.0 | 16076.2 | 16090.4 | 16104.6 | 16118.8 |
| .5 | 16133.0 | 16147.1 | 16161.0 | 16175.5 | 16189.6 | 16203.8 | 16217.9 | 16232.0 | 16246.1 | 16260.2 |
| .6 | 16274.4 | 16288.4 | 16302.3 | 16316.6 | 16330.7 | 16344.8 | 16358.8 | 16372.9 | 16386.9 | 16400.9 |
| .7 | 16415.0 | 16429.0 | 16443.0 | 16457.0 | 16471.0 | 16485.0 | 16499.0 | 16512.9 | 16526.9 | 16540.9 |
| .8 | 16554.8 | 16568.8 | 16582.7 | 16596.6 | 16610.6 | 16624.5 | 16638.4 | 16653.3 | 16666.2 | 16680.1 |
| .9 | 16694.0 | 16707.8 | 16721.7 | 16735.5 | 16749.4 | 16763.2 | 16777.1 | 16790.9 | 16804.7 | 16818.5 |

Thousandths of an inch.

| | 15.0 | | 14.2 |
|---|---|---|---|
| 1 | 1.5 | 1 | 1.4 |
| 2 | 3.0 | 2 | 2.8 |
| 3 | 4.5 | 3 | 4.3 |
| 4 | 6.0 | 4 | 5.7 |
| 5 | 7.5 | 5 | 7.1 |
| 6 | 9.0 | 6 | 8.5 |
| 7 | 10.5 | 7 | 9.9 |
| 8 | 12.0 | 8 | 11.4 |
| 9 | 13.5 | 9 | 12.8 |

## TABLE II—Continued.

$$D_1 = 60384.3 \times \log H \text{ or } h.$$

Hundredths of an inch.

| Barometer in English Inches. | 0.00 | 0.01 | 0.02 | 0.03 | 0.04 | 0.05 | 0.06 | 0.07 | 0.08 | 0.09 |
|---|---|---|---|---|---|---|---|---|---|---|
| | Feet. | Feet. | Feet. | Feet. | Feet. | Feet. | Feet. | Feet. | Feet. | Feet. |
| 19.0 | 16832.3 | 16846.1 | 16859.9 | 16873.7 | 16887.5 | 16901.3 | 16915.0 | 16928.8 | 16942.5 | 16956.3 |
| .1 | 16970.0 | 16983.7 | 16997.4 | 17011.2 | 17024.9 | 17038.6 | 17052.3 | 17065.9 | 17079.6 | 17093.3 |
| .2 | 17106.9 | 17120.6 | 17134.2 | 17147.9 | 17161.5 | 17175.0 | 17188.8 | 17202.4 | 17216.0 | 17229.6 |
| .3 | 17243.2 | 17256.8 | 17270.3 | 17283.9 | 17297.5 | 17311.0 | 17324.6 | 17338.1 | 17351.7 | 17365.2 |
| .4 | 17378.7 | 17392.2 | 17405.7 | 17419.2 | 17432.7 | 17446.2 | 17459.7 | 17473.2 | 17486.6 | 17500.1 |
| .5 | 17513.5 | 17527.0 | 17540.4 | 17553.9 | 17567.3 | 17580.7 | 17594.1 | 17607.5 | 17620.9 | 17634.3 |
| .6 | 17647.7 | 17661.1 | 17674.4 | 17687.8 | 17701.1 | 17714.5 | 17727.8 | 17741.2 | 17754.5 | 17767.8 |
| .7 | 17781.1 | 17794.4 | 17807.7 | 17821.0 | 17834.3 | 17847.6 | 17860.9 | 17874.2 | 17887.4 | 17900.7 |
| .8 | 17913.9 | 17927.2 | 17940.4 | 17953.6 | 17966.8 | 17980.1 | 17993.3 | 18006.5 | 18019.7 | 18032.9 |
| .9 | 18046.0 | 18059.2 | 18072.4 | 18085.5 | 18098.7 | 18111.8 | 18125.0 | 18138.1 | 18151.2 | 18164.4 |
| 20.0 | 18177.5 | 18190.6 | 18203.7 | 18216.8 | 18229.9 | 18243.0 | 18256.0 | 18269.1 | 18282.2 | 18295.2 |
| .1 | 18308.3 | 18321.3 | 18334.4 | 18347.4 | 18360.4 | 18373.4 | 18386.4 | 18399.5 | 18412.5 | 18425.4 |
| .2 | 18438.4 | 18451.4 | 18464.4 | 18477.3 | 18490.3 | 18503.3 | 18516.2 | 18529.1 | 18542.1 | 18555.0 |
| .3 | 18567.9 | 18580.9 | 18593.8 | 18606.7 | 18619.6 | 18632.4 | 18645.3 | 18658.2 | 18671.1 | 18683.9 |
| .4 | 18696.8 | 18709.7 | 18722.5 | 18735.3 | 18748.2 | 18761.0 | 18773.8 | 18786.6 | 18799.4 | 18812.2 |
| .5 | 18825.0 | 18837.8 | 18850.6 | 18863.4 | 18876.2 | 18888.9 | 18901.7 | 18914.4 | 18927.2 | 18939.9 |
| .6 | 18952.7 | 18965.4 | 18978.1 | 18990.8 | 19003.5 | 19016.2 | 19028.9 | 19041.6 | 19054.3 | 19067.0 |
| .7 | 19079.6 | 19092.3 | 19105.0 | 19117.6 | 19130.3 | 19142.9 | 19155.6 | 19168.2 | 19180.8 | 19193.4 |
| .8 | 19206.0 | 19218.6 | 19231.2 | 19243.8 | 19256.4 | 19269.0 | 19281.6 | 19294.1 | 19306.7 | 19319.3 |
| .9 | 19331.8 | 19344.4 | 19356.9 | 19369.4 | 19382.0 | 19394.5 | 19407.0 | 19419.5 | 19432.0 | 19444.5 |

Thousandths of an inch.

| | 13.5 | | 12.8 |
|---|---|---|---|
| 1 | 1.3 | 1 | 1.3 |
| 2 | 2.7 | 2 | 2.6 |
| 3 | 4.0 | 3 | 3.8 |
| 4 | 5.4 | 4 | 5.1 |
| 5 | 6.7 | 5 | 6.4 |
| 6 | 8.1 | 6 | 7.7 |
| 7 | 9.4 | 7 | 9.0 |
| 8 | 10.8 | 8 | 10.2 |
| 9 | 12.1 | 9 | 11.5 |

TABLE II—Continued.

$$D_1 = 60384.3 \times \log H \text{ or } h.$$

Hundredths of an Inch.

| Barometer in English inches. | 0.00 Feet. | 0.01 Feet. | 0.02 Feet. | 0.03 Feet. | 0.04 Feet. | 0.05 Feet. | 0.06 Feet. | 0.07 Feet. | 0.08 Feet. | 0.09 Feet. |
|---|---|---|---|---|---|---|---|---|---|---|
| 21.0 | 19457.0 | 19469.5 | 19482.0 | 19494.4 | 19506.9 | 19519.4 | 19531.8 | 19544.3 | 19556.7 | 19569.1 |
| .1 | 19581.6 | 19594.0 | 19606.4 | 19618.8 | 19631.2 | 19643.6 | 19656.0 | 19668.4 | 19680.8 | 19693.2 |
| .2 | 19705.6 | 19717.9 | 19730.3 | 19742.6 | 19755.0 | 19767.3 | 19779.7 | 19792.0 | 19804.3 | 19816.7 |
| .3 | 19829.0 | 19841.3 | 19853.6 | 19865.9 | 19878.2 | 19890.5 | 19902.7 | 19915.0 | 19927.3 | 19939.5 |
| .4 | 19951.8 | 19964.1 | 19976.3 | 19988.5 | 20000.8 | 20013.0 | 20025.2 | 20037.4 | 20049.7 | 20061.9 |
| .5 | 20074.1 | 20086.3 | 20098.4 | 20110.6 | 20122.8 | 20135.0 | 20147.1 | 20159.3 | 20171.5 | 20183.6 |
| .6 | 20195.8 | 20207.9 | 20220.1 | 20232.2 | 20244.3 | 20256.4 | 20268.5 | 20280.6 | 20292.7 | 20304.8 |
| .7 | 20316.9 | 20329.0 | 20341.0 | 20353.1 | 20365.2 | 20377.2 | 20389.3 | 20401.3 | 20413.4 | 20425.4 |
| .8 | 20437.5 | 20449.5 | 20461.5 | 20473.5 | 20485.5 | 20497.5 | 20509.5 | 20521.5 | 20533.5 | 20545.5 |
| .9 | 20557.5 | 20569.5 | 20581.4 | 20593.4 | 20605.3 | 20617.3 | 20629.2 | 20641.2 | 20653.1 | 20665.0 |
| 22.0 | 20677.0 | 20688.9 | 20700.8 | 20712.7 | 20724.6 | 20736.5 | 20748.4 | 20760.3 | 20772.2 | 20784.0 |
| .1 | 20795.9 | 20807.7 | 20819.6 | 20831.5 | 20843.3 | 20855.2 | 20867.0 | 20878.8 | 20890.6 | 20902.5 |
| .2 | 20914.3 | 20926.1 | 20937.9 | 20949.7 | 20961.5 | 20973.3 | 20985.1 | 20996.8 | 21008.6 | 21020.4 |
| .3 | 21032.1 | 21043.9 | 21055.7 | 21067.4 | 21079.1 | 21090.9 | 21102.6 | 21114.3 | 21126.1 | 21137.8 |
| .4 | 21149.5 | 21161.2 | 21172.9 | 21184.6 | 21196.3 | 21208.0 | 21219.6 | 21231.3 | 21243.0 | 21254.6 |
| .5 | 21266.3 | 21277.9 | 21289.6 | 21301.2 | 21312.9 | 21324.5 | 21336.1 | 21347.8 | 21359.4 | 21371.0 |
| .6 | 21382.6 | 21394.2 | 21405.8 | 21417.4 | 21429.0 | 21440.5 | 21452.1 | 21463.7 | 21475.3 | 21486.8 |
| .7 | 21498.4 | 21509.9 | 21521.5 | 21533.0 | 21544.5 | 21556.1 | 21567.6 | 21579.1 | 21590.6 | 21602.1 |
| .8 | 21613.6 | 21625.1 | 21636.6 | 21648.1 | 21659.6 | 21671.1 | 21682.6 | 21694.0 | 21705.5 | 21717.0 |
| .9 | 21728.4 | 21739.9 | 21751.3 | 21762.7 | 21774.2 | 21785.6 | 21797.0 | 21808.5 | 21819.9 | 21831.3 |

Thousandths of an inch.

| 12.2 | |
|---|---|
| 1 | 1.2 |
| 2 | 2.4 |
| 3 | 3.7 |
| 4 | 4.9 |
| 5 | 6.1 |
| 6 | 7.3 |
| 7 | 8.5 |
| 8 | 9.8 |
| 9 | 11.0 |

| 11.7 | |
|---|---|
| 1 | 1.2 |
| 2 | 2.3 |
| 3 | 3.5 |
| 4 | 4.7 |
| 5 | 5.8 |
| 6 | 7.0 |
| 7 | 8.2 |
| 8 | 9.4 |
| 9 | 10.5 |

TABLE II—Continued.

$$D_1 = 60384.3 \times \log H \text{ or } h.$$

Hundredths of an inch.

| Barometer in English inches. | 0.00 | 0.01 | 0.02 | 0.03 | 0.04 | 0.05 | 0.06 | 0.07 | 0.08 | 0.09 |
|---|---|---|---|---|---|---|---|---|---|---|
| | *Feet.* | *Feet.* | *Feet.* | *Feet.* | *Feet.* | *Feet.* | *Feet.* | *Feet.* | *Feet.* | *Feet.* |
| 23.0 | 21842.7 | 21854.1 | 21865.5 | 21876.9 | 21888.3 | 21899.6 | 21911.0 | 21922.4 | 21933.7 | 21945.1 |
| .1 | 21956.5 | 21967.8 | 21979.2 | 21990.5 | 22001.8 | 22013.2 | 22024.5 | 22035.8 | 22047.1 | 22058.4 |
| .2 | 22069.7 | 22081.0 | 22092.3 | 22103.6 | 22114.9 | 22126.2 | 22137.5 | 22148.7 | 22160.0 | 22171.3 |
| .3 | 22182.5 | 22193.8 | 22205.0 | 22216.3 | 22227.6 | 22238.7 | 22250.0 | 22261.2 | 22272.4 | 22283.6 |
| .4 | 22294.8 | 22306.0 | 22317.2 | 22328.4 | 22339.6 | 22350.8 | 22362.0 | 22373.2 | 22384.3 | 22395.5 |
| .5 | 22406.7 | 22417.8 | 22429.0 | 22440.1 | 22451.3 | 22462.4 | 22473.5 | 22484.7 | 22495.8 | 22506.9 |
| .6 | 22518.0 | 22529.1 | 22540.2 | 22551.3 | 22562.4 | 22573.5 | 22584.6 | 22595.7 | 22606.8 | 22617.8 |
| .7 | 22628.9 | 22640.0 | 22651.0 | 22662.1 | 22673.1 | 22684.2 | 22695.2 | 22706.3 | 22717.3 | 22728.3 |
| .8 | 22739.3 | 22750.4 | 22761.4 | 22772.4 | 22783.4 | 22794.4 | 22805.4 | 22816.4 | 22827.3 | 22838.3 |
| .9 | 22849.3 | 22860.3 | 22871.2 | 22882.2 | 22893.1 | 22904.1 | 22915.0 | 22926.0 | 22936.9 | 22947.9 |
| 24.0 | 22958.8 | 22969.7 | 22980.6 | 22991.6 | 23002.5 | 23013.4 | 23024.3 | 23035.2 | 23046.1 | 23056.9 |
| .1 | 23067.8 | 23078.7 | 23089.6 | 23100.5 | 23111.3 | 23122.2 | 23133.0 | 23143.9 | 23154.7 | 23165.6 |
| .2 | 23176.4 | 23187.3 | 23198.1 | 23208.9 | 23219.7 | 23230.5 | 23241.4 | 23252.2 | 23263.0 | 23273.8 |
| .3 | 23284.6 | 23295.4 | 23306.1 | 23316.9 | 23327.7 | 23338.5 | 23349.2 | 23360.0 | 23370.8 | 23381.5 |
| .4 | 23392.3 | 23403.0 | 23413.7 | 23424.5 | 23435.2 | 23445.9 | 23456.7 | 23467.4 | 23478.1 | 23488.8 |
| .5 | 23499.5 | 23510.2 | 23520.9 | 23531.6 | 23542.3 | 23553.0 | 23563.7 | 23574.3 | 23585.0 | 23595.7 |
| .6 | 23606.0 | 23617.0 | 23627.7 | 23638.3 | 23648.9 | 23659.6 | 23670.2 | 23680.9 | 23691.5 | 23702.1 |
| .7 | 23712.7 | 23723.3 | 23734.0 | 23744.6 | 23755.2 | 23765.8 | 23776.4 | 23786.9 | 23797.5 | 23808.1 |
| .8 | 23818.7 | 23829.8 | 23839.8 | 23850.4 | 23861.0 | 23871.5 | 23882.1 | 23892.8 | 23903.1 | 23913.7 |
| .9 | 23924.2 | 23934.7 | 23945.3 | 23955.8 | 23966.3 | 23976.8 | 23987.3 | 23997.8 | 24008.3 | 24018.8 |

Thousandths of an inch.

| | 11.2 | | 10.7 |
|---|---|---|---|
| 1 | 1.1 | 1 | 1.1 |
| 2 | 2.2 | 2 | 2.1 |
| 3 | 3.4 | 3 | 3.2 |
| 4 | 4.5 | 4 | 4.3 |
| 5 | 5.6 | 5 | 5.4 |
| 6 | 6.7 | 6 | 6.4 |
| 7 | 7.8 | 7 | 7.5 |
| 8 | 9.0 | 8 | 8.6 |
| 9 | 10.1 | 9 | 9.6 |

## TABLE II—Continued.

$$D_1 = 60384.3 \times \log H \text{ or } h.$$

Hundredths of an inch.

| Barometer in English inches. | 0.00 | 0.01 | 0.02 | 0.03 | 0.04 | 0.05 | 0.06 | 0.07 | 0.08 | 0.09 |
|---|---|---|---|---|---|---|---|---|---|---|
| | Feet. | Feet. | Feet. | Feet. | Feet. | Feet. | Feet. | Feet. | Feet. | Feet. |
| 25.0 | 24029.3 | 24039.8 | 24050.3 | 24060.8 | 24071.3 | 24081.7 | 24092.2 | 24102.7 | 24113.1 | 24123.6 |
| .1 | 24134.0 | 24144.5 | 24154.9 | 24165.3 | 24175.8 | 24186.2 | 24196.6 | 24207.1 | 42217.5 | 24227.9 |
| .2 | 24238.3 | 24248.7 | 24259.1 | 24269.5 | 24279.9 | 24290.3 | 24300.7 | 24311.0 | 42321.4 | 24331.8 |
| .3 | 24342.2 | 24352.5 | 24362.9 | 24373.2 | 24383.6 | 24393.9 | 24404.3 | 24414.6 | 24424.9 | 24435.3 |
| .4 | 24445.6 | 24455.9 | 24466.2 | 24476.4 | 24486.9 | 24497.2 | 24507.5 | 24517.8 | 24528.1 | 24538.4 |
| .5 | 24548.6 | 24558.9 | 24569.2 | 24579.5 | 24589.7 | 24600.0 | 24610.3 | 24620.5 | 24630.8 | 24641.0 |
| .6 | 24651.3 | 24661.5 | 24671.8 | 24682.0 | 24692.2 | 24702.5 | 24712.7 | 24722.9 | 24733.1 | 24743.3 |
| .7 | 24753.5 | 24763.7 | 24773.9 | 24784.1 | 24794.3 | 24804.5 | 24814.7 | 24824.9 | 24835.0 | 24845.2 |
| .8 | 24855.4 | 24865.5 | 24875.5 | 24885.8 | 24896.0 | 24906.1 | 24916.3 | 24926.4 | 24936.6 | 24946.7 |
| .9 | 24956.8 | 24966.9 | 24977.1 | 24987.2 | 24997.3 | 25007.4 | 25017.5 | 25027.6 | 25037.7 | 25047.8 |
| 26.0 | 25057.9 | 25068.0 | 25078.0 | 25088.1 | 25098.2 | 25108.3 | 25118.3 | 25128.4 | 25138.4 | 25148.5 |
| .1 | 25158.8 | 25168.6 | 25178.6 | 25188.7 | 25198.7 | 25208.7 | 25218.8 | 25228.8 | 25238.8 | 25248.8 |
| .2 | 25258.8 | 25268.8 | 25278.8 | 25288.8 | 25298.8 | 25308.8 | 25318.8 | 25328.8 | 25338.8 | 25348.8 |
| .3 | 25358.7 | 25368.7 | 25378.7 | 25388.6 | 25398.6 | 25408.5 | 25418.5 | 25428.4 | 25438.4 | 25448.3 |
| .4 | 25458.3 | 25468.2 | 25478.1 | 25488.0 | 25498.0 | 25507.9 | 25517.8 | 25527.7 | 25537.6 | 25547.5 |
| .5 | 25557.4 | 25567.3 | 25577.2 | 25587.1 | 25597.0 | 25606.8 | 25616.7 | 25626.6 | 25636.5 | 25646.3 |
| .6 | 25656.0 | 25666.0 | 25675.9 | 25685.7 | 25695.6 | 25705.4 | 25715.3 | 25725.1 | 25734.9 | 25744.8 |
| .7 | 25754.6 | 25764.4 | 25774.2 | 25784.0 | 25793.8 | 25803.6 | 25813.4 | 25823.2 | 25833.0 | 25842.8 |
| .8 | 25852.6 | 25862.4 | 25872.2 | 25882.0 | 25891.7 | 25901.5 | 25911.3 | 25921.0 | 25930.8 | 25940.5 |
| .9 | 25950.3 | 25960.0 | 25969.8 | 25979.5 | 25989.3 | 25999.0 | 26008.7 | 26018.4 | 26028.2 | 26037.9 |

Thousandths of an inch.

| 10.3 | | 9.9 | |
|---|---|---|---|
| 1 | 1.0 | 1 | 1.0 |
| 2 | 2.1 | 2 | 2.0 |
| 3 | 3.1 | 3 | 3.0 |
| 4 | 4.1 | 4 | 4.0 |
| 5 | 5.1 | 5 | 4.9 |
| 6 | 6.2 | 6 | 5.9 |
| 7 | 7.2 | 7 | 6.9 |
| 8 | 8.2 | 8 | 7.9 |
| 9 | 9.3 | 9 | 8.9 |

## TABLE II—Continued.

$$D_1 = 60384.3 \times \log H \text{ or } h.$$

Hundredths of an inch.

| Barometer in English inches. | 0.00 | 0.01 | 0.02 | 0.03 | 0.04 | 0.05 | 0.06 | 0.07 | 0.08 | 0.09 |
|---|---|---|---|---|---|---|---|---|---|---|
| | Feet. | Feet. | Feet. | Feet. | Feet. | Feet. | Feet. | Feet. | Feet. | Feet. |
| 27.0 | 26047.6 | 26057.3 | 26067.0 | 26076.7 | 26086.4 | 26096.1 | 26105.8 | 26115.5 | 26125.2 | 26134.9 |
| .1 | 26144.5 | 26154.2 | 26163.9 | 26173.6 | 26183.2 | 26192.9 | 26202.5 | 26212.2 | 26221.8 | 26231.5 |
| .2 | 26241.1 | 26250.8 | 26260.4 | 26270.0 | 26279.7 | 26289.3 | 26298.9 | 26308.5 | 26318.2 | 26327.8 |
| .3 | 26337.4 | 26347.0 | 26356.6 | 26366.2 | 26375.8 | 26385.4 | 26394.9 | 26404.5 | 26414.1 | 26423.7 |
| .4 | 26433.3 | 26442.8 | 26452.4 | 26462.0 | 26471.5 | 26481.1 | 26490.6 | 26500.2 | 26509.7 | 26519.3 |
| .5 | 26528.8 | 26538.3 | 26547.9 | 26557.4 | 26566.9 | 26576.4 | 26586.0 | 26595.5 | 26605.0 | 26614.5 |
| .6 | 26624.0 | 26633.5 | 26643.0 | 26652.5 | 26662.0 | 26671.5 | 26680.9 | 26690.4 | 26699.9 | 26709.4 |
| .7 | 26718.8 | 26728.3 | 26737.8 | 26747.2 | 26756.7 | 26766.1 | 26775.6 | 26785.0 | 26794.5 | 26803.9 |
| .8 | 26813.3 | 26822.8 | 26832.2 | 26841.6 | 26851.0 | 26860.5 | 26869.9 | 26879.3 | 26888.7 | 26898.1 |
| .9 | 26907.5 | 26916.9 | 26926.3 | 26935.7 | 26945.1 | 26954.5 | 26963.8 | 26973.2 | 26982.6 | 26992.0 |
| 28.0 | 27001.3 | 27010.7 | 27020.0 | 27029.4 | 27038.8 | 27048.1 | 27057.5 | 27066.8 | 27076.1 | 27085.5 |
| .1 | 27094.8 | 27104.1 | 27113.5 | 27122.8 | 27132.1 | 27141.4 | 27150.8 | 27160.1 | 27169.4 | 27178.7 |
| .2 | 27188.0 | 27197.3 | 27206.6 | 27215.9 | 27225.1 | 27234.4 | 27243.7 | 27253.0 | 27262.3 | 27271.5 |
| .3 | 27280.8 | 27290.1 | 27299.3 | 27308.6 | 27317.8 | 27327.1 | 27336.3 | 27345.6 | 27354.8 | 27364.1 |
| .4 | 27373.3 | 27382.5 | 27391.8 | 27401.0 | 27410.2 | 27419.4 | 27428.7 | 27437.9 | 27447.1 | 27456.3 |
| .5 | 27465.5 | 27474.7 | 27483.9 | 27493.1 | 27502.3 | 27511.5 | 27520.6 | 27529.8 | 27539.0 | 27548.2 |
| .6 | 27557.4 | 27566.5 | 27575.7 | 27584.8 | 27594.0 | 27603.2 | 27612.3 | 27621.5 | 27630.6 | 27639.7 |
| .7 | 27648.9 | 27658.0 | 27667.2 | 27676.3 | 27685.4 | 27694.5 | 27703.6 | 27712.8 | 27721.9 | 27731.0 |
| .8 | 27740.1 | 27749.2 | 27758.3 | 27767.1 | 27776.0 | 27785.6 | 27794.7 | 27803.8 | 27812.8 | 27821.9 |
| .9 | 27831.0 | 27840.1 | 27849.1 | 27858.2 | 27867.3 | 27876.3 | 27885.4 | 27894.4 | 27903.5 | 27912.5 |

Thousandths of an inch.

| Thousandths of an inch. | 9.5 | | 9.2 |
|---|---|---|---|
| 1 | 1.0 | 1 | 0.9 |
| 2 | 1.9 | 2 | 1.8 |
| 3 | 2.9 | 3 | 2.8 |
| 4 | 3.8 | 4 | 3.7 |
| 5 | 4.8 | 5 | 4.6 |
| 6 | 5.7 | 6 | 5.5 |
| 7 | 6.7 | 7 | 6.4 |
| 8 | 7.6 | 8 | 7.4 |
| 9 | 8.6 | 9 | 8.3 |

TABLE II—Continued.

$$D_1 = 60384.3 \times \log H \text{ or } h.$$

Hundredths of an inch.

| Barometer in English inches. | 0.00 | 0.01 | 0.02 | 0.03 | 0.04 | 0.05 | 0.06 | 0.07 | 0.08 | 0.09 |
|---|---|---|---|---|---|---|---|---|---|---|
| | Feet. | Feet. | Feet. | Feet. | Feet. | Feet. | Feet. | Feet. | Feet. | Feet. |
| 29.0 | 27921.6 | 27930.6 | 27939.6 | 27948.7 | 27957.7 | 27966.8 | 27975.8 | 27984.8 | 27993.8 | 28002.8 |
| .1 | 28011.9 | 28020.9 | 28029.9 | 28038.9 | 28047.9 | 28056.9 | 28065.9 | 28074.9 | 28083.8 | 28092.8 |
| .2 | 28101.8 | 28110.8 | 28119.8 | 28128.7 | 28137.7 | 28146.7 | 28155.6 | 28164.6 | 28173.6 | 28182.5 |
| .3 | 28191.5 | 28200.4 | 28209.4 | 28218.3 | 28227.3 | 28236.2 | 28245.1 | 28254.1 | 28263.0 | 28271.9 |
| .4 | 28280.8 | 28289.7 | 28298.7 | 28307.6 | 28316.5 | 28325.4 | 28334.3 | 28343.2 | 28352.1 | 28361.0 |
| .5 | 28369.9 | 28378.8 | 28387.6 | 28396.5 | 28405.4 | 28414.3 | 28423.2 | 28432.0 | 28440.9 | 28449.8 |
| .6 | 28458.6 | 28467.5 | 28476.3 | 28485.2 | 28494.0 | 28502.9 | 28511.7 | 28520.6 | 28529.4 | 28538.2 |
| .7 | 28547.1 | 28555.9 | 28564.7 | 28573.5 | 28582.3 | 28591.2 | 28600.0 | 28608.8 | 28617.6 | 28626.4 |
| .8 | 28635.2 | 28644.0 | 28652.8 | 28661.6 | 28670.4 | 28679.2 | 28688.0 | 28696.7 | 28705.5 | 28714.3 |
| .9 | 28723.1 | 28731.8 | 28740.6 | 28749.4 | 28758.1 | 28766.9 | 28775.6 | 28784.4 | 28793.1 | 28801.9 |
| 30.0 | 28810.6 | 28819.4 | 28828.1 | 28836.8 | 28845.6 | 28854.3 | 28863.0 | 28871.8 | 28880.5 | 28889.2 |
| .1 | 28897.9 | 28906.6 | 28915.3 | 28924.0 | 28932.7 | 28941.4 | 28950.1 | 28958.8 | 28967.5 | 28976.2 |
| .2 | 28984.9 | 28993.6 | 29002.2 | 29010.9 | 29019.6 | 29028.3 | 29036.9 | 29045.6 | 29054.3 | 29062.9 |
| .3 | 29071.6 | 29080.2 | 29088.9 | 29097.5 | 29106.2 | 29114.8 | 29123.5 | 29132.1 | 29140.7 | 29149.4 |
| .4 | 29158.0 | 29166.6 | 29175.2 | 29183.8 | 29192.5 | 29201.1 | 29209.7 | 29218.3 | 29226.9 | 29235.5 |
| .5 | 29244.1 | 29252.7 | 29261.3 | 29269.9 | 29278.5 | 29287.1 | 29295.6 | 29304.2 | 29312.8 | 29321.4 |
| .6 | 29329.9 | 29338.5 | 29347.1 | 29355.6 | 29364.2 | 29372.8 | 29381.3 | 29389.9 | 29398.4 | 29407.0 |
| .7 | 29415.5 | 29424.1 | 29432.6 | 29441.1 | 29449.7 | 29458.2 | 29466.8 | 29475.2 | 29483.7 | 29492.3 |
| .8 | 29500.8 | 29509.3 | 29517.8 | 29526.3 | 29534.8 | 29543.3 | 29551.8 | 29560.3 | 29568.8 | 29577.3 |
| .9 | 29585.8 | 29594.3 | 29602.8 | 29611.2 | 29619.7 | 29628.2 | 29636.7 | 29645.1 | 29653.6 | 29662.1 |

Thousandths of an inch.

| 8.9 | | 8.6 | |
|---|---|---|---|
| 1 | 0.9 | 1 | 0.9 |
| 2 | 1.8 | 2 | 1.7 |
| 3 | 2.7 | 3 | 2.6 |
| 4 | 3.6 | 4 | 3.4 |
| 5 | 4.4 | 5 | 4.3 |
| 6 | 5.3 | 6 | 5.2 |
| 7 | 6.2 | 7 | 6.0 |
| 8 | 7.1 | 8 | 6.9 |
| 9 | 8.0 | 9 | 7.7 |

## TABLE III.—*Correction for Temperature.*

$$D_{11} = A \log \frac{h}{H} \times B \text{ or } C.$$

| $t + t'$ | B | C | $t + t'$ | B | C |
|---|---|---|---|---|---|
| | $\dfrac{t + t' - 64}{982.2647}$ | $\dfrac{t + t' - 64}{900}$ | | $\dfrac{t + t' - 64}{982.2647}$ | $\dfrac{t + t' - 64}{900}$ |
| ° | | | ° | | |
| 30 | −0.0346 | −0.0378 | 70 | +0.0061 | +0.0067 |
| 31 | .0336 | .0367 | 71 | .0071 | .0078 |
| 32 | .0326 | .0356 | 72 | .0081 | .0089 |
| 33 | .0316 | .0344 | 73 | .0092 | .0100 |
| 34 | .0305 | .0333 | 74 | .0102 | .0111 |
| 35 | .0295 | .0322 | 75 | .0112 | .0122 |
| 36 | .0285 | .0311 | 76 | .0122 | .0133 |
| 37 | .0275 | .0300 | 77 | .0132 | .0144 |
| 38 | .0265 | .0289 | 78 | .0142 | .0156 |
| 39 | .0255 | .0278 | 79 | .0153 | .0167 |
| 40 | .0244 | .0267 | 80 | .0163 | .0178 |
| 41 | .0234 | .0256 | 81 | .0173 | .0189 |
| 42 | .0224 | .0244 | 82 | .0183 | .0200 |
| 43 | .0214 | .0233 | 83 | .0193 | .0211 |
| 44 | .0204 | .0222 | 84 | .0204 | .0222 |
| 45 | .0193 | .0211 | 85 | .0214 | .0233 |
| 46 | .0183 | .0200 | 86 | .0224 | .0244 |
| 47 | .0173 | .0189 | 87 | .0234 | .0256 |
| 48 | .0163 | .0178 | 88 | .0244 | .0267 |
| 49 | .0153 | .0167 | 89 | .0255 | .0278 |
| 50 | .0143 | .0156 | 90 | .0265 | .0289 |
| 51 | .0132 | .0144 | 91 | .0275 | .0300 |
| 52 | .0122 | .0133 | 92 | .0285 | .0311 |
| 53 | .0112 | .0122 | 93 | .0295 | .0322 |
| 54 | .0102 | .0111 | 94 | .0305 | .0333 |
| 55 | .0092 | .0100 | 95 | .0316 | .0344 |
| 56 | .0081 | .0089 | 96 | .0326 | .0356 |
| 57 | .0071 | .0078 | 97 | .0336 | .0367 |
| 58 | .0061 | .0067 | 98 | .0346 | .0378 |
| 59 | .0051 | .0056 | 99 | .0356 | .0389 |
| 60 | .0041 | .0044 | 100 | .0366 | .0400 |
| 61 | .0030 | .0033 | 101 | .0377 | .0411 |
| 62 | .0020 | .0022 | 102 | .0387 | .0422 |
| 63 | −0.0010 | −0.0011 | 103 | .0397 | .0433 |
| 64 | .0000 | .0000 | 104 | .0407 | .0444 |
| 65 | +0.0010 | +0.0011 | 105 | .0417 | .0456 |
| 66 | .0020 | .0022 | 106 | .0428 | .0467 |
| 67 | .0030 | .0033 | 107 | .0438 | .0478 |
| 68 | .0041 | .0044 | 108 | .0448 | .0489 |
| 69 | +0.0051 | +0.0056 | 109 | +0.0458 | +0.0500 |

## TABLE III.—*Correction for Temperature*—Continued.

$$D_{II} = A \log \frac{h}{H} \times B \text{ or } C.$$

| t + t' | B $\dfrac{t + t' - 64}{982.2647}$ | C $\dfrac{t + t' - 64}{900}$ | t + t' | B $\dfrac{t + t' - 64}{982.2647}$ | C $\dfrac{t + t' - 64}{900}$ |
|---|---|---|---|---|---|
| ° | | | ° | | |
| 110 | +0.0468 | +0.0511 | 150 | +0.0875 | +0.0958 |
| 111 | .0478 | .0522 | 151 | .0886 | .0967 |
| 112 | .0489 | .0533 | 152 | .0896 | .0978 |
| 113 | .0499 | .0544 | 153 | .0906 | .0989 |
| 114 | .0509 | .0556 | 154 | .0916 | .1000 |
| 115 | .0519 | .0567 | 155 | .0926 | .1011 |
| 116 | .0529 | .0578 | 156 | .0937 | .1022 |
| 117 | .0540 | .0589 | 157 | .0947 | .1033 |
| 118 | .0550 | .0600 | 158 | .0957 | .1044 |
| 119 | .0560 | .0611 | 159 | .9967 | .1056 |
| 120 | .0570 | .0622 | 160 | .0977 | .1067 |
| 121 | .0580 | .0633 | 161 | .0987 | .1078 |
| 122 | .0590 | .0644 | 162 | .0998 | .1089 |
| 123 | .0601 | .0656 | 163 | .1008 | .1100 |
| 124 | .0611 | .0667 | 164 | .1018 | .1111 |
| 125 | .0621 | .0678 | 165 | .1028 | .1122 |
| 126 | .0631 | .0689 | 166 | .1038 | .1133 |
| 127 | .0641 | .0700 | 167 | .1049 | .1144 |
| 128 | .0651 | .0711 | 168 | .1059 | .1156 |
| 129 | .0662 | .0722 | 169 | .1069 | .1167 |
| 130 | .0672 | .0733 | 170 | .1079 | .1178 |
| 131 | .0682 | .0744 | 171 | .1089 | .1189 |
| 132 | .0692 | .0756 | 172 | .1099 | .1200 |
| 133 | .0702 | .0767 | 173 | .1109 | .1211 |
| 134 | .0713 | .0778 | 174 | .1120 | .1222 |
| 135 | .0723 | .0789 | 175 | .1130 | .1233 |
| 136 | .0733 | .0800 | 176 | .1140 | .1244 |
| 137 | .0743 | .0811 | 177 | .1150 | .1256 |
| 138 | .0753 | .0822 | 178 | .1160 | .1267 |
| 139 | .0763 | .0833 | 179 | .1171 | .1278 |
| 140 | .0774 | .0844 | 180 | .1181 | .1289 |
| 141 | .0784 | .0856 | 181 | .1191 | .1300 |
| 142 | .0794 | .0867 | 182 | .1201 | .1311 |
| 143 | .0804 | .0878 | 183 | .1211 | .1322 |
| 144 | .0814 | .0889 | 184 | .1222 | .1333 |
| 145 | .0825 | .0900 | 185 | .1232 | .1344 |
| 146 | .0835 | .0911 | 186 | .1242 | .1356 |
| 147 | .0845 | .0922 | 187 | .1252 | .1367 |
| 148 | .0855 | .0933 | 188 | .1262 | .1378 |
| 149 | +0.0865 | +0.0944 | 189 | +0.1272 | +0.1389 |

TABLE IV.—*Correction for the Difference of Gravity in Various Latitudes.*

$$D_{111} = 60384.3 \log \frac{h}{H} (0.002657 \times \cos 2 \text{ lat.})$$

Correction, additive from 0° to 45°; substractive from 45° to 90°.

| Approx. diff. of level | 0°/90° | 2°/88° | 4°/86° | 6°/84° | 8°/82° | 10°/80° | 12°/78° | 14°/76° | 16°/74° | 18°/72° | 20°/70° | 22°/68° | 24°/66° | 26°/64° | 28°/62° | 30°/60° | 32°/58° | 34°/56° | 36°/54° | 38°/52° | 40°/50° | 42°/48° | 44°/46° | 45°/45° |
|---|---|---|---|---|---|---|---|---|---|---|---|---|---|---|---|---|---|---|---|---|---|---|---|---|
| | Feet. | Feet. | Feet. | Feet. | Feet. | Feet. | Feet. | Feet. | Feet. | Feet. | Feet. | Feet. | Feet. | Feet. | Feet. | Feet. | Feet. | Feet. | Feet. | Feet. | Feet. | Feet. | Feet. | |
| 1000 | 2.6 | 2.6 | 2.6 | 2.6 | 2.5 | 2.5 | 2.4 | 2.3 | 2.2 | 2.1 | 2.0 | 1.9 | 1.8 | 1.6 | 1.5 | 1.3 | 1.2 | 1.0 | 0.8 | 0.6 | 0.5 | 0.3 | 0.1 | 0 |
| 2000 | 5.3 | 5.2 | 5.2 | 5.1 | 5.0 | 4.9 | 4.8 | 4.6 | 4.5 | 4.2 | 4.0 | 3.8 | 3.5 | 3.2 | 2.9 | 2.6 | 2.3 | 2.0 | 1.6 | 1.3 | 0.9 | 0.5 | 0.2 | 0 |
| 3000 | 7.9 | 7.9 | 7.8 | 7.7 | 7.6 | 7.4 | 7.2 | 7.0 | 6.7 | 6.4 | 6.0 | 5.7 | 5.3 | 4.8 | 4.4 | 3.9 | 3.5 | 3.0 | 2.4 | 1.9 | 1.4 | 0.8 | 0.3 | 0 |
| 4000 | 10.5 | 10.5 | 10.4 | 10.3 | 10.1 | 9.9 | 9.6 | 9.3 | 9.0 | 8.5 | 8.0 | 7.6 | 7.0 | 6.5 | 5.9 | 5.3 | 4.6 | 3.9 | 3.2 | 2.5 | 1.8 | 1.1 | 0.4 | 0 |
| 5000 | 13.1 | 13.1 | 13.0 | 12.8 | 12.6 | 12.3 | 12.0 | 11.6 | 11.1 | 10.6 | 10.1 | 9.4 | 8.8 | 8.1 | 7.3 | 6.6 | 5.8 | 4.9 | 4.1 | 3.2 | 2.3 | 1.4 | 0.5 | 0 |
| 6000 | 15.8 | 15.7 | 15.6 | 15.4 | 15.1 | 14.8 | 14.4 | 13.9 | 13.4 | 12.7 | 12.1 | 11.3 | 10.5 | 9.7 | 8.8 | 7.9 | 6.9 | 5.9 | 4.9 | 3.8 | 2.7 | 1.6 | 0.5 | 0 |
| 7000 | 18.4 | 18.3 | 18.2 | 18.0 | 17.7 | 17.3 | 16.8 | 16.2 | 15.6 | 14.9 | 14.1 | 13.2 | 12.3 | 11.3 | 10.3 | 9.2 | 8.1 | 6.9 | 5.7 | 4.4 | 3.2 | 1.9 | 0.6 | 0 |
| 8000 | 21.0 | 21.0 | 20.8 | 20.5 | 20.2 | 19.7 | 19.2 | 18.5 | 17.8 | 17.0 | 16.1 | 15.1 | 14.1 | 12.9 | 11.7 | 10.5 | 9.2 | 7.9 | 6.5 | 5.1 | 3.6 | 2.2 | 0.7 | 0 |
| 9000 | 23.6 | 23.6 | 23.4 | 23.1 | 22.7 | 22.2 | 21.6 | 20.9 | 20.0 | 19.1 | 18.1 | 17.0 | 15.8 | 14.5 | 13.2 | 11.8 | 10.4 | 8.9 | 7.3 | 5.7 | 4.1 | 2.5 | 0.8 | 0 |
| 10000 | 26.3 | 26.2 | 26.0 | 25.7 | 25.2 | 24.7 | 24.0 | 23.2 | 22.3 | 21.2 | 20.1 | 18.9 | 17.6 | 16.2 | 14.7 | 13.1 | 11.5 | 9.8 | 8.1 | 6.4 | 4.6 | 2.7 | 0.9 | 0 |
| 11000 | 28.8 | 28.8 | 28.6 | 28.2 | 27.8 | 27.1 | 26.4 | 25.5 | 24.5 | 23.4 | 22.1 | 20.8 | 19.3 | 17.8 | 16.2 | 14.4 | 12.7 | 10.8 | 8.9 | 7.0 | 5.0 | 3.0 | 1.0 | 0 |
| 12000 | 31.5 | 31.4 | 31.2 | 30.8 | 30.3 | 29.6 | 28.8 | 27.8 | 26.7 | 25.5 | 24.1 | 22.7 | 21.1 | 19.4 | 17.6 | 15.8 | 13.8 | 11.8 | 9.7 | 7.6 | 5.5 | 3.3 | 1.1 | 0 |
| 13000 | 34.1 | 34.1 | 33.8 | 33.4 | 32.8 | 32.1 | 31.2 | 30.1 | 28.9 | 27.6 | 26.1 | 24.6 | 22.8 | 21.0 | 19.1 | 17.1 | 15.0 | 12.8 | 10.5 | 8.3 | 5.9 | 3.6 | 1.2 | 0 |
| 14000 | 36.8 | 36.7 | 36.4 | 36.0 | 35.3 | 34.5 | 33.6 | 32.5 | 31.2 | 29.7 | 28.1 | 26.4 | 24.6 | 22.6 | 20.6 | 18.4 | 16.1 | 13.8 | 11.4 | 8.9 | 6.4 | 3.8 | 1.3 | 0 |
| 15000 | 39.4 | 39.3 | 39.0 | 38.5 | 37.9 | 37.0 | 36.0 | 34.8 | 33.4 | 31.9 | 30.2 | 28.3 | 26.3 | 24.2 | 22.0 | 19.7 | 17.3 | 14.8 | 12.2 | 9.5 | 6.8 | 4.1 | 1.4 | 0 |
| 16000 | 42.0 | 41.9 | 41.6 | 41.1 | 40.4 | 39.5 | 38.4 | 37.1 | 35.6 | 34.0 | 32.2 | 30.2 | 28.1 | 25.7 | 23.5 | 21.0 | 18.4 | 15.7 | 13.0 | 10.2 | 7.3 | 4.4 | 1.5 | 0 |
| 17000 | 44.6 | 44.5 | 44.2 | 43.7 | 42.9 | 41.9 | 40.8 | 39.4 | 37.9 | 36.1 | 34.2 | 32.1 | 29.9 | 27.5 | 25.0 | 22.3 | 19.6 | 16.7 | 13.8 | 10.8 | 7.8 | 4.7 | 1.5 | 0 |
| 18000 | 47.3 | 47.1 | 46.8 | 46.2 | 45.4 | 44.4 | 43.2 | 41.7 | 40.1 | 38.2 | 36.2 | 34.0 | 31.6 | 29.1 | 26.4 | 23.6 | 20.7 | 17.7 | 14.6 | 11.4 | 8.2 | 4.9 | 1.6 | 0 |
| 19000 | 49.9 | 49.8 | 49.4 | 48.8 | 48.0 | 46.9 | 45.6 | 44.0 | 42.3 | 40.4 | 38.2 | 35.9 | 33.4 | 30.7 | 27.9 | 24.9 | 21.9 | 18.7 | 15.4 | 12.1 | 8.7 | 5.2 | 1.7 | 0 |
| 20000 | 52.5 | 52.4 | 52.0 | 51.4 | 50.5 | 49.3 | 48.0 | 46.4 | 44.5 | 42.5 | 40.2 | 37.8 | 35.1 | 32.3 | 29.4 | 26.3 | 23.0 | 19.7 | 16.2 | 12.7 | 9.1 | 5.5 | 1.8 | 0 |
| 21000 | 55.1 | 55.0 | 54.6 | 53.9 | 53.0 | 51.8 | 50.4 | 48.7 | 46.8 | 44.6 | 42.2 | 39.7 | 36.9 | 33.9 | 30.8 | 27.6 | 24.2 | 20.7 | 17.0 | 13.3 | 9.6 | 5.8 | 1.9 | 0 |
| 22000 | 57.8 | 57.6 | 57.2 | 56.5 | 55.5 | 54.3 | 52.8 | 51.0 | 49.0 | 46.7 | 44.3 | 41.6 | 38.7 | 35.6 | 32.3 | 28.9 | 25.3 | 21.6 | 17.9 | 14.0 | 10.0 | 6.0 | 2.0 | 0 |
| 23000 | 60.4 | 60.2 | 59.8 | 59.1 | 58.1 | 56.7 | 55.2 | 53.3 | 51.2 | 48.9 | 46.3 | 43.4 | 40.4 | 37.2 | 33.8 | 30.2 | 26.5 | 22.6 | 18.7 | 14.6 | 10.5 | 6.3 | 2.1 | 0 |
| 24000 | 63.0 | 62.9 | 62.4 | 61.6 | 60.6 | 59.2 | 57.6 | 55.6 | 53.4 | 51.0 | 48.3 | 45.3 | 42.2 | 38.8 | 35.2 | 31.5 | 27.6 | 23.6 | 19.5 | 15.2 | 10.9 | 6.6 | 2.2 | 0 |

TABLE ·V.—*Correction for Decrease of Gravity on a Vertical acting on the Density of the Mercurial Column.*

$$D_{IV} = 60384.3 \log \frac{h}{H} + \frac{60384.3 \log \frac{h}{H} + 52252}{20886860}$$

| Approx. diff. of alt. | 000 | 100 | 200 | 300 | 400 | 500 | 600 | 700 | 800 | 900 |
|---|---|---|---|---|---|---|---|---|---|---|
| | Feet. | Feet. | Feet. | Feet. | Feet. | Feet. | Feet. | Feet. | Feet. | Feet. |
| 1000 | 2.5 | 2.8 | 3.1 | 3.3 | 3.6 | 3.9 | 4.1 | 4.4 | 4.7 | 4.9 |
| 2000 | 5.2 | 5.5 | 5.7 | 6.0 | 6.3 | 6.6 | 6.8 | 7.1 | 7.4 | 7.7 |
| 3000 | 7.9 | 8.2 | 8.5 | 8.8 | 9.1 | 9.3 | 9.6 | 9.9 | 10.2 | 10.5 |
| 4000 | 10.8 | 11.1 | 11.4 | 11.6 | 11.9 | 12.2 | 12.5 | 12.8 | 13.1 | 13.4 |
| 5000 | 13.7 | 14.0 | 14.3 | 14.6 | 14.9 | 15.2 | 15.5 | 15.8 | 16.1 | 16.4 |
| 6000 | 16.7 | 17.0 | 17.4 | 17.7 | 18.0 | 18.3 | 18.6 | 18.9 | 19.2 | 19.5 |
| 7000 | 19.9 | 20.2 | 20.5 | 20.8 | 21.1 | 21.5 | 21.8 | 22.1 | 22.4 | 22.8 |
| 8000 | 23.1 | 23.4 | 23.7 | 24.1 | 24.4 | 24.7 | 25.1 | 25.4 | 25.7 | 26.1 |
| 9000 | 26.4 | 26.7 | 27.1 | 27.4 | 27.7 | 28.1 | 28.4 | 28.8 | 29.1 | 29.5 |
| 10000 | 29.8 | 30.2 | 30.5 | 30.8 | 31.2 | 31.5 | 31.9 | 32.2 | 32.6 | 33.0 |
| 11000 | 33.3 | 33.7 | 34.0 | 34.4 | 34.7 | 35.1 | 35.5 | 35.8 | 36.2 | 36.5 |
| 12000 | 36.9 | 37.3 | 37.6 | 38.0 | 38.4 | 38.8 | 39.1 | 39.5 | 39.9 | 40.2 |
| 13000 | 40.6 | 41.0 | 41.4 | 41.7 | 42.1 | 42.5 | 42.8 | 43.3 | 43.6 | 44.0 |
| 14000 | 44.4 | 44.8 | 45.2 | 45.6 | 46.0 | 46.3 | 46.7 | 47.1 | 47.5 | 47.9 |
| 15000 | 48.3 | 48.7 | 49.1 | 49.5 | 49.9 | 50.3 | 50.7 | 51.1 | 51.5 | 51.9 |
| 16000 | 52.3 | 52.7 | 53.1 | 53.5 | 53.9 | 54.3 | 54.7 | 55.1 | 55.5 | 56.0 |
| 17000 | 56.4 | 56.8 | 57.2 | 57.6 | 58.0 | 58.4 | 58.9 | 59.3 | 59.7 | 60.1 |
| 18000 | 60.5 | 61.0 | 61.4 | 61.8 | 62.2 | 62.7 | 63.1 | 63.5 | 64.0 | 64.4 |
| 19000 | 64.8 | 65.2 | 65.7 | 66.1 | 66.6 | 67.0 | 67.4 | 67.9 | 68.3 | 68.7 |
| 20000 | 69.2 | 69.6 | 70.1 | 70.5 | 71.0 | 71.4 | 71.9 | 72.3 | 72.7 | 73.2 |
| 21000 | 73.6 | 74.1 | 74.6 | 75.0 | 75.5 | 75.9 | 76.4 | 76.8 | 77.3 | 77.7 |
| 22000 | 78.2 | 78.7 | 79.1 | 79.6 | 80.1 | 80.5 | 81.0 | 81.5 | 81.9 | 82.4 |
| 23000 | 82.9 | 83.3 | 83.8 | 84.3 | 84.8 | 85.2 | 85.7 | 86.2 | 86.7 | 87.1 |
| 24000 | 87.6 | 88.1 | 88.6 | 89.1 | 89.5 | 90.0 | 90.5 | 91.0 | 91.5 | 92.0 |

NOTE.—In Table I the scale of the barometer is supposed to be of brass, extending from the cistern to the top of the mercurial column, and the difference of expansion of brass and of mercury is taken into account. The standard temperature of the yard being 62° Fahrenheit and not 32°, the difference of expansion of the scale and of the mercurial column carries the point of no correction down to 29° Fahrenheit.

TABLE VI.—*Correction for Decrease of Gravity on a Vertical acting on the Density of the Air.*

$$D_v = 60384.3 \log \frac{h}{H} \times \frac{s}{10443430}$$

Height of the barometer in English inches at the lower station.

| Approx. diff. of alt. | 12 | 13 | 14 | 15 | 16 | 17 | 18 | 19 | 20 | 21 | 22 | 23 | 24 | 25 | 26 | 27 | 28 | 29 |
|---|---|---|---|---|---|---|---|---|---|---|---|---|---|---|---|---|---|---|
| 1000 | 2.3 | 2.1 | 1.9 | 1.7 | 1.6 | 1.4 | 1.3 | 1.1 | 1.0 | 0.9 | 0.8 | 0.7 | 0.6 | 0.5 | 0.4 | 0.3 | 0.2 | 0.1 |
| 2000 | 4.6 | 4.2 | 3.8 | 3.5 | 3.1 | 2.8 | 2.6 | 2.3 | 2.0 | 1.8 | 1.5 | 1.3 | 1.1 | 0.9 | 0.7 | 0.5 | 0.3 | 0.2 |
| 3000 | 6.9 | 6.3 | 5.7 | 5.2 | 4.7 | 4.3 | 3.8 | 3.4 | 3.0 | 2.7 | 2.3 | 2.0 | 1.7 | 1.4 | 1.1 | 0.8 | 0.5 | 0.3 |
| 4000 | 9.2 | 8.4 | 7.6 | 6.9 | 6.3 | 5.7 | 5.1 | 4.6 | 4.0 | 3.6 | 3.1 | 2.6 | 2.2 | 1.8 | 1.4 | 1.0 | 0.7 | 0.4 |
| 5000 | 11.5 | 10.5 | 9.5 | 8.7 | 7.9 | 7.1 | 6.4 | 5.7 | 5.1 | 4.4 | 3.9 | 3.3 | 2.8 | 2.3 | 1.8 | 1.3 | 0.8 |  |
| 6000 | 13.8 | 12.6 | 11.4 | 10.4 | 9.4 | 8.5 | 7.7 | 6.8 | 6.1 | 5.3 | 4.6 | 4.0 | 3.3 | 2.7 | 2.1 | 1.5 | 1.0 | 0.5 |
| 7000 | 16.1 | 14.7 | 13.4 | 12.1 | 11.0 | 9.9 | 8.9 | 8.0 | 7.1 | 6.2 | 5.4 | 4.6 | 3.9 | 3.2 | 2.5 | 1.8 | 1.2 | 0.6 |
| 8000 | 18.4 | 16.7 | 15.3 | 13.9 | 12.6 | 11.4 | 10.2 | 9.1 | 8.1 | 7.1 | 6.2 | 5.3 | 4.4 | 3.6 | 2.8 | 2.1 | 1.3 | 0.7 |
| 9000 | 20.6 | 18.8 | 17.2 | 15.6 | 14.1 | 12.8 | 11.5 | 10.3 | 9.1 | 8.0 | 7.0 | 5.9 | 5.0 | 4.1 | 3.2 | 2.3 | 1.5 | 0.8 |
| 10000 | 22.9 | 20.9 | 19.1 | 17.3 | 15.7 | 14.2 | 12.8 | 11.4 | 10.1 | 8.9 | 7.7 | 6.6 | 5.5 | 4.5 | 3.5 | 2.6 | 1.7 |  |
| 11000 | 25.2 | 23.0 | 21.0 | 19.1 | 17.3 | 15.6 | 14.0 | 12.5 | 11.1 | 9.8 | 8.5 | 7.3 | 6.1 | 5.0 | 3.9 | 2.8 | 1.8 | 0.9 |
| 12000 | 27.5 | 25.1 | 22.9 | 20.8 | 18.9 | 17.0 | 15.3 | 13.7 | 12.1 | 10.7 | 9.3 | 7.9 | 6.6 | 5.4 | 4.2 | 3.1 | 2.0 | 0.9 |
| 13000 | 29.8 | 27.2 | 24.8 | 22.5 | 20.4 | 18.5 | 16.6 | 14.8 | 13.2 | 11.6 | 10.0 | 8.6 | 7.2 | 5.9 | 4.6 | 3.4 | 2.1 | 1.0 |
| 14000 | 32.1 | 29.3 | 26.7 | 24.3 | 22.0 | 19.9 | 17.9 | 16.0 | 14.2 | 12.4 | 10.8 | 9.2 | 7.8 | 6.3 | 4.9 | 3.6 | 2.3 | 1.1 |
| 15000 | 34.4 | 31.4 | 28.6 | 26.0 | 23.6 | 21.3 | 19.2 | 17.1 | 15.2 | 13.3 | 11.6 | 9.9 | 8.3 | 6.8 | 5.3 | 3.9 | 2.5 | 1.2 |
| 16000 | 36.7 | 33.5 | 30.5 | 27.7 | 25.2 | 22.7 | 20.4 | 18.2 | 16.2 | 14.2 | 12.4 | 10.6 | 8.9 | 7.2 | 5.6 | 4.1 | 2.7 | 1.3 |
| 17000 | 39.0 | 35.6 | 32.4 | 29.5 | 26.7 | 24.1 | 21.7 | 19.4 | 17.2 | 15.1 | 13.1 | 11.1 | 9.4 | 7.7 | 6.0 | 4.4 | 2.8 | 1.3 |
| 18000 | 41.3 | 37.7 | 34.3 | 31.2 | 28.3 | 25.6 | 23.0 | 20.5 | 18.2 | 16.0 | 13.9 | 11.9 | 10.0 | 8.1 | 6.4 | 4.6 | 3.0 | 1.4 |
| 19000 | 43.6 | 39.8 | 36.2 | 32.9 | 29.9 | 27.0 | 24.2 | 21.7 | 19.2 | 16.9 | 14.7 | 12.6 | 10.5 | 8.6 | 6.7 | 4.9 | 3.2 | 1.5 |
| 20000 | 45.9 | 41.9 | 38.1 | 34.7 | 31.4 | 28.4 | 25.5 | 22.8 | 20.2 | 17.8 | 15.4 | 13.2 | 11.1 | 9.0 | 7.1 | 5.2 | 3.3 | 1.6 |
| 21000 | 48.2 | 44.0 | 40.1 | 36.4 | 33.0 | 29.8 | 26.8 | 23.9 | 21.2 | 18.7 | 16.2 | 13.9 | 11.6 | 9.5 | 7.4 | 5.4 | 3.5 | 1.7 |
| 22000 | 50.5 | 46.1 | 42.0 | 38.1 | 34.6 | 31.2 | 28.1 | 25.1 | 22.3 | 19.6 | 17.0 | 14.5 | 12.2 | 9.9 | 7.8 | 5.7 | 3.7 | 1.7 |
| 23000 | 52.8 | 48.1 | 43.9 | 39.9 | 36.2 | 32.7 | 29.4 | 26.2 | 23.3 | 20.4 | 17.8 | 15.2 | 12.7 | 10.4 | 8.1 | 5.9 | 3.8 | 1.8 |
| 24000 | 55.1 | 50.2 | 45.8 | 41.6 | 37.7 | 34.1 | 30.6 | 27.4 | 24.3 | 21.3 | 18.5 | 15.9 | 13.3 | 10.8 | 8.5 | 6.2 | 4.0 | 1.9 |

TABLE VII.—*Giving the Relative Humidities in Fractions of Unity.*

| Diff. of wet and dry bulb thermometers. | Wet bulb. | | | | | | |
|---|---|---|---|---|---|---|---|
| | 10° | 15° | 20° | 22° | 24° | 26° | 28° |
| 0.0 | 1.000 | 1.000 | 1.000 | 1.000 | 1.,000 | 1.000 | 1.000 |
| .2 | .925 | .926 | .929 | .935 | .941 | .949 | .958 |
| .4 | .854 | .856 | .864 | .874 | .887 | .902 | .919 |
| .6 | .788 | .791 | .804 | .817 | .837 | .858 | .883 |
| .8 | .727 | .731 | .749 | .764 | .790 | .818 | .850 |
| 1.0 | .671 | .675 | .699 | .717 | .747 | .781 | .819 |
| .2 | .619 | .624 | .653 | .674 | .707 | .747 | .791 |
| .4 | .570 | .577 | .610 | .635 | .670 | .716 | .765 |
| .6 | .525 | .534 | .571 | .599 | .636 | .687 | .741 |
| .8 | .483 | .494 | .535 | .566 | .605 | .660 | .719 |
| 2.0 | .445 | .456 | .502 | .535 | .577 | .635 | .698 |
| .2 | .410 | .421 | .471 | .506 | .552 | .612 | .679 |
| .4 | .377 | .389 | .443 | .479 | .529 | .591 | .662 |
| .6 | .346 | .360 | .417 | .454 | .507 | .572 | .646 |
| .8 | .318 | .333 | .393 | .432 | .487 | .554 | .631 |
| 3.0 | .292 | .308 | .371 | .412 | .469 | .538 | .618 |
| .2 | .268 | .285 | .351 | .394 | .453 | .524 | .605 |
| .4 | .245 | .264 | .332 | .377 | .439 | .511 | .593 |
| .6 | .224 | .245 | .314 | .361 | .426 | .499 | .582 |
| .8 | .205 | .227 | .298 | .346 | .413 | .488 | .572 |
| 4.0 | .187 | .211 | .283 | .332 | .401 | .478 | .562 |
| .2 | .170 | .196 | .269 | .320 | .390 | .469 | .553 |
| .4 | .154 | .182 | .257 | .309 | .379 | .461 | .544 |
| .6 | .140 | .169 | .246 | .299 | .370 | .453 | .536 |
| .8 | .128 | .157 | .236 | .290 | .362 | .446 | .528 |
| 5.0 | .117 | .145 | .226 | .281 | .355 | .440 | .521 |
| .2 | | | .217 | .273 | .348 | .434 | .514 |
| .4 | | | .209 | .266 | .342 | .429 | .507 |
| .6 | | | .201 | .259 | .336 | .424 | .501 |
| .8 | | | .194 | .253 | .331 | .419 | .495 |
| 6.0 | | | .188 | .248 | .327 | .415 | .489 |
| .2 | | | .182 | .243 | .324 | .411 | .482 |
| .4 | | | .176 | .239 | .321 | .407 | .476 |
| .6 | | | .171 | .236 | .319 | .403 | .469 |
| .8 | | | .167 | .233 | .317 | .399 | .463 |
| 7.0 | | | .164 | .231 | .315 | .396 | .457 |

TABLE VII.—*Giving the Relative Humidities, &c.*—Continued.

| Diff. of wet and dry bulb thermometers. | Wet bulb. | | | | | | |
|---|---|---|---|---|---|---|---|
| | 30°. | 32° | 34° | 36° | 38° | 40° | 42° |
| 0 | 1.000 | 1.000 | 1.000 | 1.000 | 1.000 | 1.000 | 1.000 |
| 1 | .856 | .885 | .903 | .909 | .914 | .917 | .920 |
| 2 | .756 | .799 | .821 | .831 | .837 | .843 | .847 |
| 3 | .684 | .728 | .750 | .761 | .768 | .776 | .781 |
| 4 | .628 | .665 | .686 | .697 | .707 | .715 | .721 |
| 5 | .579 | .609 | .628 | .640 | .652 | .660 | .667 |
| 6 | .533 | .558 | .575 | .589 | .601 | .610 | .618 |
| 7 | .490 | .512 | .528 | .543 | .554 | .565 | .573 |
| 8 | .450 | .470 | .486 | .500 | .512 | .523 | .532 |
| 9 | .414 | .432 | .448 | .461 | .473 | .485 | .495 |
| 10 | .381 | .398 | .413 | .426 | .438 | .451 | .462 |
| 11 | .351 | .367 | .381 | .394 | .406 | .420 | .432 |
| 12 | .324 | .339 | .352 | .365 | .378 | .391 | .404 |
| 13 | .299 | .313 | .326 | .339 | .352 | .365 | .378 |
| 14 | .277 | .289 | .302 | .315 | .328 | .341 | .354 |
| 15 | .257 | .268 | .280 | .293 | .306 | .320 | .332 |
| 16 | .238 | .249 | .261 | .274 | .286 | .300 | .312 |
| 17 | .221 | .232 | .244 | .256 | .268 | .281 | .293 |
| 18 | .206 | .216 | .228 | .240 | .252 | .263 | .275 |
| 19 | .192 | .202 | .213 | .225 | .237 | .246 | .258 |
| 20 | .180 | .189 | .200 | .211 | .223 | .231 | .242 |
| 21 | .168 | .178 | .189 | .199 | .209 | .217 | .226 |
| 22 | .158 | .168 | .178 | .187 | .195 | .204 | .213 |
| 23 | .149 | .158 | .167 | .175 | .183 | .192 | .200 |
| 24 | .140 | .149 | .157 | .164 | .172 | .181 | .188 |
| 25 | .132 | .141 | .148 | .154 | .162 | .170 | .177 |
| 26 | | | | | | .160 | .168 |
| 27 | | | | | | .151 | .159 |
| 28 | | | | | | .143 | .151 |
| 29 | | | | | | .135 | .143 |
| 30 | | | | | | .128 | .135 |

TABLE VII.—*Giving the Relative Humidities,, &c.*—Continued.

| Diff. of wet and dry bulb thermometers. | Wet bulb. | | | | | | |
|---|---|---|---|---|---|---|---|
| | 44° | 46° | 48° | 50° | 52° | 54° | 56° |
| 0 | 1.000 | 1.000 | 1.000 | 1.000 | 1.000 | 1.000 | 1.000 |
| 1 | .922 | .923 | .925 | .928 | .930 | .931 | .933 |
| 2 | .851 | .853 | .858 | .862 | .866 | .869 | .872 |
| 3 | .786 | .791 | .796 | .802 | .807 | .811 | .816 |
| 4 | .727 | .734 | .740 | .747 | .753 | .758 | .764 |
| 5 | .674 | .682 | .689 | .697 | .704 | .709 | .716 |
| 6 | .626 | .635 | .643 | .651 | .658 | .664 | .671 |
| 7 | .582 | .592 | .601 | .609 | .616 | .622 | .629 |
| 8 | .542 | .552 | .562 | .570 | .577 | .583 | .590 |
| 9 | .506 | .516 | .526 | .534 | .541 | .547 | .554 |
| 10 | .473 | .483 | .493 | .501 | .507 | .513 | .521 |
| 11 | .443 | .453 | .462 | .470 | .476 | .482 | .491 |
| 12 | .415 | .425 | .433 | .441 | .448 | .454 | .463 |
| 13 | .389 | .398 | .406 | .414 | .422 | .428 | .437 |
| 14 | .365 | .373 | .381 | .389 | .397 | .404 | .413 |
| 15 | .343 | .350 | .358 | .366 | .374 | .382 | .391 |
| 16 | .322 | .329 | .337 | .345 | .353 | .361 | .370 |
| 17 | .302 | .310 | .318 | .326 | .333 | .341 | .350 |
| 18 | .284 | .292 | .300 | .308 | .315 | .322 | .331 |
| 19 | .267 | .275 | .283 | .291 | .298 | .305 | .313 |
| 20 | .251 | .259 | .267 | .275 | .282 | .289 | .296 |
| 21 | .236 | .244 | .252 | .260 | .267 | .274 | .280 |
| 22 | .222 | .230 | .238 | .246 | .253 | .260 | .265 |
| 23 | .209 | .217 | .225 | .233 | .240 | .246 | .251 |
| 24 | .197 | .205 | .213 | .221 | .227 | .233 | .238 |
| 25 | .186 | .194 | .202 | .209 | .215 | .221 | .226 |
| 26 | .176 | .184 | .192 | .198 | .204 | .209 | .214 |
| 27 | .167 | .174 | .182 | .188 | .193 | .198 | .203 |
| 28 | .158 | .165 | .172 | .178 | .183 | .188 | .193 |
| 29 | .150 | .157 | .163 | .169 | .174 | .178 | .183 |
| 30 | .142 | .148 | .154 | .160 | .165 | .169 | .174 |
| 31 | | | | .151 | .156 | .161 | .166 |
| 32 | | | | .143 | .148 | .153 | .158 |
| 33 | | | | .136 | .141 | .146 | .150 |
| 34 | | | | .129 | .134 | .139 | .143 |
| 35 | | | | .123 | .127 | .132 | .136 |

TABLE VII.—*Giving the Relative Humidities, &c.*—Continued.

| Diff. of wet and dry bulb thermometers. | Wet bulb. | | | | | | |
|---|---|---|---|---|---|---|---|
| ° | 58° | 60° | 62° | 64° | 66° | 67° | 68° |
| 0 | 1.000 | 1.000 | 1.000 | 1.000 | 1.000 | 1.000 | 1.000 |
| 1 | .935 | .936 | .937 | .938 | .940 | .940 | .941 |
| 2 | .875 | .877 | .879 | .881 | .884 | .885 | .886 |
| 3 | .819 | .822 | .825 | .828 | .832 | .834 | .835 |
| 4 | .767 | .771 | .775 | .779 | .783 | .786 | .788 |
| 5 | .719 | .724 | .729 | .734 | .739 | .741 | .743 |
| 6 | .675 | .681 | .686 | .692 | .697 | .699 | .702 |
| 7 | .634 | .641 | .647 | .653 | .658 | .660 | .663 |
| 8 | .597 | .603 | .610 | .616 | .621 | .624 | .627 |
| 9 | .562 | .568 | .575 | .582 | .587 | .590 | .593 |
| 10 | .529 | .536 | .543 | .549 | .555 | .558 | .561 |
| 11 | .499 | .506 | .513 | .519 | .526 | .528 | .531 |
| 12 | .471 | .478 | .485 | .491 | .497 | .500 | .503 |
| 13 | .445 | .452 | .459 | .465 | .471 | .474 | .476 |
| 14 | .420 | .428 | .434 | .440 | .446 | .449 | .451 |
| 15 | .397 | .405 | .411 | .417 | .422 | .426 | .428 |
| 16 | .376 | .383 | .389 | .395 | .400 | .403 | .406 |
| 17 | .356 | .362 | .368 | .374 | .379 | .382 | .385 |
| 18 | .337 | .343 | .348 | .354 | .360 | .363 | .366 |
| 19 | .319 | .325 | .330 | .336 | .341 | .344 | .447 |
| 20 | .302 | .308 | .313 | .319 | .324 | .327 | .330 |
| 21 | .286 | .292 | .297 | .303 | .308 | .311 | .314 |
| 22 | .271 | .277 | .282 | .288 | .293 | .296 | .299 |
| 23 | .257 | .263 | .268 | .273 | .279 | .282 | .284 |
| 24 | .244 | .249 | .255 | .260 | .265 | .268 | .271 |
| 25 | .231 | .236 | .242 | .247 | .252 | .255 | .258 |
| 26 | .219 | .224 | .230 | .235 | .240 | .243 | .246 |
| 27 | .208 | .213 | .219 | .224 | .229 | .232 | .235 |
| 28 | .198 | .203 | .208 | .213 | .219 | .221 | .224 |
| 29 | .188 | .193 | .198 | .203 | .209 | .211 | .214 |
| 30 | .179 | .184 | .189 | .194 | .200 | .202 | .204 |
| 31 | .170 | .175 | .180 | .185 | .190 | .193 | .195 |
| 32 | .162 | .167 | .172 | .177 | .182 | .184 | .186 |
| 33 | .154 | .159 | .164 | .169 | .173 | .176 | .178 |
| 34 | .147 | .152 | .157 | .161 | .166 | .168 | .170 |
| 35 | .141 | .145 | .150 | .154 | .159 | .161 | .163 |

TABLE VII.—*Giving the Relative Humidities, &c.*—Continued.

| Diff. of wet and dry bulb thermometers. | Wet bulb. | | | | | | |
|---|---|---|---|---|---|---|---|
| | 69° | 70° | 71° | 73° | 75° | 77° | 79° |
| 0 | 1.000 | 1.000 | 1.000 | 1.000 | 1.000 | 1.000 | 1.000 |
| 1 | .941 | .942 | .942 | .943 | .945 | .946 | .946 |
| 2 | .887 | .888 | .889 | .891 | .893 | .895 | .896 |
| 3 | .836 | .838 | .839 | .842 | .845 | .847 | .849 |
| 4 | .789 | .791 | .793 | .796 | .799 | .801 | .804 |
| 5 | .745 | .747 | .749 | .753 | .756 | .759 | .762 |
| 6 | .704 | .706 | .708 | .712 | .716 | .719 | .723 |
| 7 | .665 | .668 | .670 | .674 | .678 | .682 | .686 |
| 8 | .629 | .632 | .634 | .638 | .642 | .647 | .651 |
| 9 | .595 | .598 | .600 | .605 | .609 | .614 | .618 |
| 10 | .563 | .566 | .569 | .573 | .578 | .583 | .587 |
| 11 | .533 | .536 | .539 | .544 | .549 | .553 | .558 |
| 12 | .505 | .508 | .511 | .516 | .521 | .526 | .531 |
| 13 | .479 | .482 | .484 | .490 | .495 | .500 | .505 |
| 14 | .454 | .457 | .459 | .465 | .470 | .476 | .481 |
| 15 | .431 | .434 | .436 | .442 | .447 | .453 | .458 |
| 16 | .409 | .412 | .414 | .420 | .425 | .431 | .436 |
| 17 | .388 | .391 | .393 | .399 | .405 | .411 | .416 |
| 18 | .368 | .371 | .374 | .380 | .386 | .392 | .397 |
| 19 | .350 | .353 | .356 | .361 | .367 | .373 | .379 |
| 20 | .333 | .336 | .339 | .344 | .350 | .356 | .361 |
| 21 | .317 | .320 | .323 | .328 | .334 | .340 | .345 |
| 22 | .301 | .304 | .307 | .313 | .319 | .324 | .330 |
| 23 | .287 | .290 | .293 | .299 | .304 | .310 | .315 |
| 24 | .274 | .276 | .279 | .285 | .290 | .296 | .301 |
| 25 | .261 | .263 | .266 | .272 | .277 | .283 | .288 |
| 26 | .249 | .252 | .254 | .260 | .265 | .270 | .275 |
| 27 | .238 | .240 | .243 | .248 | .253 | .258 | .263 |
| 28 | .227 | .229 | .232 | .237 | .242 | .247 | .251 |
| 29 | .217 | .219 | .222 | .227 | .231 | .236 | .240 |
| 30 | .207 | .209 | .212 | .217 | .221 | .225 | .229 |
| 31 | .198 | .200 | .202 | .207 | .211 | .215 | .219 |
| 32 | .189 | .191 | .193 | .198 | .202 | .205 | .209 |
| 33 | .181 | .183 | .185 | .189 | .193 | .196 | .200 |
| 34 | .173 | .175 | .177 | .181 | .184 | .187 | .191 |
| 35 | .165 | .167 | .169 | .173 | .176 | .179 | .183 |

TABLE VIII.—*First Part of Correction for Atmospheric Humidity.*

$$D_{VI} = \frac{\log \frac{h}{H}}{\sqrt{hH}} \quad V_{t+t'} = 180^\circ \text{ F.}^w \quad t + t' = 180^\circ \text{ F.}$$

| Barometer at upper station. | \*Height of the barometer at lower station in English inches. | | | | | | | | | | | | | |
| --- | --- | --- | --- | --- | --- | --- | --- | --- | --- | --- | --- | --- | --- | --- |
| | 18 | 19 | 20 | 21 | 22 | 23 | 24 | 25 | 26 | 27 | 28 | 29 | 30 | 31 |
| *Eng. in.* | *Feet.* | *Feet.* | *Feet.* | *Feet.* | *Feet.* | *Feet.* | *Feet.* | *Feet.* | *Feet.* | *Feet.* | *Feet.* | *Feet.* | *Feet.* | *Feet.* |
| 17.0 | 26 | 49 | 70 | 88 | 105 | 121 | 135 | 148 | 160 | 171 | 181 | 190 | 199 | 207 |
| .5 | 13 | 36 | 56 | 75 | 92 | 108 | 122 | 135 | 147 | 158 | 168 | 177 | 186 | 194 |
| 18.0 | | 23 | 44 | 63 | 80 | 95 | 109 | 122 | 134 | 145 | 155 | 165 | 174 | 182 |
| .5 | | 11 | 32 | 51 | 68 | 83 | 98 | 111 | 123 | 134 | 144 | 153 | 162 | 170 |
| 19.0 | | | 21 | 40 | 57 | 72 | 86 | 99 | 111 | 123 | 133 | 142 | 151 | 159 |
| .5 | | | 10 | 29 | 46 | 62 | 76 | 89 | 101 | 112 | 122 | 132 | 141 | 149 |
| 20.0 | | | | 19 | 36 | 51 | 66 | 79 | 91 | 102 | 112 | 122 | 131 | 139 |
| .5 | | | | 9 | 26 | 42 | 56 | 69 | 81 | 92 | 103 | 112 | 121 | 130 |
| 21.0 | | | | | 17 | 33 | 47 | 60 | 72 | 83 | 94 | 103 | 112 | 121 |
| .5 | | | | | 8 | 24 | 38 | 51 | 63 | 75 | 85 | 95 | 104 | 112 |
| 22.0 | | | | | | 16 | 30 | 43 | 55 | 66 | 77 | 86 | 95 | 104 |
| .5 | | | | | | 8 | 22 | 35 | 47 | 58 | 69 | 78 | 87 | 96 |
| 23.0 | | | | | | | 14 | 27 | 40 | 51 | 61 | 71 | 80 | 88 |
| .5 | | | | | | | 7 | 20 | 32 | 44 | 54 | 64 | 73 | 81 |
| 24.0 | | | | | | | | 13 | 25 | 37 | 47 | 57 | 66 | 74 |
| .5 | | | | | | | | 6 | 19 | 30 | 40 | 50 | 59 | 67 |
| 25.0 | | | | | | | | | 12 | 23 | 34 | 44 | 53 | 61 |
| .5 | | | | | | | | | 6 | 17 | 28 | 37 | 46 | 55 |
| 26.0 | | | | | | | | | | 11 | 22 | 31 | 40 | 49 |
| .5 | | | | | | | | | | 6 | 16 | 26 | 35 | 43 |
| 27.0 | | | | | | | | | | | 10 | 20 | 29 | 38 |
| .5 | | | | | | | | | | | 5 | 15 | 24 | 32 |
| 28.0 | | | | | | | | | | | | 10 | 19 | 27 |
| .5 | | | | | | | | | | | | 5 | 14 | 22 |
| 29.0 | | | | | | | | | | | | | 9 | 18 |
| .5 | | | | | | | | | | | | | 4 | 13 |
| 30.0 | | | | | | | | | | | | | | 8 |
| .5 | | | | | | | | | | | | | | 4 |

TABLE IX.—*Second Part of Correc*

$$D_{VII} = \frac{\log \dfrac{h}{H}}{\sqrt{h H}}\ V.\ W.$$

| $(t + t')$ | $(t + t')$ | | | | | | |
|---|---|---|---|---|---|---|---|
| 180° F. | 170° F. | 160° F. | 150° F. | 140° F. | 130° F. | 120° F. | 110° F. |
| Feet. | Feet. | Feet. | Feet. | Feet. | Feet. | Feet. | Feet. |
| 10 | 8.5 | 7.1 | 6.0 | 5.0 | 4.2 | 3.5 | 2.9 |
| 20 | 16.9 | 14.3 | 12.0 | 10.0 | 8.4 | 6.9 | 5.8 |
| 30 | 25.4 | 21.4 | 18.0 | 15.0 | 12.5 | 10.4 | 8.6 |
| 40 | 33.8 | 28.5 | 23.9 | 20.0 | 16.7 | 13.9 | 11.5 |
| 50 | 42.3 | 35.6 | 29.9 | 25.1 | 20.9 | 17.4 | 14.4 |
| 60 | 50.7 | 42.8 | 35.9 | 30.1 | 25.1 | 20.8 | 17.3 |
| 70 | 59.2 | 49.9 | 41.9 | 35.1 | 29.3 | 24.3 | 20.1 |
| 80 | 67.6 | 57.0 | 47.9 | 40.1 | 33.4 | 27.8 | 23.0 |
| 90 | 76.1 | 64.1 | 53.9 | 45.1 | 37.6 | 31.3 | 25.9 |
| 100 | 84.5 | 71.3 | 59.9 | 50.1 | 41.8 | 34.7 | 28.8 |
| 110 | 93.0 | 78.4 | 65.8 | 55.1 | 46.0 | 38.2 | 31.6 |
| 120 | 101.5 | 85.5 | 71.8 | 60.1 | 50.1 | 41.7 | 34.5 |
| 130 | 109.9 | 92.6 | 77.8 | 65.1 | 54.3 | 45.1 | 37.4 |
| 140 | 118.4 | 99.8 | 83.8 | 70.1 | 58.5 | 48.6 | 40.3 |
| 150 | 126.8 | 106.9 | 89.8 | 75.2 | 62.7 | 52.1 | 43.1 |
| 160 | 135.3 | 114.0 | 95.8 | 80.2 | 66.9 | 55.6 | 46.0 |
| 170 | 143.7 | 121.1 | 101.8 | 85.2 | 71.0 | 59.0 | 48.9 |
| 180 | 152.2 | 128.3 | 107.7 | 90.2 | 75.2 | 62.5 | 51.8 |
| 190 | 160.6 | 135.4 | 113.7 | 95.2 | 79.4 | 66.0 | 54.6 |
| 200 | 169.1 | 142.5 | 119.7 | 100.2 | 83.6 | 69.5 | 57.5 |
| 210 | 177.6 | 149.6 | 125.7 | 105.2 | 87.8 | 72.9 | 60.4 |
| 220 | 186.0 | 156.8 | 131.7 | 110.2 | 91.9 | 76.4 | 63.3 |
| 230 | 194.5 | 163.9 | 137.7 | 115.2 | 96.1 | 79.9 | 66.1 |
| 240 | 202.9 | 171.0 | 143.7 | 120.2 | 100.3 | 83.3 | 69.0 |
| 250 | 211.4 | 178.1 | 149.6 | 125.3 | 104.5 | 86.8 | 71.9 |
| 260 | 219.8 | 185.3 | 155.6 | 130.3 | 108.6 | 90.3 | 74.8 |
| 270 | 228.3 | 192.4 | 161.6 | 135.3 | 112.8 | 93.8 | 77.6 |
| 280 | 236.7 | 199.5 | 167.6 | 140.3 | 117.0 | 97.2 | 80.5 |
| 290 | 245.2 | 206.7 | 173.6 | 145.3 | 121.2 | 100.7 | 83.4 |
| 300 | 253.6 | 213.8 | 179.6 | 150.3 | 125.4 | 104.2 | 86.3 |
| 310 | 262.1 | 220.9 | 185.6 | 155.3 | 129.5 | 107.6 | 89.1 |
| 320 | 270.6 | 228.0 | 191.5 | 160.3 | 133.7 | 111.1 | 92.0 |
| 330 | 279.0 | 235.2 | 197.5 | 165.3 | 137.9 | 114.6 | 94.9 |
| 340 | 287.5 | 242.3 | 203.5 | 170.3 | 142.1 | 118.1 | 97.8 |
| 350 | 295.9 | 249.4 | 209.5 | 175.4 | 146.3 | 121.5 | 100.6 |
| 360 | 304.4 | 256.5 | 215.5 | 180.4 | 150.4 | 125.0 | 103.5 |
| 370 | 312.8 | 263.7 | 221.5 | 185.4 | 154.6 | 128.5 | 106.4 |
| 380 | 321.3 | 270.8 | 227.5 | 190.4 | 158.8 | 132.0 | 109.3 |

*tion for Atmospheric Humidity.*

$$D_{VII} = \frac{\log \dfrac{h}{H}}{\sqrt{hH}} \; V. \; W.$$

$$(t + t')$$

| 100° F. | 90° F. | 80° F. | 70° F. | 60° F. | 50° F. | 40° F. | 30° F. | 20° F. |
|---|---|---|---|---|---|---|---|---|
| Feet. | Feet. | Feet. | Feet. | Feet. | Feet. | Feet. | Feet. | Feet. |
| 2.4 | 1.9 | 1.6 | 1.3 | 1.1 | 0.8 | 0.7 | 0.5 | 0.4 |
| 4.7 | 3.9 | 3.2 | 2.6 | 2.1 | 1.7 | 1.3 | 1.0 | 0.8 |
| 7.1 | 5.8 | 4.8 | 3.9 | 3.2 | 2.5 | 2.0 | 1.6 | 1.2 |
| 9.5 | 7.8 | 6.4 | 5.2 | 4.2 | 3.4 | 2.7 | 2.1 | 1.7 |
| 11.9 | 9.7 | 8.0 | 6.5 | 5.3 | 4.2 | 3.3 | 2.6 | 2.1 |
| 14.2 | 11.7 | 9.6 | 7.8 | 6.3 | 5.0 | 4.0 | 3.1 | 2.5 |
| 16.6 | 13.6 | 11.2 | 9.1 | 7.4 | 5.9 | 4.7 | 3.7 | 2.9 |
| 19.0 | 15.6 | 12.8 | 10.4 | 8.4 | 6.7 | 5.3 | 4.2 | 3.3 |
| 21.3 | 17.5 | 14.4 | 11.7 | 9.5 | 7.6 | 6.0 | 4.7 | 3.7 |
| 23.7 | 19.5 | 16.0 | 13.0 | 10.5 | 8.4 | 6.7 | 5.2 | 4.1 |
| 26.1 | 21.4 | 17.5 | 14.3 | 11.6 | 9.2 | 7.3 | 5.8 | 4.5 |
| 28.5 | 23.4 | 19.1 | 15.6 | 12.6 | 10.1 | 8.0 | 6.3 | 5.0 |
| 30.8 | 25.3 | 20.7 | 16.9 | 13.7 | 10.9 | 8.7 | 6.8 | 5.4 |
| 33.2 | 27.3 | 22.3 | 18.2 | 14.7 | 11.8 | 9.3 | 7.3 | 5.8 |
| 35.6 | 29.2 | 23.9 | 19.5 | 15.8 | 12.6 | 10.0 | 7.9 | 6.2 |
| 38.0 | 31.2 | 25.5 | 20.8 | 16.8 | 13.4 | 10.7 | 8.4 | 6.6 |
| 40.3 | 33.1 | 27.1 | 22.1 | 17.9 | 14.3 | 11.3 | 8.9 | 7.0 |
| 42.7 | 35.1 | 28.7 | 23.4 | 18.9 | 15.1 | 12.0 | 9.4 | 7.4 |
| 45.1 | 37.0 | 30.3 | 24.7 | 20.0 | 16.0 | 12.7 | 10.0 | 7.8 |
| 47.4 | 39.0 | 31.9 | 26.0 | 21.0 | 16.8 | 13.3 | 10.5 | 8.3 |
| 49.8 | 40.9 | 33.5 | 27.3 | 22.1 | 17.7 | 14.0 | 11.0 | 8.7 |
| 52.2 | 42.9 | 35.1 | 28.6 | 23.1 | 18.5 | 14.7 | 11.5 | 9.1 |
| 54.6 | 44.8 | 36.7 | 29.9 | 24.2 | 19.3 | 15.3 | 12.1 | 9.5 |
| 56.9 | 46.8 | 38.3 | 31.2 | 25.2 | 20.2 | 16.0 | 12.6 | 9.9 |
| 59.3 | 48.7 | 39.9 | 32.5 | 26.3 | 21.0 | 16.7 | 13.1 | 10.3 |
| 61.7 | 50.7 | 41.5 | 33.8 | 27.4 | 21.9 | 17.3 | 13.6 | 10.7 |
| 64.0 | 52.6 | 43.1 | 35.1 | 28.4 | 22.7 | 18.0 | 14.2 | 11.1 |
| 66.4 | 54.6 | 44.7 | 36.4 | 29.5 | 23.5 | 18.7 | 14.7 | 11.6 |
| 68.8 | 56.5 | 46.3 | 37.7 | 30.5 | 24.4 | 19.3 | 15.2 | 12.0 |
| 71.2 | 58.5 | 47.9 | 39.0 | 31.6 | 25.2 | 20.0 | 15.7 | 12.4 |
| 73.5 | 60.4 | 49.5 | 40.3 | 32.6 | 26.1 | 20.7 | 16.3 | 12.8 |
| 75.9 | 62.4 | 51.0 | 41.6 | 33.7 | 26.9 | 21.3 | 16.8 | 13.2 |
| 78.3 | 64.3 | 52.6 | 42.9 | 34.7 | 27.7 | 22.0 | 17.3 | 13.6 |
| 80.7 | 66.3 | 54.2 | 44.2 | 35.8 | 28.6 | 22.7 | 17.8 | 14.0 |
| 83.0 | 68.2 | 55.8 | 45.5 | 36.8 | 29.4 | 23.3 | 18.4 | 14.4 |
| 85.4 | 70.2 | 57.4 | 46.8 | 37.9 | 30.3 | 24.0 | 18.9 | 14.9 |
| 87.8 | 72.1 | 59.0 | 48.1 | 38.9 | 31.1 | 24.7 | 19.4 | 15.3 |
| 90.1 | 74.1 | 60.6 | 49.4 | 40.0 | 31.9 | 25.3 | 19.9 | 15.7 |

Table to facilitate the Calculation of the Reduction to Level, by showing at what Hours 0.001 is to be added on Account of the Fractional Part of the Correction for one Hour.

When the fractional part of the correction for one hour is—

| Hour. | No. | $\frac{1}{24}$ | $\frac{2}{24}$ | $\frac{3}{24}$ | $\frac{4}{24}$ | $\frac{5}{24}$ | $\frac{6}{24}$ | $\frac{7}{24}$ | $\frac{8}{24}$ | $\frac{9}{24}$ | $\frac{10}{24}$ | $\frac{11}{24}$ | $\frac{12}{24}$ | $\frac{13}{24}$ | $\frac{14}{24}$ | $\frac{15}{24}$ | $\frac{16}{24}$ | $\frac{17}{24}$ | $\frac{18}{24}$ | $\frac{19}{24}$ | $\frac{20}{24}$ | $\frac{21}{24}$ | $\frac{22}{24}$ | $\frac{23}{24}$ | No. | Hour. |
|---|---|---|---|---|---|---|---|---|---|---|---|---|---|---|---|---|---|---|---|---|---|---|---|---|---|---|
|  |  | Add. | Add. | Add. | Add. | Add. | Add. | Add. | Add. | Add. | Add. | Add. | Add. | Add. | Add. | Add. | Add. | Add. | Add. | Add. | Add. | Add. | Add. | Add. |  |  |
| 7 a.m. | 0 | .000 | .000 | .000 | .000 | .000 | .000 | .000 | .000 | .000 | .000 | .000 | .000 | .000 | .000 | .000 | .000 | .000 | .000 | .000 | .000 | .000 | .000 | .000 | 0 | 7 a.m. |
| 8 | 1 | 0 | 0 | 0 | 0 | 0 | .001 | .001 | .001 | .001 | .001 | .001 | .001 | .001 | .001 | .001 | .001 | .001 | .001 | .001 | .001 | .001 | .001 | .001 | 1 | 8 |
| 9 | 2 | 0 | 0 | 0 | .001 | .001 | 0 | 0 | 0 | 0 | 0 | 0 | 0 | .001 | .001 | .001 | .001 | .001 | .001 | .001 | .001 | .001 | .001 | .001 | 2 | 9 |
| 10 | 3 | 0 | .001 | .001 | 0 | 0 | 0 | 0 | 0 | .001 | .001 | .001 | .001 | .001 | .001 | 0 | 0 | 0 | .001 | .001 | .001 | .001 | .001 | .001 | 3 | 10 |
| 11 | 4 | 0 | 0 | 0 | .001 | .001 | .001 | .001 | .001 | 0 | 0 | 0 | .001 | .001 | 0 | .001 | .001 | .001 | 0 | .001 | .001 | .001 | .001 | .001 | 4 | 11 |
| Noon. | 5 | 0 | 0 | .001 | 0 | 0 | 0 | 0 | .001 | .001 | .001 | 0 | 0 | .001 | .001 | 0 | .001 | .001 | .001 | .001 | 0 | .001 | .001 | .001 | 5 | Noon. |
| 1 p.m. | 6 | .001 | .001 | 0 | 0 | .001 | .001 | .001 | 0 | 0 | .001 | .001 | 0 | 0 | .001 | .001 | 0 | .001 | .001 | .001 | .001 | 0 | .001 | .001 | 6 | 1 p.m. |
| 2 | 7 | 0 | 0 | .001 | .001 | 0 | 0 | .001 | .001 | 0 | .001 | 0 | .001 | .001 | .001 | .001 | .001 | 0 | .001 | 0 | .001 | .001 | .001 | .001 | 7 | 2 |
| 3 | 8 | 0 | 0 | 0 | .001 | 0 | .001 | 0 | 0 | .001 | 0 | .001 | .001 | 0 | 0 | .001 | .001 | .001 | .001 | .001 | .001 | .001 | 0 | .001 | 8 | 3 |
| 4 | 9 | 0 | 0 | .001 | 0 | .001 | .001 | 0 | .001 | .001 | .001 | .001 | 0 | .001 | .001 | 0 | .001 | .001 | .001 | 0 | .001 | .001 | .001 | .001 | 9 | 4 |
| 5 | 10 | 0 | .001 | 0 | 0 | .001 | 0 | .001 | .001 | 0 | 0 | 0 | .001 | .001 | .001 | .001 | 0 | .001 | .001 | .001 | .001 | .001 | .001 | .001 | 10 | 5 |
| 6 | 11 | 0 | 0 | .001 | .001 | 0 | .001 | 0 | .001 | .001 | .001 | .001 | 0 | 0 | .001 | .001 | .001 | .001 | .001 | .001 | .001 | 0 | .001 | .001 | 11 | 6 |
| 7 | 12 | 0 | .001 | 0 | .001 | .001 | .001 | .001 | 0 | .001 | .001 | 0 | .001 | .001 | 0 | .001 | .001 | 0 | .001 | .001 | .001 | .001 | .001 | 0 | 12 | 7 |
| 8 | 13 | 0 | 0 | .001 | 0 | .001 | 0 | .001 | .001 | .001 | 0 | .001 | .001 | .001 | .001 | 0 | .001 | .001 | .001 | .001 | .001 | .001 | .001 | .001 | 13 | 8 |
| 9 | 14 | 0 | .001 | 0 | .001 | 0 | .001 | .001 | .001 | 0 | .001 | .001 | 0 | .001 | .001 | .001 | .001 | .001 | 0 | .001 | .001 | .001 | .001 | .001 | 14 | 9 |
| 10 | 15 | .001 | 0 | .001 | 0 | .001 | .001 | 0 | 0 | .001 | .001 | .001 | .001 | .001 | 0 | .001 | .001 | .001 | .001 | .001 | .001 | .001 | .001 | .001 | 15 | 10 |
| 11 | 16 | 0 | 0 | 0 | .001 | .001 | 0 | .001 | .001 | .001 | 0 | .001 | .001 | 0 | .001 | .001 | .001 | .001 | .001 | .001 | .001 | .001 | .001 | 0 | 16 | 11 |
| Midnight. | 17 | .001 | 0 | .001 | 0 | 0 | .001 | .001 | .001 | .001 | .001 | 0 | .001 | .001 | .001 | .001 | .001 | .001 | .001 | .001 | .001 | .001 | .001 | .001 | 17 | Midnight. |
| 1 a.m. | 18 | 0 | .001 | 0 | .001 | .001 | 0 | .001 | .001 | .001 | .001 | .001 | 0 | .001 | .001 | .001 | .001 | .001 | .001 | .001 | .001 | .001 | .001 | .001 | 18 | 1 a.m. |
| 2 | 19 | 0 | 0 | .001 | .001 | .001 | .001 | .001 | .001 | .001 | .001 | .001 | .001 | .001 | .001 | .001 | .001 | .001 | .001 | .001 | .001 | .001 | .001 | .001 | 19 | 2 |
| 3 | 20 | 0 | .001 | .001 | 0 | .001 | .001 | .001 | .001 | .001 | .001 | .001 | .001 | .001 | .001 | .001 | .001 | .001 | .001 | .001 | .001 | .001 | .001 | .001 | 20 | 3 |
| 4 | 21 | .001 | 0 | 0 | .001 | .001 | .001 | .001 | .001 | .001 | .001 | .001 | .001 | .001 | .001 | .001 | .001 | .001 | .001 | .001 | .001 | .001 | .001 | .001 | 21 | 4 |
| 5 | 22 | 0 | .001 | .001 | .001 | .001 | .001 | .001 | .001 | .001 | .001 | .001 | .001 | .001 | .001 | .001 | .001 | .001 | .001 | .001 | .001 | .001 | .001 | .001 | 22 | 5 |
| 6 | 23 | 0 | .001 | .001 | .001 | .001 | .001 | .001 | .001 | .001 | .001 | .001 | .001 | .001 | .001 | .001 | .001 | .001 | .001 | .001 | .001 | .001 | .001 | .001 | 23 | 6 |
| 7 | 24 | .000 | .000 | .000 | .000 | .000 | .000 | .000 | .000 | .000 | .000 | .000 | .001 | .001 | .001 | .001 | .001 | .001 | .001 | .001 | .001 | .001 | .001 | .001 | 24 | 7 |

## XLVI.—*Thermometrical Measurement of Heights.*

BAROMETRIC PRESSURES CORRESPONDING TO TEMPERATURES OF THE

BOILING-POINT OF WATER.

| Degrees of Fahrenheit. | Tenths of a degree of Fahrenheit. | | | | |
|---|---|---|---|---|---|
| | 0 | 2 | 4 | 6 | 8 |
| 185 | 17.048 | 17.122 | 17.197 | 17.272 | 17.348 |
| 186 | .423 | .499 | .575 | .652 | .729 |
| 187 | .806 | .883 | .961 | 18.039 | 18.117 |
| 188 | 18.195 | 18.274 | 18.353 | .432 | .512 |
| 189 | .592 | .672 | .753 | .833 | .914 |
| 190 | .996 | 19.077 | 19.159 | 19.241 | 19.324 |
| 191 | 19.407 | .490 | .573 | .657 | .741 |
| 192 | .825 | .910 | .995 | 20.080 | 20.166 |
| 193 | 20.251 | 20.338 | 20.424 | .511 | .598 |
| 194 | .685 | .773 | .861 | .949 | 21.038 |
| 195 | 21.126 | 21.216 | 21.305 | 21.395 | .485 |
| 196 | .576 | .666 | .758 | .849 | .941 |
| 197 | 22.033 | 22.125 | 22.218 | 22.311 | 22.404 |
| 198 | .498 | .592 | .686 | .781 | .876 |
| 199 | .971 | 23.067 | 23.163 | 23.259 | 23.356 |
| 200 | 23.453 | .550 | .648 | .746 | .845 |
| 201 | .943 | 24.042 | 24.142 | 24.241 | 24.341 |
| 202 | 24.442 | .542 | .644 | .745 | .847 |
| 203 | .949 | 25.051 | 25.154 | 25.257 | 25.361 |
| 204 | 25.465 | .569 | .674 | .779 | .884 |
| 205 | .990 | 26.096 | 26.202 | 26.309 | 26.416 |
| 206 | 26.523 | .631 | .740 | .848 | .957 |
| 207 | 27.066 | 27.176 | 27.286 | 27.397 | 27.507 |
| 208 | .618 | .730 | .842 | .954 | 28.067 |
| 209 | 28.180 | 28.293 | 28.407 | 28.521 | .636 |
| 210 | .751 | .866 | .982 | 29.098 | 29.215 |
| 211 | 29.331 | 29.449 | 29.566 | .684 | .803 |
| 212 | .922 | 30.041 | 30.161 | 30.281 | 30.401 |

### Table of Comparison of Fahrenheit's Thermometer with Reaumur's and the Centesimal.

| Fah. | Reaum. | Centes. | Fah. | Reaum. | Centes. | Fah. | Reaum. | Centes. |
|---|---|---|---|---|---|---|---|---|
| ° | ° | ° | ° | ° | ° | ° | ° | ° |
|  |  |  | 33 | + 0.4 | + 0.6 | 67 | +15.6 | +19.4 |
| 0 | −14.2 | −17.8 | 34 | 0.9 | 1.1 | 68 | 16.0 | 20.0 |
| 1 | 13.8 | 17.2 | 35 | 1.3 | 1.7 | 69 | 16.4 | 20.6 |
| 2 | 13.3 | 16.7 | 36 | 1.8 | 2.2 | 70 | 16.9 | 21.1 |
| 3 | 12.9 | 16.1 | 37 | 2.2 | 2.8 | 71 | 17.3 | 21.7 |
| 4 | 12.4 | 15.6 | 38 | 2.7 | 3.3 | 72 | 17.8 | 22.2 |
| 5 | 12.0 | ·15.0 | 39 | 3.1 | 3.9 | 73 | 18.2 | 22.8 |
| 6 | 11.6 | 14.4 | 40 | 3.6 | 4.4 | 74 | 18.7 | 23.3 |
| 7 | 11.1 | 13.9 | 41 | 4.0 | 5.0 | 75 | 19.1 | 23.9 |
| 8 | 10.7 | 13.3 | 42 | 4.4 | 5.6 | 76 | 19.6 | 24.4 |
| 9 | 10.2 | 12.8 | 43 | 4.9 | 6.1 | 77 | 20.0 | 25.0 |
| 10 | 9.8 | 12.2 | 44 | 5.3 | 6.7 | 78 | 20.4 | 25.6 |
| 11 | 9.3 | 11.7 | 45 | 5.8 | 7.2 | 79 | 20.9 | 26.1 |
| 12 | 8.9 | 11.1 | 46 | 6.2 | 7.8 | 80 | 21.3 | 26.7 |
| 13 | 8.4 | 10.6 | 47 | 6.7 | 8.3 | 81 | 21.8 | 27.2 |
| 14 | 8.0 | 10.0 | 48 | 7.1 | 8.9 | 82 | 22.2 | 27.8 |
| 15 | 7.6 | 9.4 | 49 | 7.6 | 9.4 | 83 | 22.7 | 28.3 |
| 16 | 7.1 | 8.9 | 50 | 8.0 | 10.0 | 84 | 23.1 | 28.9 |
| 17 | 6.7 | 8.3 | 51 | 8.4 | 10.6 | 85 | 23.6 | 29.4 |
| 18 | 6.2 | 7.8 | 52 | 8.9 | 11.1 | 86 | 24.0 | 30.0 |
| 19 | 5.8 | 7.2 | 53 | 9.3 | 11.7 | 87 | 24.4 | 30.6 |
| 20 | 5.3 | 6.7 | 54 | 9.8 | 12.2 | 88 | 24.9 | 31.1 |
| 21 | 4.9 | 6.1 | 55 | 10.2 | 12.8 | 89 | 25.3 | 31.7 |
| 22 | 4.4 | 5.6 | 56 | 10.7 | 13.3 | 90 | 25.8 | 32.2 |
| 23 | 4.0 | 5.0 | 57 | 11.1 | 13.9 | 91 | 26.2 | 32.8 |
| 24 | 3.6 | 4.4 | 58 | 11.6 | 14.4 | 92 | 26.7 | 33.3 |
| 25 | 3.1 | 3.9 | 59 | 12.0 | 15.0 | 93 | 27.1 | 33.9 |
| 26 | 2.7 | 3.3 | 60 | 12.4 | 15.6 | 94 | 27.6 | 34.4 |
| 27 | 2.2 | 2.8 | 61 | 12.9 | 16.1 | 95 | 28.0 | 35.0 |
| 28 | 1.8 | 2.2 | 62 | 13.3 | 16.7 | 96 | 28.4 | 35.6 |
| 29 | 1.3 | 1.7 | 63 | 13.8 | 17.2 | 97 | 28.9 | 36.1 |
| 30 | 0.9 | 1.1 | 64 | 14.2 | 17.8 | 98 | 29.3 | 36.7 |
| 31 | − 0.4 | − 0.6 | 65 | 14.7 | 18.3 | 99 | 29.8 | 37.2 |
| 32 | 0.0 | 0.0 | 66 | +15.1 | +18.9 | 100 | +30.2 | +37.8 |

$$x° \text{ Reaumur} = \left(32° + \tfrac{9}{4}x°\right) \text{ Fah.} = \tfrac{5}{4}x° \text{ Centes.}$$
$$x° \text{ Centes.} = \left(32° + \tfrac{9}{5}x°\right) \text{ Fah.} = \tfrac{4}{5}x° \text{ Reaum.}$$
$$x° \text{ Fah.} = \tfrac{4}{9}\left(x° - 32°\right) \text{ Reau.} = \tfrac{5}{9}\left(x° - 32°\right) \text{ Cen.}$$

# TABLES AND FORMULÆ.

## PART III.

# ASTRONOMY.

# ASTRONOMY.

### XLVII.—Of Sidereal and Solar Time.

*True* or *apparent solar time* is that deduced from observations of the sun, or is the same as that shown by a well-adjusted sun-dial.

*Mean solar time* is derived from the time employed by the earth in revolving on its axis, as compared with the sun, supposed to move at a mean rate in its orbit, and to make 365.242218 revolutions in a mean Gregorian year.

It cannot be immediately obtained from observation, but is always deduced from apparent time by the aid of the equation of time, which is the angular distance, in time, between the mean and true sun; or, mean solar time = apparent solar time ± equation of time.

*Sidereal time* is the portion of a sidereal day which has elapsed since the transit of the first point of Aries.

Its point of origin cannot be determined by observation, but it is known at any moment by the right ascension of whatever object may be then in the meridian; or,

Sidereal time of a star's culmination = AR. of $*$ ;

Sidereal time at mean noon = AR. mean $\odot$ at mean noon; and, generally,

Sidereal time = sidereal time at mean noon ± solar time from mean noon, (expressed in sidereal intervals;)

Solar time = sidereal time — sidereal time at mean noon, (the difference being reduced to a solar interval.)

## XLVII.—*Of Sidereal and Solar Time*—Continued.

### EXAMPLE.

To find the mean solar time of the passage of Altair over the meridian of Washington, on the 10th July, 1849:

|  | h. | m. | s. |
|---|---|---|---|
| AR. Altair, July 10, 1849.................... = | 19 | 43 | 27.39 |
| Sidereal time at mean noon at Washington ..... = | 7 | 14 | 00.96 |
| Sidereal interval past Washington mean noon... = | 13 | 29 | 26.43 |
| Retardation of mean on sidereal time.......... = − | | 02 | 02.77 |

Corresponding mean time interval past mean
noon or mean time of culmination ........ = 12 27 23.66

The nautical almanacs give the sidereal time at mean noon for each day of the year for a certain meridian.

If the sidereal day be taken equal to 24 sidereal hours, the mean solar day will be equal to $24^h$ $3^m$ $56^s.55$ of those sidereal hours; or the daily acceleration of sidereal on mean solar time (which is the mean motion of the earth in a mean solar day) is $3^m$ $56^s.5554$ of sidereal time; hence the sidereal time at mean noon under any meridian other than that of the nautical almanac used will be found by allowing the proportion of this quantity due to the difference of longitude of the two places.

If the mean solar day be taken equal to 24 mean solar hours, the sidereal day will be equal to $23^h$ $56^m$ $4^s.09$ of those solar hours, or the daily retardation of mean solar on sidereal time is $3^m$ $55^s.9093$ of solar time.

The astronomical day begins at noon. In the civil or common method of reckoning, the day is supposed to commence at the *preceding* midnight. The civil reckoning is therefore 12 hours in advance of the astronomical reckoning, and in the above example, July 10th, $12^h$ $27^m$ $23^s.66$ astronomical time, corresponds to July 11th, $0^h$ $27^m$ $23^s.66$ a. m. civil time.

XLVIII.—*To find the Time by an Altitude of the Sun or a Star.*

$$\text{Sidereal time} = \text{AR.} \ast \pm \ast\text{'s hour-angle.}$$
$$\text{Solar time} = 24^h \pm \odot\text{'s hour-angle.}$$
$$2\,m = L + \triangle + A$$

$$\sin^2 \tfrac{1}{2}\,p = \frac{\cos m \cdot \sin(m - A)}{\cos L \cdot \sin \triangle}$$

where—

L = the latitude of the place of observation;

△ = the north polar distance of the sun or the star;

A = the corrected altitude of the sun or star

= observed altitude — (refraction — parallax) ± semi-diameter; and

$p$ = the hour-angle of the sun or star.

The formula gives the arc in degrees, which must be converted into time, as in one of the following four cases:

1. When we have the *corrected* altitude of the sun's *center*, the hour-angle, $p$, in time, is the *apparent* time when the sun is in the west, or the complement of 24 hours when in the east. To reduce it to *mean* time apply the equation of time.

2. But should the *sidereal* time be required, transform the mean time thus obtained to sidereal time, as previously explained.

3. When the altitude is that of a star, the sidereal time is at once deduced from the hour-angle, $p$.

4. And if, in this last instance, solar time should be required, convert this sidereal time into solar time by means of the equation—

$$\text{Solar time} = \text{AR.} \ast - \text{AR.} \odot \pm p$$

in which the sign + is used if the star is observed in the west, and the sign — if in the east; or,

Mean solar time = the mean solar equivalent of (sidereal time of observation — sidereal time of preceding mean noon at place.)

### XLIX.—*Sun-Dial.*

The most common dial is that in which the plane of the dial is horizontal, and the *style*, placed in the meridian, is inclined to the plane of the dial at an angle equal to the latitude of the place.

Hour-lines are drawn from the center, or point where the style intersects the plane, to the exterior limit of the surface of the dial. Their positions are calculated from the formula—

$$\tan x = \tan p \, \sin L$$

in which—

$x =$ hour-angle on the horizontal plane ;

$p = 15°, 30°, 45°$, etc., the hour-angle on the equatorial plane ; and

$L =$ latitude of the place.

The geometrical determination of these lines will be readily seen from the following figure.

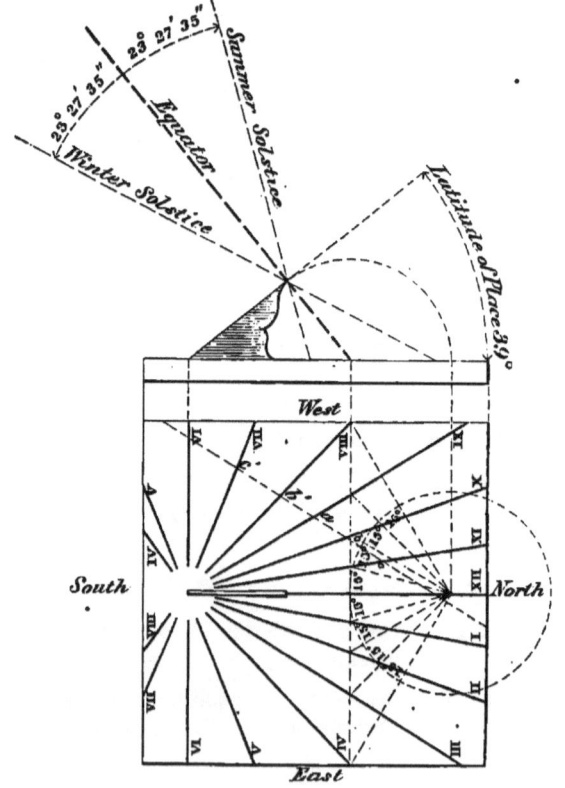

## XLIX.—*Sun-Dial*—Continued.

As the lines of IV, V, VIII, and VII cannot, generally, be directly drawn, owing to want of space upon the surface of the dial, draw, from any point of the line of IX hours, a line parallel to that of III hours, and take $ab' = ab$, and $ac' = ac$; $b'$ and $c'$ will be points in the lines VIII and VII. The lines IV and V will make the same angles on the opposite side. The line of VI is perpendicular to that of XII.

### *To determine the Meridian Line.*

Take a point in the plane of the dial through which it is intended the meridian plane shall pass. With this point as a center describe several concentric circles. Fix a straight pin in the center, perpendicular to the plane of the dial, of such a length that the extremity of the shadow cast by it shall fall within the circles at XII M. Mark the points where the extremity of the shadow passes over these circles in the forenoon and again the same in the afternoon. The line drawn from the middle of these arcs, contained between the points of passage, to the center of the circles will be the meridian.

### SUN-DIAL CORRECTION.

#### *Mean Time at Apparent Noon.*

| Day. | January. | | February. | | March. | | April. | | May. | | June. | |
|---|---|---|---|---|---|---|---|---|---|---|---|---|
| | *h.* | *m.* | *h.* | *m.* | *h.* | *m.* | *h.* | *m.* | *h.* | *m.* | *h.* | *m.* |
| 1 | 12 | 4 | 12 | 14 | 12 | 12 | 12 | 4 | 11 | 57 | 11 | 58 |
| 8 | 12 | 7 | 12 | 14 | 12 | 11 | 12 | 2 | 11 | 56 | 11 | 59 |
| 16 | 12 | 10 | 12 | 14 | 12 | 9 | 12 | 0 | 11 | 56 | 12 | 0 |
| 24 | 12 | 12 | 12 | 13 | 12 | 6 | 11 | 58 | 11 · | 57 | 12 | 2 |

| Day. | July. | | August. | | September. | | October. | | November. | | December. | |
|---|---|---|---|---|---|---|---|---|---|---|---|---|
| | *h.* | *m.* | *h.* | *m.* | *h.* | *m.* | *h.* | *m.* | *h.* | *m.* | *h.* | *m.* |
| 1 | 12 | 3 | 12 | 6 | 12 | 0 | 11 | 50 | 11 | 44 | 11 | 50 |
| 8 | 12 | 5 | 12 | 5 | 11 | 58 | 11 | 48 | 11 | 44 | 11 | 53 |
| 16 | 12 | .6 | 12 | 4 | 11 | 55 | 11 | 46 | 11 | 45 | 11 | 56 |
| 24 | 12 | 6 | 12 | 2 | 11 | 52 | 11 | 45 | 11 | 47 | 12 | 0 |

*For converting Intervals of* SIDEREAL *into Corresponding Intervals of* MEAN SOLAR *Time.*

| Hours. | | Minutes. | | | | Seconds. | | | |
|---|---|---|---|---|---|---|---|---|---|
| *h.* | *m. s.* | *m.* | *s.* | *m.* | *s.* | *s.* | *s.* | *s.* | *s.* |
| 1 | 0  9.830 | 1 | 0.164 | 31 | 5.079 | 1 | 0.003 | 31 | 0.085 |
| 2 | 0 19.659 | 2 | 0.328 | 32 | 5.242 | 2 | 0.005 | 32 | 0.087 |
| 3 | 0 29.489 | 3 | 0.491 | 33 | 5.406 | 3 | 0.008 | 33 | 0.090 |
| 4 | 0 39.318 | 4 | 0.655 | 34 | 5.570 | 4 | 0.011 | 34 | 0.093 |
| 5 | 0 49.148 | 5 | 0.819 | 35 | 5.734 | 5 | 0.014 | 35 | 0.096 |
| 6 | 0 58.977 | 6 | 0.983 | 36 | 5.898 | 6 | 0.016 | 36 | 0.098 |
| 7 | 1  8.807 | 7 | 1.147 | 37 | 6.062 | 7 | 0.019 | 37 | 0.101 |
| 8 | 1 18.636 | 8 | 1.311 | 38 | 6.225 | 8 | 0.022 | 38 | 0.104 |
| 9 | 1 28.466 | 9 | 1.474 | 39 | 6.389 | 9 | 0.025 | 39 | 0.106 |
| 10 | 1 38.296 | 10 | 1.638 | 40 | 6.553 | 10 | 0.027 | 40 | 0.109 |
| 11 | 1 48.125 | 11 | 1.802 | 41 | 6.717 | 11 | 0.030 | 41 | 0.112 |
| 12 | 1 57.955 | 12 | 1.966 | 42 | 6.881 | 12 | 0.033 | 42 | 0.115 |
| 13 | 2  7.784 | 13 | 2.130 | 43 | 7.044 | 13 | 0.036 | 43 | 0.118 |
| 14 | 2 17.614 | 14 | 2.294 | 44 | 7.208 | 14 | 0.038 | 44 | 0.120 |
| 15 | 2 27.443 | 15 | 2.457 | 45 | 7.372 | 15 | 0.041 | 45 | 0.123 |
| 16 | 2 37.273 | 16 | 2.621 | 46 | 7.536 | 16 | 0.044 | 46 | 0.126 |
| 17 | 2 47.103 | 17 | 2.785 | 47 | 7.700 | 17 | 0.047 | 47 | 0.128 |
| 18 | 2 56.932 | 18 | 2.949 | 48 | 7.864 | 18 | 0.049 | 48 | 0.131 |
| 19 | 3  6.762 | 19 | 3.113 | 49 | 8.027 | 19 | 0.052 | 49 | 0.134 |
| 20 | 3 16.591 | 20 | 3.277 | 50 | 8.191 | 20 | 0.055 | 50 | 0.137 |
| 21 | 3 26.421 | 21 | 3.440 | 51 | 8.355 | 21 | 0.057 | 51 | 0.140 |
| 22 | 3 36.250 | 22 | 3.604 | 52 | 8.519 | 22 | 0.060 | 52 | 0.142 |
| 23 | 3 46.080 | 23 | 3.768 | 53 | 8.683 | 23 | 0.063 | 53 | 0.145 |
| 24 | 3 55.909 | 24 | 3.932 | 54 | 8.847 | 24 | 0.066 | 54 | 0.148 |
| | | 25 | 4.096 | 55 | 9.010 | 25 | 0.068 | 55 | 0.150 |
| | | 26 | 4.259 | 56 | 9.174 | 26 | 0.071 | 56 | 0.153 |
| | | 27 | 4.423 | 57 | 9.338 | 27 | 0.074 | 57 | 0.156 |
| | | 28 | 4.587 | 58 | 9.502 | 28 | 0.076 | 58 | 0.159 |
| | | 29 | 4.751 | 59 | 9.666 | 29 | 0.079 | 59 | 0.161 |
| | | 30 | 4.915 | 60 | 9.830 | 30 | 0.082 | 60 | 0.164 |

The quantities taken from this table must be *subtracted* from a sidereal interval to obtain the corresponding interval in mean solar time.

*For converting Intervals of* MEAN SOLAR *Time into Corresponding Intervals of* SIDEREAL *Time.*

| Hours. | | Minutes. | | | | Seconds. | | | |
|---|---|---|---|---|---|---|---|---|---|
| h. | m. s. | m. | s. | m. | s. | s. | s. | s. | s. |
| 1 | 0 9.856 | 1 | 0.164 | 31 | 5.092 | 1 | 0.003 | 31 | 0.085 |
| 2 | 0 19.713 | 2 | 0.329 | 32 | 5.257 | 2 | 0.005 | 32 | 0.088 |
| 3 | 0 29.569 | 3 | 0.493 | 33 | 5.421 | 3 | 0.008 | 33 | 0.090 |
| 4 | 0 39.426 | 4 | 0.657 | 34 | 5.585 | 4 | 0.011 | 34 | 0.093 |
| 5 | 0 49.282 | 5 | 0.821 | 35 | 5.750 | 5 | 0.014 | 35 | 0.096 |
| 6 | 0 59.139 | 6 | 0.986 | 36 | 5.914 | 6 | 0.016 | 36 | 0.098 |
| 7 | 1 8.995 | 7 | 1.150 | 37 | 6.078 | 7 | 0.019 | 37 | 0.101 |
| 8 | 1 18.852 | 8 | 1.314 | 38 | 6.242 | 8 | 0.022 | 38 | 0.104 |
| 9 | 1 28.708 | 9 | 1.478 | 39 | 6.407 | 9 | 0.025 | 39 | 0.106 |
| 10 | 1 38.565 | 10 | 1.643 | 40 | 6.571 | 10 | 0.027 | 40 | 0.109 |
| 11 | 1 48.421 | 11 | 1.807 | 41 | 6.735 | 11 | 0.030 | 41 | 0.112 |
| 12 | 1 58.278 | 12 | 1.971 | 42 | 6.900 | 12 | 0.033 | 42 | 0.115 |
| 13 | 2 8.134 | 13 | 2.136 | 43 | 7.064 | 13 | 0.036 | 43 | 0.118 |
| 14 | 2 17.991 | 14 | 2.300 | 44 | 7.228 | 14 | 0.038 | 44 | 0.120 |
| 15 | 2 27.847 | 15 | 2.464 | 45 | 7.392 | 15 | 0.041 | 45 | 0.123 |
| 16 | 2 37.704 | 16 | 2.628 | 46 | 7.557 | 16 | 0.044 | 46 | 0.126 |
| 17 | 2 47.560 | 17 | 2.793 | 47 | 7.721 | 17 | 0.047 | 47 | 0.129 |
| 18 | 2 57.416 | 18 | 2.957 | 48 | 7.885 | 18 | 0.049 | 48 | 0.131 |
| 19 | 3 7.273 | 19 | 3.121 | 49 | 8.050 | 19 | 0.052 | 49 | 0.134 |
| 20 | 3 17.129 | 20 | 3.285 | 50 | 8.214 | 20 | 0.055 | 50 | 0.137 |
| 21 | 3 26.986 | 21 | 3.450 | 51 | 8.378 | 21 | 0.057 | 51 | 0.140 |
| 22 | 3 36.842 | 22 | 3.614 | 52 | 8.542 | 22 | 0.060 | 52 | 0.142 |
| 23 | 3 46.699 | 23 | 3.778 | 53 | 8.707 | 23 | 0.063 | 53 | 0.145 |
| 24 | 3 56.555 | 24 | 3.943 | 54 | 8.871 | 24 | 0.066 | 54 | 0.148 |
| | | 25 | 4.107 | 55 | 9.035 | 25 | 0.068 | 55 | 0.151 |
| | | 26 | 4.271 | 56 | 9.199 | 26 | 0.071 | 56 | 0.153 |
| | | 27 | 4.436 | 57 | 9.364 | 27 | 0.074 | 57 | 0.156 |
| | | 28 | 4.600 | 58 | 9.528 | 28 | 0.077 | 58 | 0.159 |
| | | 29 | 4.764 | 59 | 9.692 | 29 | 0.079 | 59 | 0.161 |
| | | 30 | 4.928 | 60 | 9.856 | 30 | 0.082 | 60 | 0.164 |

The quantities taken from this table must be *added* to a mean interval to obtain the corresponding interval in sidereal time.

*To convert Parts of the Equator in Arc into Sidereal Time, or to convert Terrestrial Longitude in Arc into Time.*

DEGREES.

| Arc. | Time. | Arc. | Time. | Arc. | Time. | Arc. | Time. |
|---|---|---|---|---|---|---|---|
| ° | *h. m.* | ° | *h. m.* | ° | *h. m.* | ° | *h. m.* |
| 1 | 0  4 | 31 | 2  4 | 61 | 4  4 | 91 | 6  4 |
| 2 | 0  8 | 32 | 2  8 | 62 | 4  8 | 92 | 6  8 |
| 3 | 0  12 | 33 | 2  12 | 63 | 4  12 | 93 | 6  12 |
| 4 | 0  16 | 34 | 2  16 | 64 | 4  16 | 94 | 6  16 |
| 5 | 0  20 | 35 | 2  20 | 65 | 4  20 | 95 | 6  20 |
| 6 | 0  24 | 36 | 2  24 | 66 | 4  24 | 96 | 6  24 |
| 7 | 0  28 | 37 | 2  28 | 67 | 4  28 | 97 | 6  28 |
| 8 | 0  32 | 38 | 2  32 | 68 | 4  32 | 98 | 6  32 |
| 9 | 0  36 | 39 | 2  36 | 69 | 4  36 | 99 | 6  36 |
| 10 | 0  40 | 40 | 2  40 | 70 | 4  40 | 100 | 6  40 |
| 11 | 0  44 | 41 | 2  44 | 71 | 4  44 | 101 | 6  44 |
| 12 | 0  48 | 42 | 2  48 | 72 | 4  48 | 102 | 6  48 |
| 13 | 0  52 | 43 | 2  52 | 73 | 4  52 | 103 | 6  52 |
| 14 | 0  56 | 44 | 2  56 | 74 | 4  56 | 104 | 6  56 |
| 15 | 1  0 | 45 | 3  0 | 75 | 5  0 | 105 | 7  0 |
| 16 | 1  4 | 46 | 3  4 | 76 | 5  4 | 106 | 7  4 |
| 17 | 1  8 | 47 | 3  8 | 77 | 5  8 | 107 | 7  8 |
| 18 | 1  12 | 48 | 3  12 | 78 | 5  12 | 108 | 7  12 |
| 19 | 1  16 | 49 | 3  16 | 79 | 5  16 | 109 | 7  16 |
| 20 | 1  20 | 50 | 3  20 | 80 | 5  20 | 110 | 7  20 |
| 21 | 1  24 | 51 | 3  24 | 81 | 5  24 | 111 | 7  24 |
| 22 | 1  28 | 52 | 3  28 | 82 | 5  28 | 112 | 7  28 |
| 23 | 1  32 | 53 | 3  32 | 83 | 5  32 | 113 | 7  32 |
| 24 | 1  36 | 54 | 3  36 | 84 | 5  36 | 114 | 7  36 |
| 25 | 1  40 | 55 | 3  40 | 85 | 5  40 | 115 | 7  40 |
| 26 | 1  44 | 56 | 3  44 | 86 | 5  44 | 116 | 7  44 |
| 27 | 1  48 | 57 | 3  48 | 87 | 5  48 | 117 | 7  48 |
| 28 | 1  52 | 58 | 3  52 | 88 | 5  52 | 118 | 7  52 |
| 29 | 1  56 | 59 | 3  56 | 89 | 5  56 | 119 | 7  56 |
| 30 | 2  0 | 60 | 4  0 | 90 | 6  0 | 120 | 8  0 |

*To convert Parts of the Equator in Arc into Sidereal Time, or to convert Terrestrial Longitude in Arc into Time—*Continued.

DEGREES.

| Arc. | Time. | | Arc. | Time. | | Arc. | Time. | | Arc. | Time. | |
|---|---|---|---|---|---|---|---|---|---|---|---|
| ° | h. | m. | ° | h. | m. | ° | h. | m. | ° | h. | m. |
| 121 | 8 | 4 | 151 | 10 | 4 | 181 | 12 | 4 | 211 | 14 | 4 |
| 122 | 8 | 8 | 152 | 10 | 8 | 182 | 12 | 8 | 212 | 14 | 8 |
| 123 | 8 | 12 | 153 | 10 | 12 | 183 | 12 | 12 | 213 | 14 | 12 |
| 124 | 8 | 16 | 154 | 10 | 16 | 184 | 12 | 16 | 214 | 14 | 16 |
| 125 | 8 | 20 | 155 | 10 | 20 | 185 | 12 | 20 | 215 | 14 | 20 |
| 126 | 8 | 24 | 156 | 10 | 24 | 186 | 12 | 24 | 216 | 14 | 24 |
| 127 | 8 | 28 | 157 | 10 | 28 | 187 | 12 | 28 | 217 | 14 | 28 |
| 128 | 8 | 32 | 158 | 10 | 32 | 188 | 12 | 32 | 218 | 14 | 32 |
| 129 | 8 | 36 | 159 | 10 | 36 | 189 | 12 | 36 | 219 | 14 | 36 |
| 130 | 8 | 40 | 160 | 10 | 40 | 190 | 12 | 40 | 220 | 14 | 40 |
| 131 | 8 | 44 | 161 | 10 | 44 | 191 | 12 | 44 | 221 | 14 | 44 |
| 132 | 8 | 48 | 162 | 10 | 48 | 192 | 12 | 48 | 222 | 14 | 48 |
| 133 | 8 | 52 | 163 | 10 | 52 | 193 | 12 | 52 | 223 | 14 | 52 |
| 134 | 8 | 56 | 164 | 10 | 56 | 194 | 12 | 56 | 224 | 14 | 56 |
| 135 | 9 | 0 | 165 | 11 | 0 | 195 | 13 | 0 | 225 | 15 | 0 |
| 136 | 9 | 4 | 166 | 11 | 4 | 196 | 13 | 4 | 226 | 15 | 4 |
| 137 | 9 | 8 | 167 | 11 | 8 | 197 | 13 | 8 | 227 | 15 | 8 |
| 138 | 9 | 12 | 168 | 11 | 12 | 198 | 13 | 12 | 228 | 15 | 12 |
| 139 | 9 | 16 | 169 | 11 | 16 | 199 | 13 | 16 | 229 | 15 | 16 |
| 140 | 9 | 20 | 170 | 11 | 20 | 200 | 13 | 20 | 230 | 15 | 20 |
| 141 | 9 | 24 | 171 | 11 | 24 | 201 | 13 | 24 | 231 | 15 | 24 |
| 142 | 9 | 28 | 172 | 11 | 28 | 202 | 13 | 28 | 232 | 15 | 28 |
| 143 | 9 | 32 | 173 | 11 | 32 | 203 | 13 | 32 | 233 | 15 | 32 |
| 144 | 9 | 36 | 174 | 11 | 36 | 204 | 13 | 36 | 234 | 15 | 36 |
| 145 | 9 | 40 | 175 | 11 | 40 | 205 | 13 | 40 | 235 | 15 | 40 |
| 146 | 9 | 44 | 176 | 11 | 44 | 206 | 13 | 44 | 236 | 15 | 44 |
| 147 | 9 | 48 | 177 | 11 | 48 | 207 | 13 | 48 | 237 | 15 | 48 |
| 148 | 9 | 52 | 178 | 11 | 52 | 208 | 13 | 52 | 238 | 15 | 52 |
| 149 | 9 | 56 | 179 | 11 | 56 | 209 | 13 | 56 | 239 | 15 | 56 |
| 150 | 10 | 0 | 180 | 12 | 0 | 210 | 14 | 0 | 240 | 16 | 0 |

*To convert Parts of the Equator in Arc into Sidereal Time, or to convert Terrestrial Longitude in Arc into Time—Continued.*

DEGREES.

| Arc. | Time. | | Arc. | Time. | | Arc. | Time. | | Arc. | Time. | |
|---|---|---|---|---|---|---|---|---|---|---|---|
| ° | h. | m. | ° | h. | m. | ° | h. | m. | ° | h. | m. |
| 241 | 16 | 4 | 271 | 18 | 4 | 301 | 20 | 4 | 331 | 22 | 4 |
| 242 | 16 | 8 | 272 | 18 | 8 | 302 | 20 | 8 | 332 | 22 | 8 |
| 243 | 16 | 12 | 273 | 18 | 12 | 303 | 20 | 12 | 333 | 22 | 12 |
| 244 | 16 | 16 | 274 | 18 | 16 | 304 | 20 | 16 | 334 | 22 | 16 |
| 245 | 16 | 20 | 275 | 18 | 20 | 305 | 20 | 20 | 335 | 22 | 20 |
| 246 | 16 | 24 | 276 | 18 | 24 | 306 | 20 | 24 | 336 | 22 | 24 |
| 247 | 16 | 28 | 277 | 18 | 28 | 307 | 20 | 28 | 337 | 22 | 28 |
| 248 | 16 | 32 | 278 | 18 | 32 | 308 | 20 | 32 | 338 | 22 | 32 |
| 249 | 16 | 36 | 279 | 18 | 36 | 309 | 20 | 36 | 339 | 22 | 36 |
| 250 | 16 | 40 | 280 | 18 | 40 | 310 | 20 | 40 | 340 | 22 | 40 |
| 251 | 16 | 44 | 281 | 18 | 44 | 311 | 20 | 44 | 341 | 22 | 44 |
| 252 | 16 | 48 | 282 | 18 | 48 | 312 | 20 | 48 | 342 | 22 | 48 |
| 253 | 16 | 52 | 283 | 18 | 52 | 313 | 20 | 52 | 343 | 22 | 52 |
| 254 | 16 | 56 | 284 | 18 | 56 | 314 | 20 | 56 | 344 | 22 | 56 |
| 255 | 17 | 0 | 285 | 19 | 0 | 315 | 21 | 0 | 345 | 23 | 0 |
| 256 | 17 | 4 | 286 | 19 | 4 | 316 | 21 | 4 | 346 | 23 | 4 |
| 257 | 17 | 8 | 287 | 19 | 8 | 317 | 21 | 8 | 347 | 23 | 8 |
| 258 | 17 | 12 | 288 | 19 | 12 | 318 | 21 | 12 | 348 | 23 | 12 |
| 259 | 17 | 16 | 289 | 19 | 16 | 319 | 21 | 16 | 349 | 23 | 16 |
| 260 | 17 | 20 | 290 | 19 | 20 | 320 | 21 | 20 | 350 | 23 | 20 |
| 261 | 17 | 24 | 291 | 19 | 24 | 321 | 21 | 24 | 351 | 23 | 24 |
| 262 | 17 | 28 | 292 | 19 | 28 | 322 | 21 | 28 | 352 | 23 | 28 |
| 263 | 17 | 32 | 293 | 19 | 32 | 323 | 21 | 32 | 353 | 23 | 32 |
| 264 | 17 | 36 | 294 | 19 | 36 | 324 | 21 | 36 | 354 | 23 | 36 |
| 265 | 17 | 40 | 295 | 19 | 40 | 325 | 21 | 40 | 355 | 23 | 40 |
| 266 | 17 | 44 | 296 | 19 | 44 | 326 | 21 | 44 | 356 | 23 | 44 |
| 267 | 17 | 48 | 297 | 19 | 48 | 327 | 21 | 48 | 357 | 23 | 48 |
| 268 | 17 | 52 | 298 | 19 | 52 | 328 | 21 | 52 | 358 | 23 | 52 |
| 269 | 17 | 56 | 299 | 19 | 56 | 329 | 21 | 56 | 359 | 23 | 56 |
| 270 | 18 | 0 | 300 | 20 | 0 | 330 | 22 | 0 | 360 | 24 | 0 |

*To convert Parts of the Equator in Arc into Sidereal Time, or to convert Terrestrial Longitude in Arc into Time*—Continued.

| | MINUTES. | | | | SECONDS. | | |
|---|---|---|---|---|---|---|---|
| Arc. | Time. | Arc. | Time. | Arc. | Time. | Arc. | Time. |
| ′ | m. s. | ′ | m. s. | ″ | s. | ″ | s. |
| 1 | 0 4 | 31 | 2 4 | 1 | 0.067 | 31 | 2.067 |
| 2 | 0 8 | 32 | 2 8 | 2 | 0.133 | 32 | 2.133 |
| 3 | 0 12 | 33 | 2 12 | 3 | 0.200 | 33 | 2.200 |
| 4 | 0 16 | 34 | 2 16 | 4 | 0.267 | 34 | 2.267 |
| 5 | 0 20 | 35 | 2 20 | 5 | 0.333 | 35 | 2.333 |
| 6 | 0 24 | 36 | 2 24 | 6 | 0.400 | 36 | 2.400 |
| 7 | 0 28 | 37 | 2 28 | 7 | 0.467 | 37 | 2.467 |
| 8 | 0 32 | 38 | 2 32 | 8 | 0.533 | 38 | 2.533 |
| 9 | 0 36 | 39 | 2 36 | 9 | 0.600 | 39 | 2.600 |
| 10 | 0 40 | 40 | 2 40 | 10 | 0.667 | 40 | 2.667 |
| 11 | 0 44 | 41 | 2 44 | 11 | 0.733 | 41 | 2.733 |
| 12 | 0 48 | 42 | 2 48 | 12 | 0.800 | 42 | 2.800 |
| 13 | 0 52 | 43 | 2 52 | 13 | 0.867 | 43 | 2.867 |
| 14 | 0 56 | 44 | 2 56 | 14 | 0.933 | 44 | 2.933 |
| 15 | 1 0 | 45 | 3 0 | 15 | 1.000 | 45 | 3.000 |
| 16 | 1 4 | 46 | 3 4 | 16 | 1.067 | 46 | 3.067 |
| 17 | 1 8 | 47 | 3 8 | 17 | 1.133 | 47 | 3.133 |
| 18 | 1 12 | 48 | 3 12 | 18 | 1.200 | 48 | 3.200 |
| 19 | 1 16 | 49 | 3 16 | 19 | 1.267 | 49 | 3.267 |
| 20 | 1 20 | 50 | 3 20 | 20 | 1.333 | 50 | 3.333 |
| 21 | 1 24 | 51 | 3 24 | 21 | 1.400 | 51 | 3.400 |
| 22 | 1 28 | 52 | 3 28 | 22 | 1.467 | 52 | 3.467 |
| 23 | 1 32 | 53 | 3 32 | 23 | 1.533 | 53 | 3.533 |
| 24 | 1 36 | 54 | 3 36 | 24 | 1.600 | 54 | 3.600 |
| 25 | 1 40 | 55 | 3 40 | 25 | 1.667 | 55 | 3.667 |
| 26 | 1 44 | 56 | 3 44 | 26 | 1.733 | 56 | 3.733 |
| 27 | 1 48 | 57 | 3 48 | 27 | 1.800 | 57 | 3.800 |
| 28 | 1 52 | 58 | 3 52 | 28 | 1.867 | 58 | 3.867 |
| 29 | 1 56 | 59 | 3 56 | 29 | 1.933 | 59 | 3.933 |
| 30 | 2 0 | 60 | 4 0 | 30 | 2.000 | 60 | 4.000 |

*To convert Sidereal Time into Parts of the Equator in Arc, or to convert Time into Terrestrial Longitude in Arc.*

| HOURS. | | MINUTES. | | | | SECONDS. | | | |
|---|---|---|---|---|---|---|---|---|---|
| Time. | Arc. | Time. | Arc. | Time. | Arc. | Time. | Arc. | Time. | Arc. |
| *h.* | ° | *m.* | ° ′ | *m.* | ° ′ | *s.* | ′ ″ | *s.* | ′ ″ |
| 1 | 15 | 1 | 0 15 | 31 | 7 45 | 1 | 0 15 | 31 | 7 45 |
| 2 | 30 | 2 | 0 30 | 32 | 8 0 | 2 | 0 30 | 32 | 8 0 |
| 3 | 45 | 3 | 0 45 | 33 | 8 15 | 3 | 0 45 | 33 | 8 15 |
| 4 | 60 | 4 | 1 0 | 34 | 8 30 | 4 | 1 0 | 34 | 8 30 |
| 5 | 75 | 5 | 1 15 | 35 | 8 45 | 5 | 1 15 | 35 | 8 45 |
| 6 | 90 | 6 | 1 30 | 36 | 9 0 | 6 | 1 30 | 36 | 9 0 |
| 7 | 105 | 7 | 1 45 | 37 | 9 15 | 7 | 1 45 | 37 | 9 15 |
| 8 | 120 | 8 | 2 0 | 38 | 9 30 | 8 | 2 0 | 38 | 9 30 |
| 9 | 135 | 9 | 2 15 | 39 | 9 45 | 9 | 2 15 | 39 | 9 45 |
| 10 | 150 | 10 | 2 30 | 40 | 10 0 | 10 | 2 30 | 40 | 10 0 |
| 11 | 165 | 11 | 2 45 | 41 | 10 15 | 11 | 2 45 | 41 | 10 15 |
| 12 | 180 | 12 | 3 0 | 42 | 10 30 | 12 | 3 0 | 42 | 10 30 |
| 13 | 195 | 13 | 3 15 | 43 | 10 45 | 13 | 3 15 | 43 | 10 45 |
| 14 | 210 | 14 | 3 30 | 44 | 11 0 | 14 | 3 30 | 44 | 11 0 |
| 15 | 225 | 15 | 3 45 | 45 | 11 15 | 15 | 3 45 | 45 | 11 15 |
| 16 | 240 | 16 | 4 0 | 46 | 11 30 | 16 | 4 0 | 46 | 11 30 |
| 17 | 255 | 17 | 4 15 | 47 | 11 45 | 17 | 4 15 | 47 | 11 45 |
| 18 | 270 | 18 | 4 30 | 48 | 12 0 | 18 | 4 30 | 48 | 12 0 |
| 19 | 285 | 19 | 4 45 | 49 | 12 15 | 19 | 4 45 | 49 | 12 15 |
| 20 | 300 | 20 | 5 0 | 50 | 12 30 | 20 | 5 0 | 50 | 12 30 |
| 21 | 315 | 21 | 5 15 | 51 | 12 45 | 21 | 5 15 | 51 | 12 45 |
| 22 | 330 | 22 | 5 30 | 52 | 13 0 | 22 | 5 30 | 52 | 13 0 |
| 23 | 345 | 23 | 5 45 | 53 | 13 15 | 23 | 5 45 | 53 | 13 15 |
| 24 | 360 | 24 | 6 0 | 54 | 13 30 | 24 | 6 0 | 54 | 13 30 |
| | | 25 | 6 15 | 55 | 13 45 | 25 | 6 15 | 55 | 13 45 |
| | | 26 | 6 30 | 56 | 14 0 | 26 | 6 30 | 56 | 14 0 |
| | | 27 | 6 45 | 57 | 14 15 | 27 | 6 45 | 57 | 14 15 |
| | | 28 | 7 0 | 58 | 14 30 | 28 | 7 0 | 58 | 14 30 |
| | | 29 | 7 15 | 59 | 14 45 | 29 | 7 15 | 59 | 14 45 |
| | | 30 | 7 30 | 60 | 15 0 | 30 | 7 30 | 60 | 15 0 |

TENTHS OF SECONDS.

| Time. | Arc. | Time. | Arc. | Time. | Arc. | Time. | Arc. |
|---|---|---|---|---|---|---|---|
| s. | '' | s. | '' | s. | '' | s. | '' |
| 0.01 | 0.15 | 0.31 | 4.65 | 0.61 | 9.15 | 0.91 | 13.65 |
| 0.02 | 0.30 | 0.32 | 4.80 | 0.62 | 9.30 | 0.92 | 13.80 |
| 0.03 | 0.45 | 0.33 | 4.95 | 0.63 | 9.45 | 0.93 | 13.95 |
| 0.04 | 0.60 | 0.34 | 5.10 | 0.64 | 9.60 | 0.94 | 14.10 |
| 0.05 | .0.75 | 0.35 | 5.25 | 0.65 | 9.75 | 0.95 | 14.25 |
| 0.06 | 0.90 | 0.36 | 5.40 | 0.66 | 9.90 | 0.96 | 14.40 |
| 0.07 | 1.05 | 0.37 | 5.55 | 0.67 | 10.05 | 0.97 | 14.55 |
| 0.08 | 1.20 | 0.38 | 5.70 | 0.68 | 10.20 | 0.98 | 14.70 |
| 0.09 | 1.35 | 0.39 | 5.85 | 0.69 | 10.35 | 0.99 | 14.85 |
| 0.10 | 1.50 | 0.40 | 6.00 | 0.70 | 10.50 | 1.00 | 15.00 |
| 0.11 | 1.65 | 0.41 | 6.15 | 0.71 | 10.65 | | |
| 0.12 | 1.80 | 0.42 | 6.30 | 0.72 | 10.80 | | |
| 0.13 | 1.95 | 0.43 | 6.45 | 0.73 | 10.95 | | |
| 0.14 | 2.10 | 0.44 | 6.60 | 0.74 | 11.10 | | |
| 0.15 | 2.25 | 0.45 | 6.75 | 0.75 | 11.25 | | |
| 0.16 | 2.40 | 0.46 | 6.90 | 0.76 | 11.40 | | |
| 0.17 | 2.55 | 0.47 | 7.05 | 0.77 | 11.55 | | |
| 0.18 | 2.70 | 0.48 | 7.20 | 0.78 | 11.70 | | |
| 0.19 | 2.85 | 0.49 | 7.35 | 0.79 | 11.85 | | |
| 0.20 | 3.00 | 0.50 | 7.50 | 0.80 | 12.00 | | |

| Time. | Arc. | Time. | Arc. | Time. | Arc. | Thousandths of seconds of time. Arc. | |
|---|---|---|---|---|---|---|---|
| | | | | | | s. | '' |
| 0.21 | 3.15 | 0.51 | 7.65 | 0.81 | 12.15 | 0.001 | 0.015 |
| 0.22 | 3.30 | 0.52 | 7.80 | 0.82 | 12.30 | 0.002 | 0.030 |
| 0.23 | 3.45 | 0.53 | 7.95 | 0.83 | 12.45 | 0.003 | 0.045 |
| 0.24 | 3.60 | 0.54 | 8.10 | 0.84 | 12.60 | 0.004 | 0.060 |
| 0.25 | 3.75 | 0.55 | 8.25 | 0.85 | 12.75 | 0.005 | 0.075 |
| 0.26 | 3.90 | 0.56 | 8.40 | 0.86 | 12.90 | 0.006 | 0.090 |
| 0.27 | 4.05 | 0.57 | 8.55 | 0.87 | 13.05 | 0.007 | 0.105 |
| 0.28 | 4.20 | 6.58 | 8.70 | 0.88 | 13.20 | 9.008 | 0.120 |
| 0.29 | 4.35 | 0.59 | 8.85 | 0.89 | 13.35 | 0.009 | 0.135 |
| 0.30 | 4.50 | 0.60 | 9.00 | 0.90 | 13.50 | 0.010 | 0.150 |

*To convert Right Ascension in Arc into Mean Time.*

DEGREES.

| AR. in arc. | Mean time. | | | AR. in arc. | Mean time. | | | AR. in arc. | Mean time. | | |
|---|---|---|---|---|---|---|---|---|---|---|---|
| ° | h. | m. | s. | ° | h. | m. | s. | ° | h. | m. | s. |
| 1 | 0 | 3 | 59.345 | 31 | 2 | 3 | 39.686 | 61 | 4 | 3 | 20.027 |
| 2 | 0 | 7 | 58.689 | 32 | 2 | 7 | 39.030 | 62 | 4 | 7 | 19.371 |
| 3 | 0 | 11 | 58.034 | 33 | 2 | 11 | 38.375 | 63 | 4 | 11 | 18.716 |
| 4 | 0 | 15 | 57.379 | 34 | 2 | 15 | 37.720 | 64 | 4 | 15 | 18.061 |
| 5 | 0 | 19 | 56.724 | 35 | 2 | 19 | 37.064 | 65 | 4 | 19 | 17.405 |
| 6 | 0 | 23 | 56.068 | 36 | 2 | 23 | 36.409 | 66 | 4 | 23 | 16.750 |
| 7 | 0 | 27 | 55.413 | 37 | 2 | 27 | 35.754 | 67 | 4 | 27 | 16.095 |
| 8 | 0 | 31 | 54.758 | 38 | 2 | 31 | 35.099 | 68 | 4 | 31 | 15.639 |
| 9 | 0 | 35 | 54.102 | 39 | 2 | 35 | 34.443 | 69 | 4 | 35 | 14.784 |
| 10 | 0 | 39 | 53.447 | 40 | 2 | 39 | 33.788 | 70 | 4 | 39 | 14.129 |
| 11 | 0 | 43 | 52.792 | 41 | 2 | 43 | 33.133 | 71 | 4 | 43 | 13.474 |
| 12 | 0 | 47 | 52.136 | 42 | 2 | 47 | 32.477 | 72 | 4 | 47 | 12.818 |
| 13 | 0 | 51 | 51.481 | 43 | 2 | 51 | 31.822 | 73 | 4 | 51 | 12.163 |
| 14 | 0 | 55 | 50.826 | 44 | 2 | 55 | 31.167 | 74 | 4 | 55 | 11.508 |
| 15 | 0 | 59 | 50.170 | 45 | 2 | 59 | 30.511 | 75 | 4 | 59 | 10.852 |
| 16 | 1 | 3 | 49.515 | 46 | 3 | 3 | 29.856 | 76 | 5 | 3 | 10.197 |
| 17 | 1 | 7 | 48.860 | 47 | 3 | 7 | 29.201 | 77 | 5 | 7 | 9.542 |
| 18 | 1 | 11 | 48.205 | 48 | 3 | 11 | 28.545 | 78 | 5 | 11 | 8.886 |
| 19 | 1 | 15 | 47.549 | 49 | 3 | 15 | 27.890 | 79 | 5 | 15 | 8.231 |
| 20 | 1 | 19 | 46.894 | 50 | 3 | 19 | 27.235 | 80 | 5 | 19 | 7.576 |
| 21 | 1 | 23 | 46.239 | 51 | 3 | 23 | 26.580 | 81 | 5 | 23 | 6.920 |
| 22 | 1 | 27 | 45.583 | 52 | 3 | 27 | 25.924 | 82 | 5 | 27 | 6.265 |
| 23 | 1 | 31 | 44.928 | 53 | 3 | 31 | 25.269 | 83 | 5 | 31 | 5.610 |
| 24 | 1 | 35 | 44.273 | 54 | 3 | 35 | 24.614 | 84 | 5 | 35 | 4.955 |
| 25 | 1 | 39 | 43.617 | 55 | 3 | 39 | 23.958 | 85 | 5 | 39 | 4.299 |
| 26 | 1 | 43 | 42.962 | 56 | 3 | 43 | 23.303 | 86 | 5 | 43 | 3.644 |
| 27 | 1 | 47 | 42.307 | 57 | 3 | 47 | 22.648 | 87 | 5 | 47 | 2.989 |
| 28 | 1 | 51 | 41.652 | 58 | 3 | 51 | 21.992 | 88 | 5 | 51 | 2.333 |
| 29 | 1 | 55 | 40.996 | .59 | 3 | 55 | 21.337 | 89 | 5 | 55 | 1.678 |
| 30 | 1 | 59 | 40.341 | 60 | 3 | 59 | 20.682 | 90 | 5 | 59 | 1.023 |

*To convert Right Ascension in Arc into Mean Time*—Continued.

DEGREES.

| AR. in arc. | Mean time. | | | AR. in arc. | Mean time. | | | AR. in arc. | Mean time. | | |
|---|---|---|---|---|---|---|---|---|---|---|---|
| ° | h. | m. | s. | ° | h. | m. | s. | ° | h. | m. | s. |
| 91 | 6 | 3 | 0.367 | 121 | 8 | 2 | 40.708 | 151 | 10 | 2 | 21.049 |
| 92 | 6 | 6 | 59.712 | 122 | 8 | 6 | 40.053 | 152 | 10 | 6 | 20.394 |
| 93 | 6 | 10 | 59.057 | 123 | 8 | 10 | 39.398 | 153 | 10 | 10 | 19.738 |
| 94 | 6 | 14 | 58.401 | 124 | 8 | 14 | 38.742 | 154 | 10 | 14 | 19.083 |
| 95 | 6 | 18 | 57.746 | 125 | 8 | 18 | 38.087 | 155 | 10 | 18 | 18.428 |
| 96 | 6 | 22 | 57.091 | 126 | 8 | 22 | 37.432 | 156 | 10 | 22 | 17.773 |
| 97 | 6 | 26 | 56.436 | 127 | 8 | 26 | 36.776 | 157 | 10 | 26 | 17.117 |
| 98 | 6 | 30 | 55.780 | 128 | 8 | 30 | 36.121 | 158 | 10 | 30 | 16.462 |
| 99 | 6 | 34 | 55.125 | 129 | 8 | 34 | 35.466 | 159 | 10 | 34 | 15.807 |
| 100 | 6 | 38 | 54.470 | 130 | 8 | 38 | 34.810 | 160 | 10 | 38 | 15.151 |
| 101 | 6 | 42 | 53.814 | 131 | 8 | 42 | 34.155 | 161 | 10 | 42 | 14.496 |
| 102 | 6 | 46 | 53.159 | 132 | 8 | 46 | 33.500 | 162 | 10 | 46 | 13.841 |
| 103 | 6 | 50 | 52.504 | 133 | 8 | 50 | 32.845 | 163 | 10 | 50 | 13.185 |
| 104 | 6 | 54 | 51.848 | 134 | 8 | 54 | 32.189 | 164 | 10 | 54 | 12.530 |
| 105 | 6 | 58 | 51.193 | 135 | 8 | 58 | 31.534 | 165 | 10 | 58 | 11.875 |
| 106 | 7 | 2 | 50.538 | 136 | 9 | 2 | 30.879 | 166 | 11 | 2 | 11.220 |
| 107 | 7 | 6 | 49.883 | 137 | 9 | 6 | 30.223 | 167 | 11 | 6 | 10.564 |
| 108 | 7 | 10 | 49.227 | 138 | 9 | 10 | 29.568 | 168 | 11 | 10 | 9.909 |
| 109 | 7 | 14 | 48.572 | 139 | 9 | 14 | 28.913 | 169 | 11 | 14 | 9.254 |
| 110 | 7 | 18 | 47.917 | 140 | 9 | 18 | 28.257 | 170 | 11 | 18 | 8.598 |
| 111 | 7 | 22 | 47.261 | 141 | 9 | 22 | 27.602 | 171 | 11 | 22 | 7.943 |
| 112 | 7 | 26 | 46.606 | 142 | 9 | 26 | 26.947 | 172 | 11 | 26 | 7.288 |
| 113 | 7 | 30 | 45.951 | 143 | 9 | 30 | 26.292 | 173 | 11 | 30 | 6.632 |
| 114 | 7 | 34 | 45.295 | 144 | 9 | 34 | 25.636 | 174 | 11 | 34 | 5.977 |
| 115 | 7 | 38 | 44.640 | 145 | 9 | 38 | 24.981 | 175 | 11 | 38 | 5.322 |
| 116 | 7 | 42 | 43.985 | 146 | 9 | 42 | 24.326 | 176 | 11 | 42 | 4.666 |
| 117 | 7 | 46 | 43.329 | 147 | 9 | 46 | 23.670 | 177 | 11 | 46 | 4.011 |
| 118 | 7 | 50 | 42.674 | 148 | 9 | 50 | 23.015 | 178 | 11 | 50 | 3.356 |
| 119 | 7 | 54 | 42.019 | 149 | 9 | 54 | 22.360 | 179 | 11 | 54 | 2.701 |
| 120 | 7 | 58 | 41.364 | 150 | 9 | 58 | 21.704 | 180 | 11 | 58 | 2.045 |

*To convert Right Ascension in Arc into Mean Time*—Continued.

| MINUTES. | | | | SECONDS. | | | |
|---|---|---|---|---|---|---|---|
| AR. in arc. | Mean time. | AR. in arc. | Mean time. | AR. in arc. | Mean time. | AR. in arc. | Mean time. |
| ' | m. s. | ' | m. s. | '' | s. | '' | s. |
| 1 | 0 3.989 | 31 | 2 3.661 | 1 | 0.066 | 31 | 2.061 |
| 2 | 0 7.978 | 32 | 2 7.650 | 2 | 0.133 | 32 | 2.128 |
| 3 | 0 11.969 | 33 | 2 11.640 | 3 | 0.199 | 33 | 2.194 |
| 4 | 0 15.956 | 34 | 2 15.629 | 4 | 0.266 | 34 | 2.261 |
| 5 | 0 19.945 | 35 | 2 19.618 | 5 | 0.332 | 35 | 2.327 |
| 6 | 0 23.935 | 36 | 2 23.607 | 6 | 0.399 | 36 | 2.393 |
| 7 | 0 27.924 | 37 | 2 27.596 | 7 | 0.465 | 37 | 2.460 |
| 8 | 0 31.913 | 38 | 2 31.585 | 8 | 0.532 | 38 | 2.526 |
| 9 | 0 35.902 | 39 | 2 35.574 | 9 | 0.598 | 39 | 2.593 |
| 10 | 0 39.891 | 40 | 2 39.563 | 10 | 0.665 | 40 | 2.659 |
| 11 | 0 43.880 | 41 | 2 43.552 | 11 | 0.731 | 41 | 2.726 |
| 12 | 0 47.869 | 42 | 2 47.541 | 12 | 0.798 | 42 | 2.792 |
| 13 | 0 51.858 | 43 | 2 51.530 | 13 | 0.864 | 43 | 2.859 |
| 14 | 0 55.847 | 44 | 2 55.519 | 14 | 0.931 | 44 | 2.925 |
| 15 | 0 59.836 | 45 | 2 59.509 | 15 | 0.997 | 45 | 2.992 |
| 16 | 1 3.825 | 46 | 3 3.498 | 16 | 1.064 | 46 | 3.058 |
| 17 | 1 7.814 | 47 | 3 7.487 | 17 | 1.130 | 47 | 3.125 |
| 18 | 1 11.803 | 48 | 3 11.476 | 18 | 1.197 | 48 | 3.191 |
| 19 | 1 15.793 | 49 | 3 15.465 | 19 | 1.263 | 49 | 3.258 |
| 20 | 1 19.782 | 50 | 3 19.454 | 20 | 1.330 | 50 | 3.324 |
| 21 | 1 23.771 | 51 | 3 23.443 | 21 | 1.396 | 51 | 3.391 |
| 22 | 1 27.760 | 52 | 3 27.432 | 22 | 1.463 | 52 | 3.457 |
| 23 | 1 31.749 | 53 | 3 31.421 | 23 | 1.529 | 53 | 3.524 |
| 24 | 1 35.738 | 54 | 3 35.410 | 24 | 1.596 | 54 | 3.590 |
| 25 | 1 39.727 | 55 | 3 39.399 | 25 | 1.662 | 55 | 3.657 |
| 26 | 1 43.716 | 56 | 3 43.388 | 26 | 1.729 | 56 | 3.723 |
| 27 | 1 47.705 | 57 | 3 47.377 | 27 | 1.795 | 57 | 3.790 |
| 28 | 1 51.694 | 58 | 3 51.367 | 28 | 1.862 | 58 | 3.856 |
| 29 | 1 55.683 | 59 | 3 55.356 | 29 | 1.928 | 59 | 3.923 |
| 30 | 1 59.672 | 60 | 3 59.345 | 30 | 1.995 | 60 | 3.989 |

*To convert Mean Time into Right Ascension in Arc.*

| HOURS. | | | | MINUTES. | | | | | | |
|---|---|---|---|---|---|---|---|---|---|---|
| Mean time. | AR. in arc. | | | Mean time. | AR. in arc. | | | Mean time. | AR. in arc. | |
| h. | ° | ′ | ″ | m. | ° | ′ | ″ | m. | ° | ′ | ″ |
| 1 | 15 | 2 | 27.85 | 1 | 0 | 15 | 2.46 | 31 | 7 | 46 | 16.39 |
| 2 | 30 | 4 | 52.69 | 2 | 0 | 30 | 4.93 | 32 | 8 | 1 | 18.85 |
| 3 | 45 | 7 | 23.54 | 3 | 0 | 45 | 30.39 | 33 | 8 | 16 | 21.31 |
| 4 | 60 | 9 | 51.39 | 4 | 1 | 0 | 9.86 | 34 | 8 | 31 | 23.78 |
| 5 | 75 | 12 | 19.24 | 5 | 1 | 15 | 12.32 | 35 | 8 | 46 | 26.24 |
| 6 | 90 | 14 | 47.08 | 6 | 1 | 30 | 14.79 | 36 | 9 | 1 | 28.71 |
| 7 | 105 | 17 | 14.93 | 7 | 1 | 45 | 17.25 | 37 | 9 | 16 | 31.17 |
| 8 | 120 | 19 | 42.78 | 8 | 2 | 0 | 19.71 | 38 | 9 | 31 | 33.64 |
| 9 | 135 | 22 | 10.62 | 9 | 2 | 15 | 22.18 | 39 | 9 | 46 | 36.10 |
| 10 | 150 | 24 | 38.47 | 10 | 2 | 30 | 24.64 | 40 | 10 | 1 | 38.57 |
| 11 | 165 | 27 | 6.32 | 11 | 2 | 45 | 27.11 | 41 | 10 | 16 | 41.03 |
| 12 | 180 | 29 | 34.16 | 12 | 3 | 0 | 29.57 | 42 | 10 | 31 | 43.39 |
| 13 | 195 | 32 | 2.01 | 13 | 3 | 15 | 32.03 | 43 | 10 | 46 | 45.96 |
| 14 | 210 | 34 | 29.86 | 14 | 3 | 30 | 34.50 | 44 | 11 | 1 | 48.42 |
| 15 | 225 | 36 | 57.70 | 15 | 3 | 45 | 36.96 | 45 | 11 | 16 | 50.89 |
| 16 | 240 | 39 | 25.55 | 16 | 4 | 0 | 39.43 | 46 | 11 | 31 | 53.35 |
| 17 | 255 | 41 | 53.40 | 17 | 4 | 15 | 41.89 | 47 | 11 | 46 | 55.81 |
| 18 | 270 | 44 | 21.24 | 18 | 4 | 30 | 44.35 | 48 | 12 | 1 | 58.38 |
| 19 | 285 | 46 | 49.09 | 19 | 4 | 45 | 46.82 | 49 | 12 | 17 | 0.74 |
| 20 | 300 | 49 | 16.94 | 20 | 5 | 0 | 49.28 | 50 | 12 | 32 | 3.21 |
| 21 | 315 | 51 | 44.78 | 21 | 5 | 15 | 51.75 | 51 | 12 | 47 | 5.57 |
| 22 | 330 | 54 | 12.63 | 22 | 5 | 30 | 54.21 | 52 | 13 | 2 | 8.13 |
| 23 | 345 | 56 | 40.48 | 23 | 5 | 45 | 56.67 | 53 | 13 | 17 | 10.60 |
| 24 | 360 | 59 | 8.33 | 24 | 6 | 0 | 59.14 | 54 | 13 | 32 | 13.06 |
| | | | | 25 | 6 | 16 | 1.60 | 55 | 13 | 47 | 15.53 |
| | | | | 26 | 6 | 31 | 4.07 | 56 | 14 | 2 | 17.99 |
| | | | | 27 | 6 | 46 | 6.53 | 57 | 14 | 17 | 20.45 |
| | | | | 28 | 7 | 1 | 9.00 | 58 | 14 | 32 | 22.92 |
| | | | | 29 | 7 | 16 | 11.46 | 59 | 14 | 47 | 25.38 |
| | | | | 30 | 7 | 31 | 13.92 | 60 | 15 | 2 | 27.85 |

*To convert Mean Time into Right Ascension in Arc*—Continued.

| | SECONDS. | | | | TENTHS OF SECONDS. | | |
|---|---|---|---|---|---|---|---|
| Mean time. | AR. in arc. | Mean time. | AR. in arc. | Mean time. | AR. in arc. | Mean time. | AR. in arc. |
| *s.* | *′* *″* | *s.* | *′* *″* | *s.* | *″* | *s.* | *″* |
| 1 | 0 15.04 | 31 | 7 46.27 | 0.01 | 0.15 | 0.31 | 4.66 |
| 2 | 0 30.08 | 32 | 8 1.31 | 0.02 | 0.30 | 0.32 | 4.81 |
| 3 | 0 45.12 | 33 | 8 16.36 | 0.03 | 0.45 | 0.33 | 4.96 |
| 4 | 1 0.16 | 34 | 8 31.40 | 0.04 | 0.60 | 0.34 | 5.12 |
| 5 | 1 15.21 | 35 | 8 46.44 | 0.05 | 0.75 | 0.35 | 5.27 |
| 6 | 1 30.25 | 36 | 9 1.48 | 0.06 | 0.90 | 0.36 | 5.42 |
| 7 | 1 45.29 | 37 | 9 16.52 | 0.07 | 1.05 | 0.37 | 5.57 |
| 8 | 2 0.33 | 38 | 9 31.56 | 0.08 | 1.20 | 0.38 | 5.72 |
| 9 | 2 15.37 | 39 | 9 46.60 | 0.09 | 1.35 | 0.39 | 5.87 |
| 10 | 2 30.41 | 40 | 10 1.64 | 0.10 | 1.50 | 0.40 | 6.02 |
| 11 | 2 45.45 | 41 | 10 16.68 | 0.11 | 1.65 | 0.41 | 6.17 |
| 12 | 3 0.49 | 42 | 10 31.73 | 0.12 | 1.81 | 0.42 | 6.32 |
| 13 | 3 15.53 | 43 | 10 46.77 | 0.13 | 1.96 | 0.43 | 6.47 |
| 14 | 3 30.58 | 44 | 11 1.81 | 0.14 | 2.11 | 0.44 | 6.62 |
| 15 | 3 45.62 | 45 | 11 16.85 | 0.15 | 2.26 | 0.45 | 6.77 |
| 16 | 4 0.66 | 46 | 11 31.89 | 0.16 | 2.41 | 0.46 | 6.92 |
| 17 | 4 15.70 | 47 | 11 46.93 | 0.17 | 2.56 | 0.47 | 7.07 |
| 18 | 4 30.74 | 48 | 12 1.97 | 0.18 | 2.71 | 0.48 | 7.22 |
| 19 | 4 45.78 | 49 | 12 17.01 | 0.19 | 2.86 | 0.49 | 7.37 |
| 20 | 5 0.82 | 50 | 12 32.05 | 0.20 | 3.01 | 0.50 | 7.52 |
| 21 | 5 15.86 | 51 | 12 47.09 | 0.21 | 3.16 | 0.51 | 7.67 |
| 22 | 5 30.90 | 52 | 13 2.14 | 0.22 | 3.31 | 0.52 | 7.82 |
| 23 | 5 45.94 | 53 | 13 17.18 | 0.23 | 3.46 | 0.53 | 7.97 |
| 24 | 6 1.00 | 54 | 13 32.22 | 0.24 | 3.61 | 0.54 | 8.12 |
| 25 | 6 16.03 | 55 | 13 47.26 | 0.25 | 3.76 | 0.55 | 8.27 |
| 26 | 6 31.07 | 56 | 14 2.30 | 0.26 | 3.91 | 0.56 | 8.43 |
| 27 | 6 46.11 | 57 | 14 17.34 | 0.27 | 4.06 | 0.57 | 8.58 |
| 28 | 7 1.15 | 58 | 14 32.38 | 0.28 | 4.21 | 0.58 | 8.73 |
| 29 | 7 16.19 | 59 | 14 47.42 | 0.29 | 4.36 | 0.59 | 8.88 |
| 30 | 7 31.23 | 60 | 15 2.46 | 0.30 | 4.51 | 0.60 | 9.03 |

*To convert Mean Time into Right Ascension in Arc*—Continued.

| TENTHS OF SECONDS. | | | | | | THOUSANDTHS OF SECONDS. | |
|---|---|---|---|---|---|---|---|
| Mean time. | AR. in arc. | Mean time. | AR. in arc. | Mean time. | AR. in arc. | Mean time. | AR. in arc. |
| s. | '' | s. | '' | s. | '' | s. | '' |
| 0.61 | 9.18 | 0.76 | 11.43 | 0.91 | 13.69 | 0.001 | 0.02 |
| 0.62 | 9.33 | 0.77 | 11.58 | 0.92 | 13.84 | 0.002 | 0.03 |
| 0.63 | 9.48 | 0.78 | 11.74 | 0.93 | 13.99 | 0.003 | 0.05 |
| 0.64 | 9.63 | 0.79 | 11.89 | 0.94 | 14.14 | 0.004 | 0.06 |
| 0.65 | 9.78 | 0.80 | 12.04 | 0.95 | 14.29 | 0.005 | 0.08 |
| 0.66 | 9.93 | 0.81 | 12.19 | 0.96 | 14.44 | 0.006 | 0.09 |
| 0.67 | 10.08 | 0.82 | 12.34 | 0.97 | 14.59 | 0.007 | 0.11 |
| 0.68 | 10.23 | 0.83 | 12.49 | 0.98 | 14.74 | 0.008 | 0.12 |
| 0.69 | 10.38 | 0.84 | 12.64 | 0.99 | 14.89 | 0.009 | 0.14 |
| 0.70 | 10.53 | 0.85 | 12.79 | 1.00 | 15.05 | 0.010 | 0.15 |
| 0.71 | 10.68 | 0.86 | 12.94 | | | | |
| 0.72 | 10.83 | 0.87 | 13.09 | | | | |
| 0.73 | 10.98 | 0.88 | 13.24 | | | | |
| 0.74 | 11.13 | 0.89 | 13.39 | | | | |
| 0.75 | 11.28 | 0.90 | 13.54 | | | | |

## CONSTANT LOGARITHMS.

| | | | Logarithms. |
|---|---|---|---|
| 12 hours, expressed in seconds .......... = | 43200. | | 4.6354837 |
| Complement to the same .......... = | .00002315 | | 5.3645163 |
| 24 hours, expressed in seconds.......... = | 86400. | | 4.9365137 |
| Complement to the same .......... = | .00001157 | | 5.0634863 |
| 360 degrees, expressed in seconds ........ = | 1296000. | | 6.1126050 |
| To convert sidereal time into mean solar time .......... | | | 9.9988126 |

FORM FOR

SURVEY OF ............................ DETERMINATION OF TIME,

DATE AND STATION.—1843, *October* 13.—*Mouth of the Big Black River,*

INSTRUMENTS ... $\left\{\begin{array}{l}\text{Sextant No. 2197, by } \textit{Troughton \& Simms,} \text{ and} \\ \textit{Mean Solar} \text{ Chronometer No. 76, by } \textit{Charles}\end{array}\right.$

| Names of stars. | Observed double altitudes of star with sextant. | True altitudes of star affected by corrections for refraction and errors of sextant, =A. | *Mean solar* time of observation deduced. | Time of observation noted by chronometer. |
|---|---|---|---|---|
| | ° ′ ″ | ° ′ ″ | h. m. s. | h. m. s. |
| | 91 43 40 | 45 52 58.8 | 7 05 47.69 | 6 57 02.4 |
| | 92 18 00 | 46 10 09.3 | 7 07 28.67 | 6 58 43.2 |
| | 92 41 15 | 46 21 47.3 | 7 08 37.15 | 6 59 52.8 |
| α *Andromedæ,* | 93 04 05 | 46 33 12.6 | 7 09 44.37 | 7 00 59.6 |
| (*east.*) | 93 45 20 | 46 53 50.8 | 7 11 45.92 | 7 03 01.2 |
| | 94 13 45 | 47 08 03.7 | 7 13 09.73 | 7 04 25.6 |
| | 94 40 50 | 47 21 36.6 | 7 14 29.64 | 7 05 45. |
| | 95 07 25 | 47 34 54.5 | 7 15 48.14 | 7 07 03.6 |

Mean result of 8 observations on α *Andromedæ,* in the *east* ................

| | ° ′ ″ | ° ′ ″ | h. m. s. | h. m. s. |
|---|---|---|---|---|
| | 95 20 05 | 47 41 14.7 | 8 55 32.36 | 8 46 49.2 |
| | 95 00 00 | 47 31 11.6 | 8 56 32.06 | 8 47 50.4 |
| | 94 30 40 | 47 16 31.2 | 8 57 59.42 | 8 49 16. |
| α *Lyræ* | 94 12 20 | 47 07 21. | 8 58 54. | 8 50 10.8 |
| (*west.*) | 93 53 45 | 46 58 03.1 | 8 59 49.4 | 8 51 06.9 |
| | 93 29 20 | 46 45 50.2 | 9 01 02.1 | 8 52 19.4 |
| | 93 07 35 | 46 34 57.3 | 9 02 07. | 8 53 24.8 |
| | 92 46 50 | 46 24 34.5 | 9 03 09. | 8 54 26. |
| | 92 28 45 | 46 15 31.7 | 9 04 02.96 | 8 55 21.2 |

Mean result of *nine* observations on the star α *Lyræ,* in the *west.* ...........
Mean result of *eight* observations on the star α *Andromedæ,* in the east, as above.

CHRONOMETER ERROR.—*Slow* of *mean solar* time at 8ʰ *p. m.,* by a mean of these results from east and west stars..................................

RECORD AND COMPUTATION.

*by Observed Double Altitudes of East and West Stars.*

*a tributary to the river Saint John, Maine.*
artificial horizon of Mercury.

*Young.*

| Chronometer (C.J. 76) slow of mean solar time by each observation. | Remarks. |
|---|---|
| *h. m. s.* | Index error of sextant ............... $= + 2' 40''$ |
| 0 08 45.29 | Error of eccentricity of sextant ....... $= + 1\ 32$ |
| 8 45.47 | Thermometer, 31°.5 Fahrenheit. |
| 8 44.35 | Barometer, 29.14 inches. |
| 8 44.77 | Apparent right ascension of star ..... $=\ 0^h\ 00^m\ 21^s.72$ |
| 8 44.56 | Apparent declination of star ......... $= 28°\ 13'\ 59''.5$ N. |
| 8 44.13 | Apparent north polar distance of star . $= 61\ 46\ 00\ .5\ = \triangle$ |
| 8 44.64 | Approximate latitude of this station .. $= 46\ 57\ 00$ N. $=$ L. |
| 8 44.54 | Approximate longitude of this station . $=\ 4^h\ 37^m\ 47^s$ |
|  | Sidereal time of mean noon at station . $= 13\ 26\ 20\ .83$ |
| 0$^h$ 08$^m$ 44$^s$.74 |  |
| *h. m. s.* |  |
| 0 08 43.16 | Thermometer, 29° Fahrenheit. |
| 8 41.66 | Barometer, 29.14 inches. |
| 8 43.42 | Apparent right ascension of star ..... $= 18^h\ 31^m\ 39^s.16$ |
| 8 43.20 | Apparent declination of star north.... $= 38°\ 38'\ 46''.5$ |
| 8 42.50 | Apparent north polar distance of star . $= 51\ 21\ 13\ .5\ = \triangle$ |
| 8 42.70 | Index error of sextant...... ........ $= + 2'\ 40''$ |
| 8 42.20 | Error of eccentricity of sextant....... $= + 1\ 32$ |
| 8 43.00 |  |
| 8 41.76 |  |
| 0$^h$ 08$^m$ 42$^s$.6 |  |
| 0 08 44.7 |  |
| 0$^h$ 08$^m$ 43$^s$.6 | Observer, *Major J. D. Graham.*<br>Computer, *Private F. Herbst.* |

*Computation of the Fifth of the Preceding Altitudes of α An-
dromeda, (formula, page* 187.)

Observed double altitude................. =   93° 45′ 20″
Index error, sextant .................... = +    02 40
Eccentricity, sextant .................... = +    01 32

Double altitude corrected................ =   93 49 32
Altitude ............................... =   46 54 46
Refraction, (thermom., 31°.5; barom., 29°.1).. = −        56 .6

True altitude of ✳ = A................. =   46 53 49 .4
$2 m = L + Δ + A$
L = 46° 57′                 cos =.   9.8341894
Δ = 61 46 00″.5             sin =    9.9449899

A = 46 53 49.4         cos L sin Δ =    9.7791793

$2 m =$ 155 36 49.9
$m =$ 77 48 24.4            cos =    9.3247127
(m − A) = 30 54 35.0       sin =    9.7106984

                    cos m sin (m − A) =    9.0354111
$\sin^2 \frac{1}{2} p =$ $\dfrac{\cos m \sin (m - A)}{\cos L \sin Δ}$ =   19.2562318

                         $\sin \frac{1}{2} p =$    9.6281159
                         $\frac{1}{2} p =$    25° 08′ 01″.5
                         $p$ in arc =   50 16 03 .0
        (page 192) $p$ in time = −   3 21 04 .20
                         AR. ✳ =   24 00 21 .72

Sidereal time of observation... = AR. ± $p$ =   20 39 17 .52
Sidereal time, mean noon, at place, (nauti-
cal almanac) ......................... =   13 26 20 .83

Sidereal interval past mean noon ......... =    7 12 56 .69
Retardation of mean on sidereal interval,
(page 190) ........................... = −     01 10 .93

Mean solar interval past mean noon, or
mean time p. m. of observation.........   7 11 45 .76
Time of observation by chronometer ......   7 03 01 .20

Chronometer slow......................        8 44 .56

L.—*To Find the Time by Equal Altitudes of the Sun.*

Correction in time, to be applied as an equation to the mean of the times of observed equal altitudes of the sun, in order to obtain the time of its meridional passage:

$$x = \partial \tan D \, \frac{T}{30 \tan 7\frac{1}{2} \, T} - \partial \tan L \, \frac{T}{30 \sin 7\frac{1}{2} \, T}$$

Make—

$$\frac{T}{30 \sin 7\frac{1}{2} \, T} = A; \quad \frac{T}{30 \tan 7\frac{1}{2} \, T} = B$$

$$x = \mp A \, \partial \tan L + B \, \partial \tan D$$

$$\text{Apparent noon} = \tfrac{1}{2} \, (t + t') + x$$

$t, t'$ = the times of observations;

$T = (t' - t)$ = the interval of time between the observations, expressed in hours and decimals:

$L$ = the latitude of the place of observation, (*minus* when south;)

$D$ = the sun's declination at apparent noon on the given day, (*minus* when south;)

$\partial$ = the hourly variation in the declination at noon, (*minus* when the sun is proceeding toward the south;) and

$x$ = required correction in *seconds*, where A is to be *minus* where the time of noon is required and *plus* where the time of midnight is required, *i. e.*, when the first observation is made in the afternoon and the corresponding one the morning following.

Logarithmic values of A and B are given in the following tables.

*Equations to Equal Altitudes.*

| Interval. | Log A. | Log B. | Interval. | Log A. | Log B. |
|---|---|---|---|---|---|
| h. m. | | | h. m. | | |
| 2  0 | 9.4109 | 9.3959 | 3  0 | 9.4172 | 9.3828 |
| 2 | .4111 | .3955 | 2 | .4174 | .3822 |
| 4 | .4113 | .3952 | 4 | .4177 | .3817 |
| 6 | .4114 | .3948 | 6 | .4179 | .3811 |
| 8 | .4116 | .3944 | 8 | .4182 | .3806 |
| 10 | .4118 | .3941 | 10 | .4184 | .3800 |
| 12 | .4120 | .3937 | 12 | .4187 | .3794 |
| 14 | .4121 | .3933 | 14 | .4190 | .3789 |
| 16 | .4123 | .3929 | 16 | .4193 | .3783 |
| 18 | .4125 | .3925 | 18 | .4195 | .3777 |
| 20 | .4127 | .3921 | 20 | .4198 | .3771 |
| 22 | .4129 | .3917 | 22 | .4201 | .3765 |
| 24 | .4131 | .3913 | 24 | .4204 | .3759 |
| 26 | .4133 | .3909 | 26 | .4207 | .3752 |
| 28 | .4135 | .3905 | 28 | .4209 | .3746 |
| 30 | .4137 | .3900 | 30 | .4212 | .3740 |
| 32 | .4139 | .3896 | 32 | .4215 | .3733 |
| 34 | .4141 | .3892 | 34 | .4218 | .3727 |
| 36 | .4144 | .3887 | 36 | .4221 | .3720 |
| 38 | .4146 | .3882 | 38 | .4224 | .3713 |
| 40 | .4148 | .3878 | 40 | .4227 | .3707 |
| 42 | .4150 | .3873 | 42 | .4231 | .3700 |
| 44 | .4152 | .3868 | 44 | .4234 | .3693 |
| 46 | .4155 | .3863 | 46 | .4237 | .3686 |
| 48 | .4157 | .3859 | 48 | .4240 | .3679 |
| 50 | .4159 | .3854 | 50 | .4243 | .3672 |
| 52 | .4162 | .3849 | 52 | .4246 | .3665 |
| 54 | .4164 | .3843 | 54 | .4250 | .3657 |
| 56 | .4167 | .3838 | 56 | .4253 | .3650 |
| 2  58 | 9.4169 | 9.3833 | 3  58 | 9.4256 | 9.3643 |

$$x = \mp A\, \delta \tan L + B\, \delta \tan D$$

## *Equations to Equal Altitudes*—Continued.

| Interval. | | Log A. | Log B. | Interval. | | Log A. | Log B. |
|---|---|---|---|---|---|---|---|
| *h.* | *m.* | | | *h.* | *m.* | | |
| 4 | 0 | 9.4260 | 9.3635 | 5 | 0 | 9.4374 | 9.3369 |
| | 2 | .4263 | .3627 | | 2 | .4378 | .3358 |
| | 4 | .4266 | .3620 | | 4 | .4383 | .3348 |
| | 6 | .4270 | .3612 | | 6 | .4387 | .3337 |
| | 8 | .4273 | .3604 | | 8 | .4391 | .3327 |
| | 10 | .4277 | .3596 | | 10 | .4396 | .3316 |
| | 12 | .4280 | .3588 | | 12 | .4400 | .3305 |
| | 14 | .4284 | .3580 | | 14 | .4405 | .3294 |
| | 16 | .4288 | .3572 | | 16 | .4409 | .3283 |
| | 18 | .4291 | .3564 | | 18 | .4414 | .3272 |
| | 20 | .4295 | .3555 | | 20 | .4418 | .3261 |
| | 22 | .4299 | .3547 | | 22 | .4423 | .3249 |
| | 24 | .4302 | .3538 | | 24 | .4427 | .3238 |
| | 26 | .4306 | .3530 | | 26 | .4432 | .3226 |
| | 28 | .4310 | .3521 | | 28 | .4437 | .3214 |
| | 30 | .4314 | .3512 | | 30 | .4441 | .3203 |
| | 32 | .4317 | .3503 | | 32 | .4446 | .3191 |
| | 34 | .4321 | .3494 | | 34 | .4451 | .3178 |
| | 36 | .4325 | .3485 | | 36 | .4456 | .3166 |
| | 38 | .4329 | .3476 | | 38 | .4460 | .3154 |
| | 40 | .4333 | .3467 | | 40 | .4465 | .3142 |
| | 42 | .4337 | .3457 | | 42 | .4470 | .3129 |
| | 44 | .4341 | .3448 | | 44 | .4475 | .3116 |
| | 46 | .4345 | .3438 | | 46 | .4480 | .3103 |
| | 48 | .4349 | .3429 | | 48 | .4485 | .3091 |
| | 50 | .4353 | .3419 | | 50 | .4490 | .3078 |
| | 52 | .4357 | .3409 | | 52 | .4494 | .3064 |
| | 54 | .4361 | .3399 | | 54 | .4500 | .3051 |
| | 56 | .4366 | .3389 | | 56 | .4505 | .3038 |
| 4 | 58 | 9.4370 | 9.3379 | 5 | 58 | 9.4510 | 9.3024 |

$$x = \mp \, A \, \delta \, \tan L + B \, \delta \, \tan D$$

## *Equations to Equal Altitudes*—Continued.

| Interval. | Log A. | Log B. | Interval. | Log A. | Log B. |
|---|---|---|---|---|---|
| h. m. | | | h. m. | | |
| 6 0 | 9.4515 | 9.3010 | 7 0 | 9.4685 | 9.2530 |
| 2 | .4521 | .2996 | 2 | .4691 | .2511 |
| 4 | .4526 | .2982 | 4 | .4697 | .2492 |
| 6 | .4531 | .2968 | 6 | .4704 | .2473 |
| 8 | .4536 | .2954 | 8 | .4710 | .2454 |
| 10 | .4542 | .2940 | 10 | .4716 | .2434 |
| 12 | .4547 | .2925 | 12 | .4723 | .2415 |
| 14 | .4552 | .2911 | 14 | .4729 | .2395 |
| 16 | .4558 | .2896 | 16 | .4735 | .2375 |
| 18 | .4563 | .2881 | 18 | .4742 | .2355 |
| 20 | .4569 | .2866 | 20 | .4748 | .2334 |
| 22 | .4574 | .2850 | 22 | .4755 | .2313 |
| 24 | .4580 | .2835 | 24 | .4761 | .2292 |
| 26 | .4585 | .2819 | 26 | .4768 | .2271 |
| 28 | .4591 | .2804 | 28 | .4774 | .2250 |
| 30 | .4597 | .2788 | 30 | .4781 | .2228 |
| 32 | .4602 | .2772 | 32 | .4788 | .2206 |
| 34 | .4608 | .2756 | 34 | .4794 | .2184 |
| 36 | .4614 | .2739 | 36 | .4801 | .2162 |
| 38 | .4620 | .2723 | 38 | .4808 | .2140 |
| 40 | .4625 | .2706 | 40 | .4815 | .2117 |
| 42 | .4631 | .2689 | 42 | .4821 | .2094 |
| 44 | .4637 | .2672 | 44 | .4828 | .2070 |
| 46 | .4643 | .2655 | 46 | .4835 | .2047 |
| 48 | .4649 | .2638 | 48 | .4842 | .2023 |
| 50 | .4655 | .2620 | 50 | .4849 | .1999 |
| 52 | .4661 | .2602 | 52 | .4856 | .1974 |
| 54 | .4667 | .2584 | 54 | .4863 | .1950 |
| 56 | .4673 | .2566 | 56 | .4870 | .1925 |
| 6 58 | 9.4679 | 9.2548 | 7 58 | 9.4877 | 9.1900 |

$$x = \mp \, A \, \delta \tan L + B \, \delta \tan D$$

*Equations to Equal Altitudes*—Continued.

| Interval. | Log A. | Log B. | Interval. | Log A. | Log B. |
|---|---|---|---|---|---|
| h.   m. | | | h.   m. | | |
| 8   0 | 9. 4884 | 9. 1874 | 9   0. | 9. 5115 | 9. 0943 |
| 2 | . . 4892 | . 1848 | 2 | . . 5123 | . 0906 |
| 4 | . 4899 | . 1822 | 4 | . 5132 | . 0867 |
| 6 | . 4906 | . 1796 | 6 | . 5140 | . 0828 |
| 8 | . 4913 | . 1769 | 8 | . 5148 | . 0789 |
| 10 | . 4921 | . 1742 | 10 | . 5157 | . 0749 |
| 12 | . 4928 | . 1715 | 12 | . 5165 | . 0708 |
| 14 | . 4935 | . 1687 | 14 | . 5174 | . 0667 |
| 16 | . 4943 | . 1659 | 16 | . 5182 | . 0625 |
| 18 | . 4950 | . 1630 | 18 | . 5191 | . 0583 |
| 20 | . 4958 | . 1602 | 20 | . 5199 | . 0540 |
| 22 | . 4965 | . 1573 | 22 | . 5208 | . 0496 |
| 24 | . 4973 | . 1543 | 24 | . 5217 | . 0452 |
| 26 | . 4980 | . 1513 | 26 | . 5225 | . 0406 |
| 28 | . 4988 | . 1483 | 28 | . 5234 | . 0360 |
| 30 | . 4996 | . 1453 | 30 | . 5243 | . 0314 |
| 32 | . 5003 | . 1422 | 32 | . 5252 | . 0266 |
| 34 | . 5011 | . 1390 | 34 | . 5261 | . 0218 |
| 36 | . 5019 | . 1359 | 36 | . 5269 | . 0169 |
| 38 | . 5027 | . 1327 | 38 | . 5278 | . 0119 |
| 40 | . 5035 | . 1294 | 40 | . 5287 | . 0069 |
| 42 | . 5042 | . 1261 | 42 | . 5296 | . 0017 |
| 44 | . 5050 | . 1228 | 44 | . 5305 | 8. 9965 |
| 46 | . 5058 | . 1194 | 46 | . 5315 | . 9911 |
| 48 | . 5066 | . 1159 | 48 | . 5324 | . 9857 |
| 50 | . 5074 | . 1125 | 50 | . 5333 | . 9802 |
| 52 | . 5082 | . 1089 | 52 | . 5342 | . 9745. |
| 54 | . 5091 | . 1054 | 54 | . 5351 | . 9688 |
| 56 | . 5099 | . 1017 | 56 | . 5361 | . 9630 |
| 8   58 | 9. 5107 | 9. 0981 | 9   58 | 9. 5370 | 8. 9570 |

$$x = \mp \, A\,\delta \tan L + B\,\delta \tan D$$

*Equations to Equal Altitudes*—Continued.

| Interval. | Log A. | Log B. | Interval. | Log A. | Log B. |
|---|---|---|---|---|---|
| h. m. | | | h. m. | | |
| 14  0 | 9.6841 | —9.0971 | 15  0 | 9.7333 | —9.3162 |
| 2 | .6856 | .1057 | 2 | .7351 | .3225 |
| 4 | .6872 | .1141 | 4 | .7369 | .3287 |
| 6 | .6887 | .1224 | 6 | .7386 | .3350 |
| 8 | .6903 | .1306 | 8 | .7404 | .3411 |
| 10 | .6919 | .1387 | 10 | .7422 | .3472 |
| 12 | .6934 | .1468 | 12 | .7440 | .3533 |
| 14 | .6950 | .1547 | 14 | .7458 | .3593 |
| 16 | .6966 | .1625 | 16 | .7476 | .3653 |
| 18 | .6982 | .1703 | 18 | .7494 | .3713 |
| 20 | .6998 | .1779 | 20 | .7512 | .3772 |
| 22 | .7014 | .1855 | 22 | .7531 | .3831 |
| 24 | .7030 | .1930 | 24 | .7549 | .3889 |
| 26 | .7047 | .2004 | 26 | .7568 | .3947 |
| 28 | .7063 | .2078 | 28 | .7586 | .4005 |
| 30 | .7079 | .2150 | 30 | .7605 | .4062 |
| 32 | .7096 | .2222 | 32 | .7624 | .4119 |
| 34 | .7112 | .2293 | 34 | .7642 | .4175 |
| 36 | .7129 | .2364 | 36 | .7661 | .4232 |
| 38 | .7146 | .2434 | 38 | .7680 | .4288 |
| 40 | .7162 | .2503 | 40 | .7699 | .4343 |
| 42 | .7179 | .2571 | 42 | .7718 | .4399 |
| 44 | .7196 | .2639 | 44 | .7738 | .4454 |
| 46 | .7213 | .2706 | 46 | .7757 | .4509 |
| 48 | .7230 | .2773 | 48 | .7776 | .4563 |
| 50 | .7247 | .2839 | 50 | .7796 | .4617 |
| 52 | .7264 | .2905 | 52 | .7815 | .4671 |
| 54 | .7281 | .2970 | 54 | .7835 | .4725 |
| 56 | .7299 | .3034 | 56 | .7855 | .4778 |
| 14  58 | 9.7316 | —9.3098 | 15  58 | 9.7875 | —9.4831 |

$$x = \mp\, A\, \delta \tan L + B\, \delta \tan D$$

## *Equations to Equal Altitudes*—Continued.

| Interval. | Log A. | Log B. | Interval. | Log A. | Log B. |
|---|---|---|---|---|---|
| h.  m. | | | h.  m. | | |
| 16   0 | 9. 7895 | − 9. 4884 | 17   0 | 9. 8539 | − 9. 6383 |
| 2 | . 7915 | . 4937 | 2 | . 8562 | . 6431 |
| 4 | . 7935 | . 4990 | 4 | . 8585 | . 6478 |
| 6 | . 7955 | . 5042 | 6 | . 8608 | . 6526 |
| 8 | . 7975 | . 5094 | 8 | . 8632 | . 6573 |
| 10 | . 7996 | . 5146 | 10 | . 8655 | . 6621 |
| 12 | . 8016 | . 5197 | 12 | . 8679 | . 6668 |
| 14 | . 8037 | . 5248 | 14 | . 8703 | . 6715 |
| 16 | . 8058 | . 5300 | 16 | . 8727 | . 6762 |
| 18 | . 8078 | . 5351 | 18 | . 8751 | . 6809 |
| 20 | . 8099 | . 5401 | 20 | . 8775 | . 6856 |
| 22 | . 8120 | . 5452 | 22 | . 8799 | . 6903 |
| 24 | . 8141 | . 5502 | 24 | . 8824 | . 6949 |
| 26 | . 8162 | . 5553 | 26 | . 8848 | . 6996 |
| 28 | . 8184 | . 5603 | 28 | . 8873 | . 7043 |
| 30 | . 8205 | . 5653 | 30 | . 8898 | . 7089 . |
| 32 | . 8227 | . 5702 | 32 | . 8923 | . 7136 |
| 34 | . 8248 | . 5752 | 34 | . 8948 | . 7182 |
| 36 | . 8270 | . 5801 | 36 | . 8973 | . 7228 |
| 38 | . 8292 | . 5850 | 38 | . 8999 | . 7275 |
| 40 | . 8314 | . 5900 | 40 | . 9024 | . 7321 |
| 42 | . 8336 | . 5948 | 42 | . 9050 | . 7367 |
| 44 | . 8358 | . 5997 | 44 | . 9075 | . 7413 |
| 46 | . 8380 | . 6046 | 46 | . 9101 | . 7459 |
| 48 | . 8402 | . 6094 | 48 | . 9127 | . 7505 |
| 50 | . 8425 | . 6143 | 50 | . 9154 | . 7552 |
| 52 | . 8447 | . 6191 | 52 | . 9180 | . 7598 |
| 54 | . 8470 | . 6239 | 54 | . 9206 | . 7644 |
| 56 | . 8493 | . 6287 | 56 | . 9233 | . 7690 |
| 16   58 | 9. 8516 . | − 9. 6335 | 17   58 | 9. 9260 | − 9. 7736 |

$$x = \mp\ A\ \delta \tan L + B\ \delta \tan D$$

## Equations to Equal Altitudes—Continued.

| Interval. | | Log A. | Log B. | Interval. | | Log A. | Log B. |
|---|---|---|---|---|---|---|---|
| h. | m. | | | h. | m. | | |
| 18 | 0 | 9. 9287 | −9. 7782 | 19 | 0 | 0. 0172 | −9. 9167 |
| | 2 | . 9314 | . 7827 | | 2 | . 0204 | . 9213 |
| | 4 | . 9341 | . 7873 | | 4 | . 0237 | . 9260 |
| | 6 | . 9368 | . 7919 | | 6 | . 0270 | . 9307 |
| | 8 | . 9396 | . 7965 | | 8 | . 0303 | . 9355 |
| | 10 | . 9424 | . 8011 | | 10 | . 0336 | . 9402 |
| | 12 | . 9451 | . 8057 | | 12 | . 0370 | . 9449 |
| | 14 | . 9479 | . 8103 | | 14 | . 0403 | . 9497 |
| | 16 | . 9508 | . 8149 | | 16 | . 0437 | . 9544 |
| | 18 | . 9536 | . 8195 | | 18 | . 0472 | . 9592 |
| | 20 | . 9564 | . 8241 | | 20 | . 0506 | . 9640 |
| | 22 | . 9593 | . 8287 | | 22 | . 0541 | . 9687 |
| | 24 | . 9622 | . 8333 | | 24 | . 0576 | . 9735 |
| | 26 | . 9651 | . 8379 | | 26 | . 0611 | . 9784 |
| | 28 | . 9680 | . 8425 | | 28 | . 0646 | . 9832 |
| | 30 | . 9709 | . 8471 | | 30 | . 0682 | . 9880 |
| | 32 | . 9739 | . 8517 | | 32 | . 0718 | . 9929 |
| | 34 | . 9769 | . 8563 | | 34 | . 0754 | −9. 9977 |
| | 36 | . 9798 | . 8609 | | 36 | . 0790 | −0. 0026 |
| | 38 | . 9829 | . 8655 | | 38 | . 0827 | . 0075 |
| | 40 | . 9859 | . 8701 | | 40 | . 0864 | . 0124 |
| | 42 | . 9889 | . 8748 | | 42 | . 0901 | . 0173 |
| | 44 | . 9920 | . 8794 | | 44 | . 0939 | . 0223 |
| | 46 | . 9951 | . 8840 | | 46 | . 0976 | . 0272 |
| | 48 | 9. 9982 | . 8887 | | 48 | . 1015 | . 0322 |
| | 50 | 0. 0013 | . 8933 | | 50 | . 1053 | . 0372 |
| | 52 | . 0044 | . 8980 | | 52 | . 1092 | . 0422 |
| | 54 | . 0076 | . 9026 | | 54 | . 1131 | . 0473 |
| | 56 | . 0108 | . 9073 | | 56 | . 1170 | . 0523 |
| 18 | 58 | 0. 0140 | −9. 9120 | 19 | 58 | 0. 1209 | −0. 0574 |

$$x = \mp \text{A } \partial \tan \text{L} + \text{B } \partial \tan \text{D}$$

*Equations to Equal Altitudes*—Continued.

| Interval. | Log A. | Log B. | Interval. | Log A. | Log B. |
|---|---|---|---|---|---|
| *h.* *m.* | | | *h.* *m.* | | |
| 20   0 | 0. 1249 | —0. 0625 | 21   0 | 0. 2623 | —0. 2279 |
| 2 | . 1290 | . 0676 | 2 | . 2676 | . 2339 |
| 4 | . 1330 | . 0727 | 4 | . 2729 | . 2401 |
| 6 | . 1371 | . 0779 | 6 | . 2783 | . 2462 |
| 8 | . 1412 | . 0830 | 8 | . 2838 | . 2524 |
| 10 | . 1454 | . 0882 | 10 | . 2893 | . 2587 |
| 12 | . 1496 | . 0935 | 12 | . 2949 | . 2650 |
| 14 | . 1538 | . 0987 | 14 | . 3005 | . 2714 |
| 16 | . 1581 | . 1040 | 16 | . 3063 | . 2778 |
| 18 | . 1623 | . 1093 | 18 | . 3120 | . 2843 |
| 20 | . 1667 | . 1146 | 20 | . 3179 | . 2909 |
| 22 | . 1711 | . 1200 | 22 | . 3238 | . 2975 |
| 24 | . 1755 | . 1253 | 24 | . 3298 | . 3041 |
| 26 | . 1799 | . 1308 | 26 | . 3359 | . 3109 |
| 28 | . 1844 | . 1362 | 28 | . 3420 | . 3177 |
| 30 | . 1889 | . 1417 | 30 | . 3482 | . 3245 |
| 32 | . 1935 | . 1472 | 32 | . 3545 | . 3315 |
| 34 | . 1981 | . 1527 | 34 | . 3609 | . 3385 |
| 36 | . 2028 | . 1582 | 36 | . 3674 | . 3456 |
| 38 | . 2075 | . 1638 | 38 | . 3739 | . 3527 |
| 40 | . 2122 | . 1695 | 40 | . 3805 | . 3599 |
| 42 | . 2170 | . 1751 | 42 | . 3873 | . 3673 |
| 44 | . 2218 | . 1808 | 44 | . 3941 | . 3747 |
| 46 | . 2267 | . 1866 | 46 | . 4010 | . 3822 |
| 48 | . 2316 | . 1924 | 48 | . 4080 | . 3897 |
| 50 | . 2366 | . 1982 | 50 | . 4151 | . 3974 |
| 52 | . 2416 | . 2040 | 52 | . 4223 | . 4052 |
| 54 | . 2467 | . 2099 | 54 | . 4297 | . 4130 |
| 56 | . 2518 | . 2159 | 56 | . 4371 | . 4210 |
| 20   58 | 0. 2570 | —0. 2219 | 21   58 | 0. 4446 | —0. 4291 |

$$x = \mp\, A\, \delta \tan L + B\, \delta \tan D$$

## *Equations to Equal Altitudes*—Continued.

| Interval. | Log A. | Log B. | Interval. | Log A. | Log B. |
|---|---|---|---|---|---|
| h.  m. | | | h.  m. | | |
| 22   0 | 0. 4523 | —0. 4372 | 23   0 | 0. 7689 | '—0. 7652 |
| 2 | .4601 | .4455 | 2 | .7842 | .7807 |
| 4 | .4680 | .4540 | 4 | .8000 | .7967 |
| 6 | .4761 | .4625 | 6 | .8163 | .8133 |
| 8 | .4842 | .4711 | 8 | .8333 | .8305 |
| 10 | .4926 | .4799 | 10 | .8508 | .8483 |
| 12 | .5010 | .4889 | 12 | .8691 | .8667 |
| 14 | .5097 | .4980 | 14 | .8882 | .8860 |
| 16 | .5184 | .5072 | 16 | .9080 | .9060 |
| 18 | .5274 | .5165 | 18 | .9288 | .9270 |
| 20 | .5365 | .5261 | 20 | .9506 | .9489 |
| 22 | .5458 | .5358 | 22 | .9734 | .9719 |
| 24 | .5553 | .5457 | 24 | 0. 9975 | —0. 9961 |
| 26 | .5649 | .5557 | 26 | 1. 0228 | —1. 0216 |
| 28 | .5748 | .5660 | 28 | .0497 | .0487 |
| 30 | .5848 | .5764 | 30 | .0783 | .0774 |
| 32 | .5951 | .5871 | 32 | .1089 | .1081 |
| 34 | .6056 | .5979 | 34 | .1416 | .1409 |
| 36 | .6164 | .6090 | 36 | .1770 | .1764 |
| 38 | .6273 | .6204 | 38 | .2154 | .2149 |
| 40 | .6386 | .6319 | 40 | .2573 | .2569 |
| 42 | .6501 | .6438 | 42 | .3037 | .3033 |
| 44 | .6619 | .6559 | 44 | .3554 | .3552 |
| 46 | .6740 | .6684 | 46 | .4140 | .4138 |
| 48 | .6865 | .6811 | 48 | .4815 | .4814 |
| 50 | .6993 | .6942 | 50 | .5613 | .5612 |
| 52 | .7124 | .7076 | 52 | .6588 | .6587 |
| 54 | .7259 | .7214 | 54 | .7844 | .7843 |
| 56 | .7398 | .7355 | 56 | 1. 9610 | —1. 9610 |
| 22   58 | 0. 7541 | —0. 7501 | 23   58 | 2. 2627 | —2. 2627 |

$$x = \mp A \, \delta \tan L + B \, \delta \tan D$$

*Computation of the Equation of Equal Altitudes to correct the Chronometer for Noon, August 9, 1844, by the First of the following Equal Altitudes of the Sun's Limbs.*

$$x = (- A \, \delta \, \tan L) + (B \, \delta \, \tan D)$$

| | | | |
|---|---|---|---|
| T = 6$^h$ 33$^m$ | log A = $-$ 9.4605 | | log B = 9.2764 |
| $\delta$ = 43$''$.63 | log $\delta$ = $-$ 1.6397 | | log $\delta$ = $-$ 1.6397 |
| L = 45° 48$'$ | log tan L = 0.0121 | | log tan D = 9.4493 |

| | | | |
|---|---|---|---|
| 1st term | 12$^s$.95 = | 1.1123 | $-$ 2$^s$.32 = $-$ 0.3654 |
| 2d term | $-$ 2 .32 | | |

$+$ 10 .63 = equation of equal altitudes.

*Computation of the First Two of the following Pairs of Equal Altitudes of the Sun's Limbs.*

| | | 1st pair. | 2d pair. |
|---|---|---|---|
| A. M. | = $t$ = | 1$^h$ 28$^m$ 23$^s$.0 | 1$^h$ 29$^m$ 52$^s$.8 |
| P. M. | = $t'$ = | 8 03 16 .5 | 8 01 46 .5 |
| | $t + t'$ = | 9 31 39 .5 | 9 31 39 .3 |
| | $\dfrac{t + t'}{2}$ = | 4 45 49 .75 | 4 45 49 .65 |
| Equat'n of equal alts. = | $x$ = | + 10 .63 | 10 .63 |
| Time by chron. of appt. noon = | | 4 46 00 .38 | 4 46 00 .28 |
| Correct mean time at apparent noon (Naut. Alm.) | = | 0 05 09 .09 | 0 05 09 .09 |
| Chron. fast of mean time at app't noon, August 9, 1844 | = | 4 40 51 .29 | 4 40 51 .19 |

SURVEY OF                          DETERMINATION OF THE TIME,

                                                        *Chronometer*


DATE AND STATION.—1844, *August* 9—*American Camp, near Tasche*

INSTRUMENTS $\Big\{$ Sextant No. 2197, by *Troughton &*
                   *Mean Solar* Chronometer, No. 2440,

| Observed double altitudes of the sun's upper and lower limbs. | Times, by chronometer, of observed equal altitudes. | | $t' - t = $ the elapsed time, $= T.$ | Equation of equal altitudes $= x.$ |
| --- | --- | --- | --- | --- |
| | *August 9th.* | | | |
| | A. M. $= t$ | P. M. $= t'$ | | |
| *Upper Limb.* | h. m. s. | h. m. s. | h. m. | s. |
| 78° 50′ 00″ | 1 28 23 | 8 03 16.5 $\Big\}$ | 6 33 | +10.63 |
| 79 9 30 | 1 29 52.8 | 8 01 46.5 | | |
| *Lower Limb.* | | | | |
| 83° 10′ 00″ | 1 45 01 | 7 46 40.5 | | |
| 83 40 00 | 1 46 34.5 | 7 45 06.2 $\Big\}$ | 5 59½ | +10.24 |
| 84 00 00 | 1 47 38 | 7 44 04 | | |
| *Upper Limb.* | | | | |
| 85° 36′ 00″ | 1 49 23 | 7 42 18 $\Big\}$ | 5 48 | +10.1 |
| 87 02 10 | 1 53 55.5 | 7 37 46.2 | | |

CHRONOMETER ERROR.—*Fast* of mean solar time at apparent noon of August 9, 1844, by a mean of 7 pairs of equal altitudes of the sun.........

*by Observed Equal Altitudes of the Sun's Limbs, to Correct the at Noon.*

*reau's house, on the highland boundary between Maine and Canada.*

*Simms,* and Artificial Horizon of Mercury.
by *Parkinson & Frodsham.*

| Chronometer No. 2440 *fast* of mean time at apparent noon by each pair of equal altitudes. | Remarks. |
|---|---|
| *h. m. s.* | |
| $\begin{cases} 4 \ 40 \ 51.29 \\ 4 \ 40 \ 51.19 \end{cases}$ | Index error of sextant ................... |
| | Error of eccentricity of sextant.......... |
| | Thermometer (a. m.) 70° Fahr. barom... |
| | Thermometer (p. m.) 69° Fahr. barom... |
| $\begin{cases} 4 \ 40 \ 51.9 \\ 4 \ 40 \ 51.5 \\ 4 \ 40 \ 52.15 \end{cases}$ | Sun's appar't declination at appar't noon (D) $= 15° \ 43' \ 12'' \ $N. |
| | Hourly variation of sun's declination...$(\delta) = \quad 43''.63$ |
| | Equation of time at apparent noon....... $+ \quad 5^{m} \ 09^{s}.09$ |
| | Latitude of station (approximate) ....... $+ 45° \ 48' = $(L.) |
| $\begin{cases} 4 \ 40 \ 51.51 \\ 4 \ 40 \ 51.86 \end{cases}$ | |
| | Observer, *Major J. D. Graham.* |
| | Computer, *Do.* |
| 4 40 51.6 | |

## Sun's Parallax in Altitude.

| Sun's altitude. | Sun's horizontal parallax. | | | | Sun's altitude. | Sun's horizontal parallax. | | | |
|---|---|---|---|---|---|---|---|---|---|
| | 8″.7 | 8″.8 | 8″.9 | 9″.0 | | 8″.7 | 8″.8 | 8″.9 | 9″.0 |
| ° | ″ | ″ | ″ | ″ | ° | ″ | ″ | ″ | ″ |
| 0 | 8.70 | 8.80 | 8.90 | 9.00 | 45 | 6.15 | 6.22 | 6.29 | 6.36 |
| 5 | 8.67 | 8.77 | 8.87 | 8.97 | 50 | 5.59 | 5.66. | 5.72 | 5.79 |
| 10 | 8.57 | 8.67 | 8.76 | 8.86 | 55 | 4.99 | 5.05 | 5.12 | 5.16 |
| 15 | 8.40 | 8.50 | 8.60 | 8.70 | 60 | 4.35 | 4.40 | 4.45 | 4.50 |
| 20 | 8.18 | 8.27 | 8.36 | 8.46 | 65 | 3.68 | 3.72 | 3.76 | 3.80 |
| 25 | 7.88 | 7.98 | 8.06 | 8.16 | 70 | 2.98 | 3.01 | 3.04 | 3.08 |
| 30 | 7.53 | 7.62 | 7.70 | 7.79 | 75 | 2.25 | 2.28 | 2.31 | 2.33 |
| 35 | 7.13 | 7.21 | 7.29 | 7.37 | 80 | 1.51 | 1.53 | 1.55 | 1.56 |
| 40 | 6.66 | 6.74 | 6.82 | 6.90 | 85 | 0.76 | 0.77 | 0.77 | 0.78 |
| 45 | 6.15 | 6.22 | 6.29 | 6.36 | 90 | 0.00 | 0.00 | 0.00 | 0.00 |

Parallax in altitude = horizontal parallax × cosine of altitude.

## Decimals of an Hour.

| Minutes. | | | | | | Seconds. | | | | | |
|---|---|---|---|---|---|---|---|---|---|---|---|
| m. | decm. | m. | decm. | m. | decm. | s. | decm. | s. | decm. | s. | decm. |
| 1 | .01667 | 21 | .35000 | 41 | .68333 | 1 | .00028 | 21 | .00583 | 41 | .01139 |
| 2 | .03333 | 22 | .36667 | 42 | .70000 | 2 | .00056 | 22 | .00611 | 42 | .01167 |
| 3 | .05000 | 23 | .38333 | 43 | .71667 | 3 | .00083 | 23 | .00639 | 43 | .01194 |
| 4 | .06667 | 24 | .40000 | 44 | .73333 | 4 | .00111 | 24 | .00667 | 44 | .01222 |
| 5 | .08333 | 25 | .41667 | 45 | .75000 | 5 | .00139 | 25 | .00694 | 45 | .01250 |
| 6 | .10000 | 26 | .43333 | 46 | .76667 | 6 | .00167 | 26 | .00722 | 46 | .01278 |
| 7 | .11667 | 27 | .45000 | 47 | .78333 | 7 | .00194 | 27 | .00750 | 47 | .01306 |
| 8 | .13333 | 28 | .46667 | 48 | .80000 | 8 | .00222 | 28 | .00778 | 48 | .01333 |
| 9 | .15000 | 29 | .48333 | 49 | .81667 | 9 | .00250 | 29 | .00806 | 49 | .01361 |
| 10 | .16667 | 30 | .50000 | 50 | .83333 | 10 | .00278 | 30 | .00833 | 50 | .01389 |
| 11 | .18333 | 31 | .51667 | 51 | .85000 | 11 | .00306 | 31 | .00861 | 51 | .01417 |
| 12 | .20000 | 32 | .53333 | 52 | .86667 | 12 | .00333 | 32 | .00889 | 52 | .01444 |
| 13 | .21667 | 33 | .55000 | 53 | .88333 | 13 | .00361 | 33 | .00917 | 53 | .01472 |
| 14 | .23333 | 34 | .56667 | 54 | .90000 | 14 | .00389 | 34 | .00944 | 54 | .01500 |
| 15 | .25000 | 35 | .58333 | 55 | .91667 | 15 | .00417 | 35 | .00972 | 55 | .01528 |
| 16 | .26667 | 36 | .60000 | 56 | .93333 | 16 | .00444 | 36 | .01000 | 56 | .01556 |
| 17 | .28333 | 37 | .61667 | 57 | .95000 | 17 | .00472 | 37 | .01028 | 57 | .01583 |
| 18 | .30000 | 38 | .63333 | 58 | .96667 | 18 | .00500 | 38 | .01056 | 58 | .01611 |
| 19 | .31667 | 39 | .65000 | 59 | .98333 | 19 | .00528 | 39 | .01083 | 59 | .01639 |
| 20 | .33333 | 40 | .66667 | 60 | 1.00000 | 20 | .00556 | 40 | .01111 | 60 | .01667 |

## LI.—*The Transit Instrument.*

Knowing the apparent right ascension of a star, to compute the corrections to its observed transit on account of the three principal errors of the transit instrument—in azimuth, in the inclination of the axis, and in collimation—in order to obtain the correct clock error:

$$E = T + a\,\frac{\sin(L - D)}{\cos D} + b\,\frac{\cos(L - D)}{\cos D} + \frac{c}{\cos D} - AR.$$

where—

E denotes the error of the clock, *minus* when slow;

T, the observed time of transit;

L, the latitude of the place;

D, the declination of the star, *plus* when north, and *minus* when south, for the upper culminations, and *vice versa* for the lower culminations;

a, the deviation in the telescope in azimuth, *plus* when (pointing to the south) the vertical which it describes falls to the east, and *minus* when it falls to the west, and *vice versa* when pointing to the north;

b, the bias or inclination of the axis of the telescope, *plus* when the west end of the axis is too high;

c, the error in collimation, *plus* when the circle described by the line of collimation of the telescope falls to the east, and *minus* when it falls to the west, for upper culminations, and *vice versa* for lower culminations; and

AR., the right ascension of the star.

When the clock marks mean solar time, the mean time of transit of the object over the meridian must be substituted for AR.

### LI.—*The Transit Instrument*—Continued.

1. To determine the value (*in time*) of the co-efficients $a$, $b$, $c$, in the preceding formula :

*For inclination of the axis of the telescope:*

$$b = \frac{d}{60} \left\{ (w + w') - (e + e') \right\}$$

where—

> $w'$ and $e'$ denote respectively the values of $w$ and $e$ after *reversing* the *level*;
>
> $d$, the value of each division of the level in seconds of space;
>
> $w$, the inclination of the level to the west; and
>
> $e$, the inclination of the level to the east.

*For collimation :*

$$c = \tfrac{1}{2}(t' - t)\cos D + \tfrac{1}{2}(b' - b)\cos(L - D)$$

where—

> $t'$ and $b'$ denote respectively the values of $t$ and $b$, after *reversing* the *instrument ;*
>
> D, the declination of a circumpolar star; and
>
> $t$, the time of the transit of the circumpolar star deduced from an observation at a given *side* wire of the instrument.

*For the deviation in azimuth :*

By observations of a circumpolar star :

$$a = \frac{12^{\mathrm{h}} - (T' - T)}{2\cos L \tan D} + \frac{b\cos(L - D) - b'\cos(L + D) + 2c}{2\cos L \sin D}$$

where—

> T' and $b'$ denote respectively the values of T and $b$ at the *lower* culmination.

LI.—*The Transit Instrument*—Continued.

Deviation in azimuth by transits of a high and low star :

$$a = \left\{ (AR.' - AR.) - (T' - T) \right\} \times \frac{\cos D' \cos D}{\cos L \sin (D - D')}$$

where—

T', AR.', and D' denote respectively the values of T, AR., and D of the *second* star observed.

Or—

$$a = \frac{(AR.' - AR.) - (T' - T)}{\cos L (\tan D - \tan D')}$$

If one of the stars is observed at its lower culmination, use $180° - D'$ and $12^h + AR.'$ for its declination and right ascension.

Or make—

$$\frac{\sin (L - D)}{\cos D} \text{ for the } first \text{ star} = n$$

and—

$$\frac{\sin (L - D')}{\cos D'} \text{ for the } second \text{ star} = n$$

then—

$$a = \frac{(AR.' - AR.) - (T' - T)}{n' - n}$$

$n$ is negative for a star north of the zenith.

2. To find the equatorial interval of each wire from the central wire, observe the transit of a star of any declination D ; then—

Equatorial interval = observed interval × cos D.

3. When the intervals on each side of the central wire are equal, the mean of the times of transit over each wire will denote the transit over the middle wire. But should they not be equal, a correction must be applied to obtain a correct mean.

Call I. II; IV. V, the equatorial intervals of each wire from the central wire, the instrument having say 5 wires ; then—

$$\text{Reduction to middle wire} = \frac{(I + II) - (IV + V)}{5 \cos D}$$

FORM FOR RECORD AND COMPUTATION.

SURVEY OF                                    STATION.

*Transits of Stars .................. with ........................... Inch transit No. ............. ;*
*Sidereal Chronometer Hardy No. 50,.*

Illuminated end of axis, west.

| Date (1847) ......... | October 6th. | October 6th. | October 6th. |
|---|---|---|---|
| Observer ............. | T. J. L. | T. J. L. | T. J. L. |
| Object .............. | π Capricorni. | 14 Capricorni. | a Cygni. |
| Level............... | E. 32.2 W. 33.0 | E. 32.7 W. 32.5 | E. 32.7 W. 32.5 |
| Value of 1 division of scale = 7″.5 ....... | E. 32.2 W. 33.0 | E. 32.5 W. 33.3 | E. 33.0 W. 32.5 |

| | h. m. s. | h. m. s. | h. m. s. |
|---|---|---|---|
| Wires ........... I | 20 17 33.0 | 20 29 43.7 | 20 35 00.0 |
| II | 17 53.5 | 30 02.7 | 35 26.0 |
| III | 18 12.7 | 30 22.0 | 35 52.0 |
| IV | 18 32.7 | 30 41.7 | 36 18.7 |
| V | 20 18 52.5 | 20 31 00.7 | 20 36 45.5 |
| Sum................ .... | 184.4 | 110.8 | 142.2 |
| Mean ...... ......... | 20 18 12.88 | 20 30 22.16 | 20 35 52.44 |
| Reduc'n to middle wire | — .07 | — .07 | — .10 |
| Transit on instrument. | 12.81 | 22.09 | 52.39 |
| Corr'ns in time { for collimation. | | | |
| for level ...... | + .10 | + .04 | — .12 |
| for dev'n in az'h | + .17 | + .18 | — .01 |
| Transit by chronom'r. | 20 18 13.08 | 20 30 22.31 | 20 35 52.21 |
| AR. of star .......... | 20 18 36.66 | 20 30 45.89 | 20 36 15.80 |
| Error of chronometer. | 23.58 | 23.58 | 23.59 |

Chronometer ............................................. slow of ..................................... time
at .................................... p. m., October 6th, 1847.

*Computation of the Corrections a and b, in the Preceding Transits.*

Declination of $\pi$ Capricorni $= 18°$ 42 S.

14 Capricorni $= 15°$ 29 S.

$a$ Cygni $= 44°$ 44 N.

Latitude of Station $= L = 43°13'$

*Level Correction of $\pi$ Capricorni.*

| | | E. 32.2 | W. 33 |
|---|---|---|---|
| L $=$ | 43° 13' | 32.2 | 33 |
| D $= -$ | 18° 42' | | |
| (L $-$ D) $=$ | 61° 55' | 64.4 | 66 |

$$\cos \frac{(L-D)}{\cos D} = 0.50$$

$$66 - 64.4 = 1.6$$

$$b = \frac{7.5}{60} \times 1.6 = 0^{s}.20$$

$$\text{Level correction} = b \frac{\cos (L-D)}{\cos D} = 0^{s}.20 \times 0.50 = 0^{s}.10$$

*Deviation in Azimuth.*

$$a = \frac{(AR.' - AR.) - (T' - T)}{n' - n}$$

T' and T being the times of transit corrected for level and collimation.

Combining $\pi$ Capricorni and $a$ Cygni,

| | h. m. s. | | h. m. s. |
|---|---|---|---|
| AR.' $=$ | 20 36 15.80 | T' $=$ | 20 35 52.22 |
| AR. $=$ | 20 18 36.66 | T $=$ | 20 18 12.91 |
| | 17 39.14 | | 17 39.31 |
| | 17 39.31 | | |

$$(AR.' - AR.) - (T' - T) = -0.17$$

$$n' = \frac{(\sin L - D')}{\cos D'} = \frac{\sin (-1° \ 31')}{\cos 44° 44'} = -0.03$$

$$n = \frac{\sin (L - D)}{\cos D} = \frac{\sin 61° 55'}{\cos 18° 42'} = +0.93$$

$$a = \frac{-0^{s}.17}{-0.03 - 0.93} = \frac{0.17}{0.96} = +0^{s}.18$$

Combining 14 Capricorni and $a$ Cygni, $a = +0^{s}.19$

Correction for deviation in azimuth of $\pi$ Capricorni,

$$^{s} = a \frac{\sin (L - D)}{\cos D} = 0^{s}.18 \times 0.93 = 0^{s}.17$$

*Numerical Values of Factors* $\dfrac{\sin\ (L - D)}{\cos\ D},\quad \dfrac{\cos\ (L - D)}{\cos\ D},$

| For deviation. | Star's declination $= \pm$ D | | | | | | | For level. |
|---|---|---|---|---|---|---|---|---|
| Star's Z.-D. $=(L-D)$ | 0° | 10° | 20° | 25° | 30° | 35° | 40° | Star's Z.-D. $=(L-D)$ |
| o | | | | | | | | o |
| 1 | .02 | .02 | .02 | .02 | .02 | .02 | .02 | 89 |
| 2 | .04 | .04 | .04 | .04 | .04 | .04 | .05 | 88 |
| 4 | .07 | .07 | .07 | .08 | .08 | .08 | .09 | 86 |
| 6 | .11 | .11 | .11 | .11 | .12 | .13 | .14 | 84 |
| 8 | .14 | .14 | .15 | .15 | .16 | .17 | .18 | 82 |
| 10 | .17 | .18 | .19 | .19 | .20 | .21 | .23 | 80 |
| 12 | .21 | .21 | .22 | .23 | .24 | .25 | .27 | 78 |
| 14 | .24 | .25 | .26 | .27 | .28 | .29 | .32 | 76 |
| 16 | .28 | .28 | .29 | .30 | .32 | .34 | .36 | 74 |
| 18 | .31 | .31 | .33 | .34 | .36 | .38 | .40 | 72 |
| 20 | .34 | .35 | .36 | .38 | .40 | .42 | .45 | 70 |
| 22 | .37 | .38 | .40 | .42 | .44 | .46 | .49 | 68 |
| 24 | .41 | .41 | .43 | .45 | .47 | .49 | .53 | 66 |
| 26 | .44 | .45 | .47 | .49 | .51 | .54 | .57 | 64 |
| 28 | .47 | .48 | .50 | .52 | .54 | .57 | .61 | 62 |
| 30 | .50 | .51 | .53 | .55 | .58 | .61 | .65 | 60 |
| 32 | .53 | .54 | .56 | .58 | .61 | .65 | .69 | 58 |
| 34 | .56 | .57 | .59 | .61 | .65 | .69 | .73 | 56 |
| 36 | .59 | .60 | .63 | .65 | .68 | .72 | .77 | 54 |
| 38 | .62 | .63 | .66 | .68 | .71 | .75 | .80 | 52 |
| 40 | .64 | .65 | .68 | .71 | .74 | .78 | .84 | 50 |
| 45 | .71 | .72 | .75 | .78 | .82 | .86 | .92 | 45 |
| 50 | .77 | .78 | .82 | .84 | .89 | .93 | 1.00 | 40 |
| 55 | .82 | .83 | .87 | .90 | .95 | .98 | 1.07 | 35 |
| 60 | .87 | .88 | .92 | .95 | 1.00 | 1.06 | 1.13 | 30 |
| 65 | .91 | .92 | .96 | 1.00 | 1.05 | 1.10 | 1.18 | 25 |
| 70 | .94 | .95 | 1.00 | 1.04 | 1.09 | 1.14 | 1.23 | 20 |
| 75 | .97 | .98 | 1.03 | 1.07 | 1.12 | 1.17 | 1.26 | 15 |
| 80 | .98 | 1.00 | 1.05 | 1.09 | 1.14 | 1.20 | 1.29 | 10 |
| 89 | 1.00 | 1.02 | 1.06 | 1.10 | 1.15 | 1.22 | 1.31 | 1 |
| For colli- mation | 1.00 | 1.02 | 1.06 | 1.10 | 1.15 | 1.22 | 1.31 | $=\dfrac{1}{\cos D}$ |

$\dfrac{1}{\cos D}$, *for facilitating the Reduction of Transit Observations.*

| For deviation. | Star's declination $= \pm$ D | | | | | | | For level. |
|---|---|---|---|---|---|---|---|---|
| Star's Z.-D. $=(L-D)$ | 45° | 50° | 55° | 60° | 65° | 70° | 75° | Star's Z.-D. $=(L-D)$ |
| ° | | | | | | | | ° |
| 1 | .02 | .03 | .03 | .03 | .04 | .05 | .07 | 89 |
| 2 | .05 | .05 | .06 | .07 | .08 | .10 | .13 | 88 |
| 4 | .10 | .11 | .12 | .14 | .17 | .20 | .27 | 86 |
| 6 | .15 | .16 | .18 | .21 | .25 | .31 | .40 | 84 |
| 8 | .20 | .22 | .24 | .28 | .33 | .41 | .54 | 82 |
| 10 | .25 | .27 | .30 | .35 | .41 | .51 | .67 | 80 |
| 12 | .29 | .32 | .36 | .42 | .49 | .61 | .80 | 78 |
| 14 | .34 | .38 | .42 | .48 | .57 | .71 | .94 | 76 |
| 16 | .39 | .43 | .48 | .55 | .65 | .81 | 1.06 | 74 |
| 18 | .44 | .48 | .54 | .62 | .73 | .90 | 1.19 | 72 |
| 20 | .48 | .53 | .60 | .68 | .81 | 1.00 | 1.32 | 70 |
| 22 | .53 | .58 | .65 | .75 | .89 | 1.09 | 1.45 | 68 |
| 24 | .58 | .63 | .71 | .81 | .96 | 1.19 | 1.57 | 66 |
| 26 | .62 | .68 | .76 | .88 | 1.04 | 1.28 | 1.69 | 64 |
| 28 | .66 | .73 | .82 | .94 | 1.11 | 1.37 | 1.81 | 62 |
| 30 | .71 | .78 | .87 | 1.00 | 1.18 | 1.46 | 1.93 | 60 |
| 32 | .75 | .82 | .92 | 1.06 | 1.25 | 1.55 | 2.05 | 58 |
| 34 | .79 | .87 | .97 | 1.12 | 1.32 | 1.65 | 2.16 | 56 |
| 36 | .83 | .91 | 1.03 | 1.18 | 1.39 | 1.74 | 2.27 | 54 |
| 38 | .87 | .96 | 1.07 | 1.23 | 1.46 | 1.80 | 2.38 | 52 |
| 40 | .91 | 1.00 | 1.12 | 1.29 | 1.52 | 1.88 | 2.48 | 50 |
| 45 | 1.00 | 1.10 | 1.23 | 1.41 | 1.67 | 2.07 | 2.73 | 45 |
| 50 | 1.08 | 1.19 | 1.34 | 1.53 | 1.81 | 2.24 | 2.96 | 40 |
| 55 | 1.16 | 1.27 | 1.43 | 1.64 | 1.94 | 2.40 | 3.16 | 35 |
| 60 | 1.22 | 1.35 | 1.51 | 1.73 | 2.05 | 2.53 | 3.35 | 30 |
| 65 | 1.28 | 1.41 | 1.58 | 1.81 | 2.14 | 2.65 | 3.50 | 25 |
| 70 | 1.33 | 1.46 | 1.64 | 1.88 | 2.22 | 2.75 | 3.63 | 20 |
| 75 | 1.37 | 1.50 | 1.68 | 1.93 | 2.29 | 2.82 | 3.67 | 15 |
| 80 | 1.39 | 1.53 | 1.72 | 1.97 | 2.33 | 2.88 | 3.81 | 10 |
| 89 | 1.41 | 1.56 | 1.74 | 2.00 | 2.37 | 2.92 | 3.86 | 1 |
| For collimation $\}$ .. | 1.41 | 1.56 | 1.74 | 2.00 | 2.37 | 2.92 | 3.86 | $= \dfrac{1}{\cos D}$ |

### LII.—*Reduction of Transits by Least Squares.*

Let—

E be the error of chronometer at an assumed time T;

$t_1, t_2, t_3$, &c., the observed times of transit (corrected for rate and level error) of stars having the right ascensions $AR_{.1}, AR_{.2}, AR_{.3}$, &c.;

$a$ and $c$, the errors of azimuth and collimation; and

$A_1, A_2, A_3$, &c., $C_1, C_2, C_3$, &c., the factors of azimuth and of collimation for the several stars;

then—

$$t_1 + E + A_1\,a + C_1\,c = AR_1$$
$$t_2 + E + A_2\,a + C_2\,c = AR_2$$
$$t_3 + E + A_3\,a + C_3\,c = AR_3$$
$$\&c., \&c.$$

Let—

$$E = E + \varepsilon$$

where $\varepsilon$ is the unknown correction to an assumed chronometer error $E$;

and let, also,

$$AR_1 - t_1 = e_1$$
$$AR_2 - t_2 = e_2$$
$$AR_3 - t_3 = e_3$$
$$\&c., \&c.$$

then—

$$E + \varepsilon + A_1\,a + C_1\,c = e_1$$
$$E + \varepsilon + A_2\,a + C_2\,c = e_2$$
$$E + \varepsilon + A_3\,a + C_3\,c = e_3$$
$$\&c., \&c.$$

Let now—

$$e_1 - E = n_1$$
$$e_2 - E = n_2$$
$$e_3 - E = n_3$$
$$\&c., \&c.$$

then—

$$\varepsilon + A_1\,a + C_1\,c = n_1$$
$$\varepsilon + A_2\,a + C_2\,c = n_2$$
$$\varepsilon + A_3\,a + C_3\,c = n_3$$
$$\&c., \&c.$$

From which form the normal equations—

$$\Sigma \varepsilon + \Sigma A\,a + \Sigma C\,c = n$$
$$\Sigma A \varepsilon + \Sigma A^2\,a + \Sigma A C\,c = A\,n \qquad (1)$$
$$\Sigma C \varepsilon + \Sigma A C\,a + \Sigma C_2\,c = C\,n$$

from which $\varepsilon$, $a$, and $c$ can be obtained.

LII.—*Reduction of Transits, &c.*—Continued.

If the errors of collimation are known, and the times $t_1$, $t_2$, $t_3$, &c., corrected for it, the azimuthal deviation and correction to assumed chronometer-error may be deduced from the equations—

$$\Sigma \epsilon + \Sigma A\, a = \Sigma n$$
$$\Sigma A\, \epsilon + \Sigma A^2\, a = \Sigma A\, n \qquad (2)$$

Equations (1) cannot be advantageously employed unless the instrument be reversed.

*Example of the Computation of Equations* (1).

Latitude, 36° 38′ N.—April 11, 1852—Assumed time, T = 11ʰ sidereal—Chronometer losing 1ˢ.83 daily—Assume E = + 3ʰ 14ᵐ 30ˢ.0.

| Illum'n star. | $\epsilon$ | $n$ | A | C | A$n$ | C$n$ | A$^2$ | AC | C$^2$ |
|---|---|---|---|---|---|---|---|---|---|
| | h. m. s. | | | | | | | | |
| E. ⎰ a Urs. Maj....... | 3 14 29.77 | −0.23 | −0.95 | +2.17 | +0.22 | −0.50 | +0.90 | − 2.06 | +4.71 |
| ⎱ δ Leonis ........ | 29.91 | −0.09 | +0.27 | +1.07 | −0.02 | −0.10 | 0.07 | + 0.29 | 1.14 |
| ⎩ δ Hydra ........ | 29.87 | −0.13 | +0.80 | +1.04 | −0.10 | −0.14 | 0.64 | + 0.83 | 1.08 |
| W. ⎧ γ Cephei sub. polo | 28.50 | −1.50 | +4.05 | +4.39 | −6.07 | −6.58 | 16.40 | −17.78 | 19.27 |
| ⎨ γ Urs. Maj....... | 30.56 | +0.56 | −0.55 | −1.72 | −0.31 | −0.96 | 0.30 | + 0.95 | 2.96 |
| ⎪ 4052 B. A. C .... | 30.19 | +0.19 | +0.46 | −1.01 | +0.09 | −0.19 | 0.21 | − 0.46 | 1.02 |
| ⎩ 4072 B. A. C .... | 30.06 | +0.06 | +0.46 | −1.02 | +0.03 | −0.06 | 0.21 | − 0.47 | 1.04 |
| | | −1.14 | +4.54 | +4.92 | −6.16 | −8.53 | +18.73 | +16.86 | +31.22 |

*Normal Equations.*

$$7\,\epsilon + \ 4.54\,a + \ 4.92\,c = -\,1.14$$
$$4.54\,\epsilon + 18.73\,a + 16.86\,c = -\,6.16$$
$$4.92\,\epsilon + 16.86\,a + 31.22\,c = -\,8.53$$

from which—

$$\epsilon = +\ 0^{s}.09$$
$$a = -\ 0^{s}.18$$
$$c = -\ 0^{s}.19$$

hence,

Azimuthal deviation of the instrument ....... = 0ˢ.18 W. of S.
Error of collimation of mean of wires, illumi-
 nation east .........................: = 0ˢ.19 W.
Error of chronometer, (slow) .............. = 3ʰ 14ᵐ 30ˢ.06

## LIII.—*Tables of Refraction.*

Table I gives the refraction when the barometer stands at 30 inches and the Fahrenheit thermometer at 50°.

Table II, to be used when greater accuracy is desired, gives the correction of the mean refraction depending upon the observed height of the barometer and thermometer.

In column A of this table, the refraction is regarded as a function of the *apparent* zenith-distance Z. The adopted form of this function is—

$$r = a \beta^A \gamma^\lambda \tan Z$$

in which $a$ varies slowly with the zenith-distance, and its logarithm is therefore readily taken from the table with the argument Z. The exponents A and $\lambda$ differ sensibly from unity only for great zenith-distances, and also vary slowly; their values are therefore readily found from the table.

The factor $\beta$ depends upon the barometer. The actual pressure indicated by the barometer depends not only upon the height of the column, but also upon its temperature. It is therefore put under the form—

$$\beta = B\,T$$

and log B and log T are given in the supplementary tables with the arguments "Height of the barometer" and "Height of the attached thermometer," respectively; so that—

$$\log \beta = \log B + \log T$$

Finally, log $\gamma$ is given directly in the supplementary table with the argument "External thermometer." This thermometer should be so exposed as to indicate truly the temperature of the atmosphere at the place of observation.

LIII.—*Tables of Refraction*—Continued.

*Example.*—Given the apparent zenith distance, Z = 78°30′0″; barometer, 29.770 inches; attached thermometer, − 0°.4 F.; external thermometer, − 2°.0 F.

From table II for 78°30′; log $a$ = 1.74981
$$A = 1.0032 ; \lambda = 1.0328$$

and from the tables for barometer and thermometer—

$$\log B = + \ 0.00253$$
$$\log T = + \ 0.00127 \qquad \log \gamma = + \ 0.04545$$

$$\log \beta = + \ 0.00380$$

Hence the refraction is computed as follows:

$$\log a = \quad 1.74981$$
$$A \log \beta = \log \beta^A = + \ 0.00381$$
$$\lambda \log \gamma = \log \gamma^\lambda = + \ 0.04694$$
$$\log \tan Z = \quad 0.69154$$

$$r = 5' \ 10''.53 = 310''.53; \ \log r = \quad 2.49210$$

The true zenith-distance is, therefore,

$$78° \ 30' + 5' \ 10''.53 = 78° \ 35' \ 10''.53$$

### TABLE I.—*Mean Refraction.*

Barometer, 30 inches—Fahrenheit thermometer, 50°.

| Apparent altitude. | | Mean refraction. | | Apparent altitude. | | Mean refraction. | | Apparent altitude. | | Mean refraction. | |
|---|---|---|---|---|---|---|---|---|---|---|---|
| ° | ′ | ′ | ″ | ° | ′ | ′ | ″ | ° | ′ | ′ | ″ |
| | | | | 5 | 30 | 9 | 7.0 | 6 | 30 | 7 | 53.9 |
| 0 | 0 | 36 | 29 | | 35 | 9 | 0.1 | | 35 | 7 | 48.7 |
| 1 | 0 | 24 | 54 | | 40 | 8 | 53.4 | | 40 | 7 | 43.5 |
| 2 | 0 | 18 | 26 | | 45 | 8 | 46.8 | | 45 | 7 | 38.4 |
| 3 | 0 | 14 | 25 | | 50 | 8 | 40.4 | | 50 | 7 | 33.5 |
| 4 | 0 | 11 | 44 | | 55 | 8 | 34.2 | | 55 | 7 | 28.6 |
| 5 | 0 | 9 | 52.0 | 6 | 0 | 8 | 28.0 | 7 | 0 | 7 | 23.8 |
| | 5 | 9 | 44.0 | | 5 | 8 | 22.1 | | 5 | 7 | 19.2 |
| | 10 | 9 | 36.2 | | 10 | 8 | 16.2 | | 10 | 7 | 14.6 |
| | 15 | 9 | 28.6 | | 15 | 8 | 10.5 | | 15 | 7 | 10.1 |
| | 20 | 9 | 21.2 | | 20 | 8 | 4.8 | | 20 | 7 | 5.7 |
| | 25 | 9 | 14.0 | | 25 | 7 | 59.3 | | 25 | 7 | 1.4 |

## Table I.—*Mean Refraction*—Continued.

Barometer, 30 inches—Fahrenheit thermometer, 50°.

| Apparent altitude. | | Mean refraction. | | Apparent altitude. | | Mean refraction. | | Apparent altitude. | | Mean refraction. | |
|---|---|---|---|---|---|---|---|---|---|---|---|
| ° | ′ | ′ | ″ | ° | ′ | ′ | ″ | ° | ′ | ′ | ″ |
| 7 | 30 | 6 | 57.1 | 10 | 30 | 5 | 4.6 | 13 | 30 | 3 | 58.1 |
|  | 35 | 6 | 53.0 |  | 35 | 5 | 2.3 |  | 35 | 3 | 56.6 |
|  | 40 | 6 | 48.9 |  | 40 | 5 | 0.0 |  | 40 | 3 | 55.2 |
|  | 45 | 6 | 44.9 |  | 45 | 4 | 57.8 |  | 45 | 3 | 53.7 |
|  | 50 | 6 | 41.0 |  | 50 | 4 | 55.6 |  | 50 | 3 | 52.3 |
|  | 55 | 6 | 37.1 |  | 55 | 4 | 53.4 |  | 55 | 3 | 50.9 |
| 8 | 0 | 6 | 33.3 | 11 | 0 | 4 | 51.2 | 14 | 0 | 3 | 49.5 |
|  | 5 | 6 | 29.6 |  | 5 | 4 | 49.1 |  | 5 | 3 | 48.1 |
|  | 10 | 6 | 25.9 |  | 10 | 4 | 47.0 |  | 10 | 3 | 46.8 |
|  | 15 | 6 | 22.3 |  | 15 | 4 | 44.9 |  | 15 | 3 | 45.5 |
|  | 20 | 6 | 18.8 |  | 20 | 4 | 42.9 |  | 20 | 3 | 44.2 |
|  | 25 | 6 | 15.3 |  | 25 | 4 | 40.9 |  | 25 | 3 | 42.9 |
|  | 30 | 6 | 11.9 |  | 30 | 4 | 38.9 |  | 30 | 3 | 41.6 |
|  | 35 | 6 | 8.5 |  | 35 | 4 | 36.9 |  | 35 | 3 | 40.3 |
|  | 40 | 6 | 5.2 |  | 40 | 4 | 35.0 |  | 40 | 3 | 39.0 |
|  | 45 | 6 | 2.0 |  | 45 | 4 | 33.1 |  | 45 | 3 | 37.7 |
|  | 50 | 5 | 58.8 |  | 50 | 4 | 31.2 |  | 50 | 3 | 36.5 |
|  | 55 | 5 | 55.7 |  | 55 | 4 | 29.4 |  | 55 | 3 | 35.3 |
| 9 | 0 | 5 | 52.6 | 12 | 0 | 4 | 27.5 | 15 | 0 | 3 | 34.1 |
|  | 5 | 5 | 49.6 |  | 5 | 4 | 25.7 |  | 5 | 3 | 32.9 |
|  | 10 | 5 | 46.6 |  | 10 | 4 | 23.9 |  | 10 | 3 | 31.7 |
|  | 15 | 5 | 43.6 |  | 15 | 4 | 22.2 |  | 15 | 3 | 30.5 |
|  | 20 | 5 | 40.7 |  | 20 | 4 | 20.4 |  | 20 | 3 | 29.4 |
|  | 25 | 5 | 37.9 |  | 25 | 4 | 18.7 |  | 25 | 3 | 28.2 |
|  | 30 | 5 | 35.1 |  | 30 | 4 | 17.0 |  | 30 | 3 | 27.1 |
|  | 35 | 5 | 32.4 |  | 35 | 4 | 15.3 |  | 35 | 3 | 25.9 |
|  | 40 | 5 | 29.6 |  | 40 | 4 | 13.6 |  | 40 | 3 | 24.8 |
|  | 45 | 5 | 27.0 |  | 45 | 4 | 12.0 |  | 45 | 3 | 23.7 |
|  | 50 | 5 | 24.3 |  | 50 | 4 | 10.4 |  | 50 | 3 | 22.6 |
|  | 55 | 5 | 21.7 |  | 55 | 4 | 8.8 |  | 55 | 3 | 21.5 |
| 10 | 0 | 5 | 19.2 | 13 | 0 | 4 | 7.2 | 16 | 0 | 3 | 20.5 |
|  | 5 | 5 | 16.7 |  | 5 | 4 | 5.6 |  | 5 | 3 | 19.4 |
|  | 10 | 5 | 14.2 |  | 10 | 4 | 4.1 |  | 10 | 3 | 18.4 |
|  | 15 | 5 | 11.7 |  | 15 | 4 | 2.6 |  | 15 | 3 | 17.3 |
|  | 20 | 5 | 9.3 |  | 20 | 4 | 1.0 |  | 20 | 3 | 16.3 |
|  | 25 | 5 | 6.9 |  | 25 | 3 | 59.6 |  | 25 | 3 | 15.2 |

## TABLE I.—*Mean Refraction*—Continued.

Barometer, 30 inches—Fahrenheit thermometer, 50°.

| Apparent altitude. | Mean refraction. | Apparent altitude. | Mean refraction. | Apparent altitude. | Mean refraction. |
|---|---|---|---|---|---|
| ° ′ | ′ ″ | ° ′ | ′ ″ | ° ′ | ′ ″ |
| 16 30 | 3 14.2 | 20 0 | 2 38.8 | 26 0 | 1 58.9 |
| 35 | 3 13.2 | 10 | 2 37.4 | 10 | 1 58.1 |
| 40 | 3 12.2 | 20 | 2 36.0 | 20 | 1 57.2 |
| 45 | 3 11.2 | 30 | 2 34.6 | 30 | 1 56.4 |
| 50 | 3 10.3 | 40 | 2 33.3 | 40 | 1 55.5 |
| 55 | 3 9.3 | 50 | 2 32.0 | 50 | 1 54.7 |
| 17 0 | 3 8.3 | 21 0 | 2 30.7 | 27 0 | 1 53.9 |
| 5 | 3 7.3 | 10 | 2 29.4 | 10 | 1 53.1 |
| 10 | 3 6.4 | 20 | 2 28.1 | 20 | 1 52.3 |
| 15 | 3 5.5 | 30 | 2 26.9 | 30 | 1 51.5 |
| 20 | 3 4.6 | 40 | 2 25.7 | 40 | 1 50.7 |
| 25 | 3 3.7 | 50 | 2 24.5 | 50 | 1 50.0 |
| 30 | 3 2.8 | 22 0 | 2 23.3 | 28 0 | 1 49.2 |
| 35 | 3 1.9 | 10 | 2 22.1 | 10 | 1 48.4 |
| 40 | 3 1.0 | 20 | 2 20.9 | 20 | 1 47.7 |
| 45 | 3 0.1 | 30 | 2 19.8 | 30 | 1 46.9 |
| 50 | 2 59.2 | 40 | 2 18.7 | 40 | 1 46.2 |
| 55 | 2 58.3 | 50 | 2 17.5 | 50 | 1 45.5 |
| 18 0 | 2 57.5 | 23 0 | 2 16.4 | 29 0 | 1 44.8 |
| 5 | 2 56.6 | 10 | 2 15.4 | 20 | 1 43.4 |
| 10 | 2 55.8 | 20 | 2 14.3 | 40 | 1 42.0 |
| 15 | 2 54.9 | 30 | 2 13.3 | 30 0 | 1 40.6 |
| 20 | 2 54.1 | 40 | 2 12.2 | 20 | 1 39.3 |
| 25 | 2 53.2 | 50 | 2 11.2 | 40 | 1 38.0 |
| 30 | 2 52.4 | 24 0 | 2 10.2 | 31 0 | 1 36.7 |
| 35 | 2 51.6 | 10 | 2 9.2 | 20 | 1 35.5 |
| 40 | 2 50.8 | 20 | 2 8.2 | 40 | 1 34.2 |
| 45 | 2 50.0 | 30 | 2 7.2 | 32 0 | 1 33.0 |
| 50 | 2 49.2 | 40 | 2 6.2 | 20 | 1 31.8 |
| 55 | 2 48.4 | 50 | 2 5.3 | 40 | 1 30.7 |
| 19 0 | 2 47.7 | 25 0 | 2 4.4 | 33 0 | 1 29.5 |
| 10 | 2 46.1 | 10 | 2 3.4 | 20 | 1 28.4 |
| 20 | 2 44.6 | 20 | 2 2.5 | 40 | 1 27.3 |
| 30 | 2 43.1 | 30 | 2 1.6 | 34 0 | 1 26.2 |
| 40 | 2 41.6 | 40 | 2 0.7 | 20 | 1 25.1 |
| 50 | 2 40.2 | 50 | 1 59.8 | 40 | 1 24.1 |

TABLE I.—*Mean Refraction*—Continued.

Barometer, 30 inches—Fahrenheit thermometer, 50°.

| Apparent altitude. | | Mean re-fraction. | | Apparent altitude. | | Mean re-fraction. | | Apparent altitude. | | Mean re-fraction. | |
|---|---|---|---|---|---|---|---|---|---|---|---|
| ° | ′ | ′ | ″ | ° | ′ | ′ | ″ | ° | ′ | ′ | ″ |
| 35 | 0 | 1 | 23.1 | 47 | 0 | 0 | 54.3 | 59 | 0 | 0 | 35.0 |
| | 20 | 1 | 22.0 | | 20 | 0 | 53.7 | | 20 | 0 | 34.5 |
| | 40 | 1 | 21.0 | | 40 | 0 | 53.1 | | 40 | 0 | 34.1 |
| 36 | 0 | 1 | 20.1 | 48 | 0 | 0 | 52.5 | 60 | 0 | 0 | 33.6 |
| | 20 | 1 | 19.1 | | 20 | 0 | 51.9 | | 20 | 0 | 33.2 |
| | 40 | 1 | 18.2 | | 40 | 0 | 51.2 | | 40 | 0 | 32.7 |
| 37 | 0 | 1 | 17.2 | 49 | 0 | 0 | 50.6 | 61 | 0 | 0 | 32.3 |
| | 20 | 1 | 16.3 | | 20 | 0 | 50.0 | 62 | 0 | 0 | 31.0 |
| | 40 | 1 | 15.4 | | 40 | 0 | 49.4 | 63 | 0 | 0 | 29.7 |
| 38 | 0 | 1 | 14.5 | 50 | 0 | 0 | 48.9 | 64 | 0 | 0 | 28.4 |
| | 20 | 1 | 13.6 | | 20 | 0 | 48.3 | 65 | 0 | 0 | 27.2 |
| | 40 | 1 | 12.7 | | 40 | 0 | 47.8 | 66 | 0 | 0 | 25.9 |
| 39 | 0 | 1 | 11.9 | 51 | 0 | 0 | 47.2 | 67 | 0 | 0 | 24.7 |
| | 20 | 1 | 11.0 | | 20 | 0 | 46.6 | 68 | 0 | 0 | 23.6 |
| | 40 | 1 | 10.2 | | 40 | 0 | 46.1 | 69 | 0 | 0 | 22.4 |
| 40 | 0 | 1 | 9.4 | 52 | 0 | 0 | 45.5 | 70 | 0 | 0 | 21.2 |
| | 20 | 1 | 8.6 | | 20 | 0 | 45.0 | 71 | 0 | 0 | 20.1 |
| | 40 | 1 | 7.8 | | 40 | 0 | 44.4 | 72 | 0 | 0 | 18.9 |
| 41 | 0 | 1 | 7.0 | 53 | 0 | 0 | 43.9 | 73 | 0 | 0 | 17.8 |
| | 20 | 1 | 6.2 | | 20 | 0 | 43.4 | 74 | 0 | 0 | 16.7 |
| | 40 | 1 | 5.4 | | 40 | 0 | 42.8 | 75 | 0 | 0 | 15.6 |
| 42 | 0 | 1 | 4.7 | 54 | 0 | 0 | 42.3 | 76 | 0 | 0 | 14.5 |
| | 20 | 1 | 3.9 | | 20 | 0 | 41.8 | 77 | 0 | 0 | 13.5 |
| | 40 | 1 | 3.2 | | 40 | 0 | 41.3 | 78 | 0 | 0 | 12.4 |
| 43 | 0 | 1 | 2.4 | 55 | 0 | 0 | 40.8 | 79 | 0 | 0 | 11.3 |
| | 20 | 1 | 1.7 | | 20 | 0 | 40.3 | 80 | 0 | 0 | 10.3 |
| | 40 | 1 | 1.0 | | 40 | 0 | 39.8 | 81 | 0 | 0 | 9.2 |
| 44 | 0 | 1 | 0.3 | 56 | 0 | 0 | 39.3 | 82 | 0 | 0 | 8.2 |
| | 20 | 0 | 59.6 | | 20 | 0 | 38.8 | 83 | 0 | 0 | 7.2 |
| | 40 | 0 | 58.9 | | 40 | 0 | 38.3 | 84 | 0 | 0 | 6.1 |
| 45 | 0 | 0 | 58.2 | 57 | 0 | 0 | 37.8 | 85 | 0 | 0 | 5.1 |
| | 20 | 0 | 57.6 | | 20 | 0 | 37.3 | 86 | 0 | 0 | 4.1 |
| | 40 | 0 | 56.9 | | 40 | 0 | 36.9 | 87 | 0 | 0 | 3.1 |
| 46 | 0 | 0 | 56.2 | 58 | 0 | 0 | 36.4 | 88 | 0 | 0 | 2.0 |
| | 20 | 0 | 55.6 | | 20 | 0 | 35.9 | 89 | 0 | 0 | 1.0 |
| | 40 | 0 | 55.0 | | 40 | 0 | 35.5 | 90 | 0 | 0 | 0.0 |

## TABLE II.—*Bessel's Refraction-Table.*

| Zenith-distance | | Log a. | | A | λ |
|---|---|---|---|---|---|
| ° | ′ | | | | |
| 0 | 0 | 1.76156 | 2 | | |
| 10 | 0 | 1.76154 | 2 | | |
| 20 | 0 | 1.76149 | 5 | | |
| 30 | 0 | 1.76139 | 10 | | |
| 35 | 0 | 1.76130 | 9 | | |
| 40 | 0 | 1.76119 | 11 | | |
| | | | 15 | | |
| 45 | 0 | 1.76104 | 4 | | 1.0018 |
| 46 | 0 | 1.76100 | 4 | | 1.0019 |
| 47 | 0 | 1.76096 | 4 | | 1.0019 |
| 48 | 0 | 1.76092 | 4 | | 1.0020 |
| 49 | 0 | 1.76087 | 5 | | 1.0021 |
| 50 | 0 | 1.76082 | 5 | | 1.0023 |
| | | | 5 | | |
| 51 | 0 | 1.76077 | 6 | | 1.0025 |
| 52 | 0 | 1.76071 | 6 | | 1.0026 |
| 53 | 0 | 1.76065 | 7 | | 1.0027 |
| 54 | 0 | 1.76058 | 8 | | 1.0029 |
| 55 | 0 | 1.76050 | 8 | | 1.0031 |
| 56 | 0 | 1.76042 | 9 | | 1.0034 |
| 57 | 0 | 1.76033 | 10 | | 1.0037 |
| 58 | 0 | 1.76023 | 11 | | 1.0040 |
| 59 | 0 | 1.76012 | 11 | | 1.0043 |
| 60 | 0 | 1.76001 | 13 | | 1.0046 |
| 61 | 0 | 1.75988 | 15 | | 1.0049 |
| 62 | 0 | 1.75973 | 16 | | 1.0054 |
| 63 | 0 | 1.75957 | 18 | | 1.0058 |
| 64 | 0 | 1.75939 | 20 | | 1.0063 |
| 65 | 0 | 1.75919 | 22 | | 1.0068 |
| 66 | 0 | 1.75897 | 26 | | 1.0075 |
| 67 | 0 | 1.75871 | 29 | | 1.0083 |
| 68 | 0 | 1.75842 | 33 | | 1.0092 |
| 69 | 0 | 1.75809 | 38 | | 1.0101 |
| 70 | 0 | 1.75771 | 45 | | 1.0111 |
| 71 | 0 | 1.75726 | 51 | | 1.0124 |
| 72 | 0 | 1.75675 | 60 | | 1.0139 |
| 73 | 0 | 1.75615 | 72 | | 1.0156 |
| 74 | 0 | 1.75543 | 86 | | 1.0175 |
| 75 | 0 | 1.75457 | 16 | | 1.0197 |
| 75 | 10 | 1.75441 | 16 | | 1.0200 |
| 75 | 20 | 1.75425 | 17 | | 1.0204 |
| 75 | 30 | 1.75408 | 17 | | 1.0208 |
| 75 | 40 | 1.75391 | 18 | | 1.0212 |
| 75 | 50 | 1.75373 | 18 | | 1.0216 |
| 76 | 0 | 1.75355 | 19 | | 1.0220 |
| 76 | 10 | 1.75336 | 20 | | 1.0225 |
| 76 | 20 | 1.75316 | 21 | | 1.0230 |
| 76 | 30 | 1.75295 | 21 | | 1.0235 |
| 76 | 40 | 1.75274 | 22 | | 1.0241 |
| 76 | 50 | 1.75252 | 23 | | 1.0246 |
| 77 | 0 | 1.75229 | | 1.0026 | 1.0252 |

| Zenith-distance | | Log a. | | A | λ |
|---|---|---|---|---|---|
| ° | ′ | | | | |
| 77 | 0 | 1.75229 | 24 | 1.0026 | 1.0252 |
| 77 | 10 | 1.75205 | 25 | 1.0026 | 1.0258 |
| 77 | 20 | 1.75180 | 25 | 1.0027 | 1.0264 |
| 77 | 30 | 1.75155 | 26 | 1.0027 | 1.0272 |
| 77 | 40 | 1.75129 | 28 | 1.0028 | 1.0281 |
| 77 | 50 | 1.75101 | 29 | 1.0029 | 1.0290 |
| 78 | 0 | 1.75072 | 29 | 1.0030 | 1.0299 |
| 78 | 10 | 1.75043 | 30 | 1.0030 | 1.0308 |
| 78 | 20 | 1.75013 | 32 | 1.0031 | 1.0318 |
| 78 | 30 | 1.74981 | 34 | 1.0032 | 1.0328 |
| 78 | 40 | 1.74947 | 35 | 1.0033 | 1.0338 |
| 78 | 50 | 1.74912 | 36 | 1.0034 | 1.0347 |
| 79 | 0 | 1.74876 | 37 | 1.0035 | 1.0357 |
| 79 | 10 | 1.74839 | 40 | 1.0036 | 1.0367 |
| 79 | 20 | 1.74799 | 42 | 1.0037 | 1.0377 |
| 79 | 30 | 1.74757 | 43 | 1.0038 | 1.0387 |
| 79 | 40 | 1.74714 | 44 | 1.0039 | 1.0398 |
| 79 | 50 | 1.74670 | 47 | 1.0040 | 1.0409 |
| 80 | 0 | 1.74623 | 50 | 1.0041 | 1.0420 |
| 80 | 10 | 1.74573 | 52 | 1.0042 | 1.0431 |
| 80 | 20 | 1.74521 | 53 | 1.0043 | 1.0442 |
| 80 | 30 | 1.74468 | 56 | 1.0045 | 1.0454 |
| 80 | 40 | 1.74412 | 60 | 1.0046 | 1.0466 |
| 80 | 50 | 1.74352 | 64 | 1.0047 | 1.0479 |
| 81 | 0 | 1.74288 | 65 | 1.0049 | 1.0493 |
| 81 | 10 | 1.74223 | 68 | 1.0050 | 1.0508 |
| 81 | 20 | 1.74155 | 72 | 1.0052 | 1.0523 |
| 81 | 30 | 1.74083 | 76 | 1.0054 | 1.0540 |
| 81 | 40 | 1.74007 | 79 | 1.0056 | 1.0559 |
| 81 | 50 | 1.73928 | 83 | 1.0058 | 1.0579 |
| 82 | 0 | 1.73845 | 88 | 1.0060 | 1.0600 |
| 82 | 10 | 1.73757 | 94 | 1.0062 | 1.0622 |
| 82 | 20 | 1.73663 | 99 | 1.0065 | 1.0646 |
| 82 | 30 | 1.73564 | 105 | 1.0067 | 1.0671 |
| 82 | 40 | 1.73459 | 112 | 1.0070 | 1.0697 |
| 82 | 50 | 1.73347 | 118 | 1.0073 | 1.0725 |
| 83 | 0 | 1.73229 | 124 | 1.0075 | 1.0754 |
| 83 | 10 | 1.73105 | 131 | 1.0078 | 1.0784 |
| 83 | 20 | 1.72974 | 142 | 1.0081 | 1.0815 |
| 83 | 30 | 1.72832 | 151 | 1.0084 | 1.0846 |
| 83 | 40 | 1.72681 | 162 | 1.0088 | 1.0879 |
| 83 | 50 | 1.72519 | 173 | 1.0092 | 1.0914 |
| 84 | 0 | 1.72346 | 186 | 1.0096 | 1.0951 |
| 84 | 10 | 1.72160 | 199 | 1.0100 | 1.0992 |
| 84 | 20 | 1.71961 | 212 | 1.0105 | 1.1036 |
| 84 | 30 | 1.71749 | 227 | 1.0110 | 1.1082 |
| 84 | 40 | 1.71522 | 243 | 1.0115 | 1.1130 |
| 84 | 50 | 1.71279 | 259 | 1.0121 | 1.1178 |
| 85 | 0 | 1.71020 | | 1.0127 | 1.1229 |

## TABLE II.—*Bessel's Refraction-Table.*

**Factor depending upon the barometer.**

| Eng. ins. | Log B. |
|---|---|
| 27.5 | −0.03191 |
| 27.6 | 0.03033 |
| 27.7 | 0.02876 |
| 27.8 | 0.02720 |
| 27.9 | 0.02564 |
| 28.0 | 0.02409 |
| 28.1 | 0.02254 |
| 28.2 | 0.02099 |
| 28.3 | 0.01946 |
| 28.4 | 0.01793 |
| 28.5 | 0.01640 |
| 28.6 | 0.01488 |
| 28.7 | 0.01336 |
| 28.8 | 0.01185 |
| 28.9 | 0.01035 |
| 29.0 | 0.00885 |
| 29.1 | 0.00735 |
| 29.2 | 0.00586 |
| 29.3 | 0.00438 |
| 29.4 | 0.00290 |
| 29.5 | −0.00142 |
| 29.6 | +0.00005 |
| 29.7 | 0.00151 |
| 29.8 | 0.00297 |
| 29.9 | 0.00443 |
| 30.0 | 0.00588 |
| 30.1 | 0.00732 |
| 30.2 | 0.00876 |
| 30.3 | 0.01020 |
| 30.4 | 0.01163 |
| 30.5 | 0.01306 |
| 30.6 | 0.01448 |
| 30.7 | 0.01589 |
| 30.8 | 0.01731 |
| 30.9 | 0.01871 |
| 31.0 | +0.02012 |

**Factor depending upon the attached thermometer.**

| F. | Log T. |
|---|---|
| −30 | +0.00242 |
| 20 | 0.00203 |
| −10 | 0.00164 |
| 0 | 0.00125 |
| +10 | 0.00086 |
| 20 | 0.00047 |
| 30 | +0.00008 |
| 40 | −0.00031 |
| 50 | 0.00070 |
| 60 | 0.00109 |
| 70 | 0.00148 |
| 80 | 0.00186 |
| 90 | 0.00225 |
| +100 | −0.00264 |

$\text{Log } \beta = \text{log } B + \text{log } T.$

**Factor depending upon the external thermometer.**

| F. | Log γ. | F. | Log γ. |
|---|---|---|---|
| −20 | +0.06279 | +35 | +0.01185 |
| 19 | 0.06181 | 36 | 0.01098 |
| 18 | 0.06083 | 37 | 0.01011 |
| 17 | 0.05985 | 38 | 0.00924 |
| 16 | 0.05887 | 39 | 0.00837 |
| 15 | 0.05790 | 40 | 0.00750 |
| 14 | 0.05693 | 41 | 0.00664 |
| 13 | 0.05596 | 42 | 0.00578 |
| 12 | 0.05500 | 43 | 0.00492 |
| 11 | 0.05403 | 44 | 0.00406 |
| 10 | 0.05307 | 45 | 0.00320 |
| 9 | 0.05211 | 46 | 0.00234 |
| 8 | 0.05115 | 47 | 0.00149 |
| 7 | 0.05020 | 48 | +0.00064 |
| 6 | 0.04924 | 49 | −0.00021 |
| 5 | 0.04829 | 50 | 0.00106 |
| 4 | 0.04734 | 51 | 0.00191 |
| 3 | 0.04640 | 52 | 0.00275 |
| 2 | 0.04545 | 53 | 0.00360 |
| −1 | 0.04451 | 54 | 0.00444 |
| 0 | 0.04357 | 55 | 0.00528 |
| +1 | 0.04263 | 56 | 0.00612 |
| 2 | 0.04169 | 57 | 0.00696 |
| 3 | 0.04076 | 58 | 0.00780 |
| 4 | 0.03982 | 59 | 0.00863 |
| 5 | 0.03889 | 60 | 0.00946 |
| 6 | 0.03796 | 61 | 0.01029 |
| 7 | 0.03704 | 62 | 0.01112 |
| 8 | 0.03611 | 63 | 0.01195 |
| 9 | 0.03519 | 64 | 0.01278 |
| 10 | 0.03427 | 65 | 0.01360 |
| 11 | 0.03335 | 66 | 0.01443 |
| *12 | 0.03243 | 67 | 0.01525 |
| 13 | 0.03152 | 68 | 0.01607 |
| 14 | 0.03060 | 69 | 0.01689 |
| 15 | 0.02969 | 70 | 0.01770 |
| 16 | 0.02878 | 71 | 0.01852 |
| 17 | 0.02787 | 72 | 0.01933 |
| 18 | 0.02697 | 73 | 0.02015 |
| 19 | 0.02606 | 74 | 0.02096 |
| 20 | 0.02516 | 75 | 0.02177 |
| 21 | 0.02426 | 76 | 0.02257 |
| 22 | 0.02336 | 77 | 0.02338 |
| 23 | 0.02247 | 78 | 0.02419 |
| 24 | 0.02157 | 79 | 0.02499 |
| 25 | 0.02068 | 80 | 0.02579 |
| 26 | 0.01979 | 81 | 0.02659 |
| 27 | 0.01890 | 82 | 0.02738 |
| 28 | 0.01801 | 83 | 0.02819 |
| 29 | 0.01713 | 84 | 0.02898 |
| 30 | 0.01624 | 85 | 0.02978 |
| 31 | 0.01536 | 86 | 0.03057 |
| 32 | 0.01448 | 87 | 0.03136 |
| 33 | 0.01360 | 88 | 0.03216 |
| 34 | 0.01273 | 89 | 0.03294 |
| +35 | +0.01185 | +90 | −0.03373 |

LIV.—*To determine the Latitude from the Meridional Altitude of an Object whose Declination is known.*

1. When the object observed is south of the zenith:

$$L = 90° + D - A = Z + D = 90° + Z - \triangle = 180° - (A + \triangle)$$

2. When the star is between the zenith and the pole:

$$L = A - \triangle = D - Z = 90° - (Z + \triangle) = A + D - 90°$$

3. When the star is between the pole and the horizon to the north:

$$L = A + \triangle = 90° + \triangle - Z = 90° + A - D = 180° - (Z + D)$$

where L = the latitude sought;

D = the declination of the object, *minus* when south;

$\triangle$ = its north-polar distance;

A = its meridional altitude; and

Z = its meridional zenith-distance.

A and Z must be corrected for refraction.

When the sun is the object observed,

A = observed altitude — (refraction — parallax) ± semi-diameter.

LV.—*Determination of the Latitude of a Place by the Method of Circum-Meridian Altitudes.*

Reduction to meridian =

$$x = k \left\{ i \frac{\cos l \cos D}{\cos a} \right\} - m \tan a \left\{ i \frac{\cos l \cos D}{\cos a} \right\}^2$$

$$k = \frac{2 \sin^2 \tfrac{1}{2} p}{\sin 1''}$$

$$m = \frac{2 \sin^4 \tfrac{1}{2} p}{\sin 1''}$$

$$a = 90° + D - l$$

Where—

A = $a + x$ = the meridional altitude of the object;

$a$ = its observed altitude — (refraction — parallax) ± semi-diameter;

$p$ = its correct hour-angle;

D = its declination;

$l$ = the assumed latitude of the place; and

$x$ = the required correction in seconds.

When a *star* is the object observed and the chronometer marks *mean* time—

$$i = 1.005473; \quad \log i = 0.0023708$$

When the *sun* is observed and the chronometer marks *sidereal* time—

$$i = 0.99455418; \quad \log i = 9.9976285$$

and, generally, when the chronometer has a large losing rate, $x$ must be multiplied by $1 + 0.00002315\ r$; when it has a gaining rate it must be divided by $1 + 00002315\ r$; $r$ being the rate in 24 hours, which must be assumed *minus* when *gaining*, and *plus* when *losing*.

LV.—*Determination of the Latitude, &c.*—Continued.

The values of $k$ and $m$ for each value of $p$ are given in the following tables.

The meridian altitude,

$$\Lambda = a + x$$

for each observation; for any number of observations, $n$,

$$\frac{a' + a'' + \cdots}{n} + \frac{x' + x'' + \cdots}{n}$$

= the mean, $a$, of all the observed altitudes + the mean, $x$, of all the corrections. Consequently,

1. Measure several successive altitudes of the object both before and after its meridional passage.

2. Note the times of each observation, and compute the time of the object's culmination; the differences between this and the times of each successive observation are the values of $p'$, $p''$, &c., in time, for which the corresponding values of $k'$, $k''$, &c., and $m'$, $m''$, &c., must be taken from the tables.

3. The means $k$ and $m$ of these results will be introduced into the equation for the value of the correction, $x$, to be applied to $a$ to obtain the meridional altitude, $\Lambda$, of the object.

4. If the final latitude differ much from the assumed, the computation should be repeated with the new value for $l$.

5. It is not necessary that the time of the object's culmination should be known with great precision, provided an equal number of altitudes be taken upon each side of the meridian, and at nearly equal distances from it.

6. The second correction, $m$, is seldom necessary, unless great accuracy is desired, and the object is observed more than ten minutes of time from the meridian.

$$\textit{Reduction to the Meridian; Values of } k = \frac{2 \sin^2 \frac{1}{2} p}{\sin 1''}$$

| Sec. | 0$^m$ | 1$^m$ | 2$^m$ | 3$^m$ | 4$^m$ | 5$^m$ | 6$^m$ | 7$^m$ |
|---|---|---|---|---|---|---|---|---|
|   | '' | '' | '' | '' | '' | '' | '' | '' |
| 0 | 0. 00 | 1. 96 | 7. 8 | 17. 7 | 31. 4 | 49. 1 | 70. 7 | 96. 2 |
| 1 | 0. 00 | 2. 03 | 8. 0 | 17. 9 | 31. 7 | 49. 4 | 71. 1 | 96. 7 |
| 2 | 0. 00 | 2. 10 | 8. 1 | 18. 1 | 31. 9 | 49. 7 | 71. 5 | 97. 1 |
| 3 | 0. 00 | 2. 16 | 8. 2 | 18. 3 | 32. 2 | 50. 1 | 71. 9 | 97. 6 |
| 4 | 0. 01 | 2. 23 | 8. 4 | 18. 5 | 32. 5 | 50. 4 | 72. 3 | 98. 0 |
| 5 | 0. 01 | 2. 31 | 8. 5 | 18. 7 | 32. 7 | 50. 7 | 72. 7 | 98. 5 |
| 6 | 0. 02 | 2. 38 | 8. 7 | 18. 9 | 33. 0 | 51. 1 | 73. 1 | 99. 0 |
| 7 | 0. 02 | 2. 45 | 8. 8 | 19. 1 | 33. 3 | 51. 4 | 73. 5 | 99. 4 |
| 8 | 0. 03 | 2. 52 | 8. 9 | 19. 3 | 33. 5 | 51. 7 | 73. 9 | 99. 9 |
| 9 | 0. 04 | 2. 60 | 9. 1 | 19. 5 | 33. 8 | 52. 1 | 74. 3 | 100. 4 |
| 10 | 0. 05 | 2. 67 | 9. 2 | 19. 7 | 34. 1 | 52. 4 | 74. 7 | 100. 8 |
| 11 | 0. 06 | 2. 75 | 9. 4 | 19. 9 | 34. 4 | 52. 7 | 75. 1 | 101. 3 |
| 12 | 0. 08 | 2. 83 | 9. 5 | 20. 1 | 34. 6 | 53. 1 | 75. 5 | 101. 8 |
| 13 | 0. 09 | 2. 91 | 9. 6 | 20. 3 | 34. 9 | 53. 4 | 75. 9 | 102. 3 |
| 14 | 0. 11 | 2. 99 | 9. 8 | 20. 5 | 35. 2 | 53. 8 | 76. 3 | 102. 7 |
| 15 | 0. 12 | 3. 07 | 9. 9 | 20. 7 | 35. 5 | 54. 1 | 76. 7 | 103. 2 |
| 16 | 0. 14 | 3. 15 | 10. 1 | 20. 9 | 35. 7 | 54. 5 | 77. 1 | 103. 7 |
| 17 | 0. 16 | 3. 23 | 10. 2 | 21. 2 | 36. 0 | 54. 8 | 77. 5 | 104. 2 |
| 18 | 0. 18 | 3. 32 | 10. 4 | 21. 4 | 36. 3 | 55. 1 | 77. 9 | 104. 6 |
| 19 | 0. 20 | 3. 40 | 10. 5 | 21. 6 | 36. 6 | 55. 5 | 78. 3 | 105. 1 |
| 20 | 0. 22 | 3. 49 | 10. 7 | 21. 8 | 36. 9 | 55. 8 | 78. 8 | 105. 6 |
| 21 | 0. 24 | 3. 58 | 10. 8 | 22. 0 | 37. 2 | 56. 2 | 79. 2 | 106. 1 |
| 22 | 0. 26 | 3. 67 | 11. 0 | 22. 3 | 37. 4 | 56. 5 | 79. 6 | 106. 6 |
| 23 | 0. 28 | 3. 76 | 11. 2 | 22. 5 | 37. 7 | 56. 9 | 80. 0 | 107. 0 |
| 24 | 0. 31 | 3. 85 | 11. 3 | 22. 7 | 38. 0 | 57. 3 | 80. 4 | 107. 5 |
| 25 | 0. 34 | 3. 94 | 11. 5 | 22. 9 | 38. 3 | 57. 6 | 80. 8 | 108. 0 |
| 26 | 0. 37 | 4. 03 | 11. 6 | 23. 1 | 38. 6 | 58. 0 | 81. 3 | 108. 5 |
| 27 | 0. 40 | 4. 12 | 11. 8 | 23. 4 | 38. 9 | 58. 3 | 81. 7 | 109. 0 |
| 28 | 0. 43 | 4. 22 | 11. 9 | 23. 6 | 39. 2 | 58. 7 | 82. 1 | 109. 5 |
| 29 | 0. 46 | 4. 32 | 12. 1 | 23. 8 | 39. 5 | 59. 0 | 82. 5 | 110. 0 |

*Reduction to the Meridian; Values of* $k = \dfrac{2 \sin^2 \frac{1}{2} p}{\sin 1''}$

| Sec. | 0ᵐ | 1ᵐ | 2ᵐ | 3ᵐ | 4ᵐ | 5ᵐ | 6ᵐ | 7ᵐ |
|---|---|---|---|---|---|---|---|---|
| | '' | '' | '' | '' | '' | '' | '' | '' |
| 30 | 0. 49 | 4. 42 | 12. 3 | 24. 0 | 39. 8 | 59. 4 | 83. 0 | 110. 4 |
| 31 | 0. 52 | 4. 52 | 12. 4 | 24. 3 | 40. 1 | 59. 8 | 83. 4 | 110. 9 |
| 32 | 0. 56 | 4. 62 | 12. 6 | 24. 5 | 40. 3 | 60. 1 | 83. 8 | 111. 4 |
| 33 | 0. 59 | 4. 72 | 12. 8 | 24. 7 | 40. 6 | 60. 5 | 84. 2 | 111. 9 |
| 34 | 0. 63 | 4. 82 | 12. 9 | 25. 0 | 40. 9 | 60. 8 | 84. 7 | 112. 4 |
| 35 | 0. 67 | 4. 92 | 13. 1 | 25. 2 | 41. 2 | 61. 2 | 85. 1 | 112. 9 |
| 36 | 0. 71 | 5. 03 | 13. 3 | 25. 4 | 41. 5 | 61. 6 | 85. 5 | 113. 4 |
| 37 | 0. 75 | 5. 13 | 13. 4 | 25. 7 | 41. 8 | 61. 9 | 86. 0 | 113. 9 |
| 38 | 0. 80 | 5. 24 | 13. 6 | 25. 9 | 42. 1 | 62. 3 | 86. 4 | 114. 4 |
| 39 | 0. 83 | 5. 34 | 13. 8 | 26. 2 | 42. 5 | 62. 7 | 86. 8 | 114. 9 |
| 40 | 0. 87 | 5. 45 | 14. 0 | 26. 4 | 42. 8 | 63. 0 | 87. 3 | 115. 4 |
| 41 | 0. 91 | 5. 56 | 14. 1 | 26. 6 | 43. 1 | 63. 4 | 87. 7 | 115. 9 |
| 42 | 0. 96 | 5. 67 | 14. 3 | 26. 9 | 43. 4 | 63. 8 | 88. 1 | 116. 4 |
| 43 | 1. 01 | 5. 78 | 14. 5 | 27. 1 | 43. 7 | 64. 2 | 88. 6 | 116. 9 |
| 44 | 1. 06 | 5. 90 | 14. 7 | 27. 4 | 44. 0 | 64. 5 | 89. 0 | 117. 4 |
| 45 | 1. 10 | 6. 01 | 14. 8 | 27. 6 | 44. 3 | 64. 9 | 89. 5 | 117. 9 |
| 46 | 1. 15 | 6. 13 | 15. 0 | 27. 9 | 44. 6 | 65. 3 | 89. 9 | 118. 4 |
| 47 | 1. 20 | 6. 24 | 15. 2 | 28. 1 | 44. 9 | 65. 7 | 90. 3 | 118. 9 |
| 48 | 1. 26 | 6. 36 | 15. 4 | 28. 3 | 45. 2 | 66. 0 | 90. 8 | 119. 5 |
| 49 | 1. 31 | 6. 48 | 15. 6 | 28. 6 | 45. 5 | 66. 4 | 91. 2 | 120. 0 |
| 50 | 1. 36 | 6. 60 | 15. 8 | 28. 8 | 45. 9 | 66; 8 | 91. 7 | 120. 5 |
| 51 | 1. 42 | 6. 72 | 15. 9 | 29. 1 | 46. 2 | 67. 2 | 92. 1 | 121. 0 |
| 52 | 1. 48 | 6. 84 | 16. 1 | 29. 4 | 46. 5 | 67. 6 | 92. 6 | 121. 5 |
| 53 | 1. 53 | 6. 96 | 16. 3 | 29. 6 | 46. 8 | 68. 0 | 93. 0 | 122. 0 |
| 54 | 1. 59 | 7. 09 | 16. 5 | 29. 9 | 47. 1 | 68. 3 | 93. 5 | 122. 5 |
| 55 | 1. 65 | 7. 21 | 16. 7 | 30. 1 | 47. 5 | 68. 7 | 93. 9 | 123. 1 |
| 56 | 1. 71 | 7. 34 | 16. 9 | 30. 4 | 47. 8 | 69. 1 | 94. 4 | 123. 6 |
| 57 | 1. 77 | 7. 46 | 17. 1 | 30. 6 | 48. 1 | 69. 5 | 94. 8 | 124. 1 |
| 58 | 1. 83 | 7. 60 | 17. 3 | 30. 9 | 48. 4 | 69. 9 | 95. 3 | 124. 6 |
| 59 | 1. 89 | 7. 72 | 17. 5 | 31. 1 | 48. 8 | 70. 3 | 95. 7 | 125. 1 |

*Reduction to the Meridian; Values of* $k = \dfrac{2 \sin^2 \frac{1}{2} p}{\sin 1''}$

| Sec. | $8^m$ | $9^m$ | $10^m$ | $11^m$ | $12^m$ | $13^m$ | $14^m$ |
|------|-------|-------|--------|--------|--------|--------|--------|
|      | $''$  | $''$  | $''$   | $''$   | $''$   | $''$   | $''$   |
| 0  | 125.7 | 159.0 | 196.3 | 237.5 | 282.7 | 331.8 | 384.7 |
| 1  | 126.2 | 159.6 | 197.0 | 238.3 | 283.5 | 332.6 | 385.6 |
| 2  | 126.7 | 160.2 | 197.6 | 239.0 | 284.2 | 333.4 | 386.6 |
| 3  | 127.2 | 160.8 | 198.3 | 239.7 | 285.0 | 334.3 | 387.5 |
| 4  | 127.8 | 161.4 | 198.9 | 240.4 | 285.8 | 335.2 | 388.4 |
| 5  | 128.3 | 162.0 | 199.6 | 241.2 | 286.6 | 336.0 | 389.3 |
| 6  | 128.8 | 162.6 | 200.3 | 241.9 | 287.4 | 336.9 | 390.2 |
| 7  | 129.3 | 163.2 | 200.9 | 242.6 | 288.2 | 337.7 | 391.1 |
| 8  | 129.9 | 163.8 | 201.6 | 243.3 | 289.0 | 338.6 | 392.1 |
| 9  | 130.4 | 164.4 | 202.2 | 244.1 | 289.8 | 339.4 | 393.0 |
| 10 | 131.0 | 165.0 | 202.9 | 244.8 | 290.6 | 340.3 | 393.9 |
| 11 | 131.5 | 165.6 | 203.6 | 245.5 | 291.4 | 341.2 | 394.8 |
| 12 | 132.0 | 166.2 | 204.2 | 246.3 | 292.2 | 342.0 | 395.8 |
| 13 | 132.6 | 166.8 | 204.9 | 247.0 | 293.0 | 342.9 | 396.7 |
| 14 | 133.1 | 167.4 | 205.6 | 247.7 | 293.8 | 343.7 | 397.6 |
| 15 | 133.6 | 168.0 | 206.3 | 248.5 | 294.6 | 344.6 | 398.6 |
| 16 | 134.2 | 168.6 | 206.9 | 249.2 | 295.4 | 345.5 | 399.5 |
| 17 | 134.7 | 169.2 | 207.6 | 249.9 | 296.2 | 346.4 | 400.5 |
| 18 | 135.3 | 169.8 | 208.3 | 250.7 | 297.0 | 347.2 | 401.4 |
| 19 | 135.8 | 170.4 | 208.9 | 251.4 | 297.8 | 348.1 | 402.3 |
| 20 | 136.3 | 171.0 | 209.6 | 252.2 | 298.6 | 349.0 | 403.3 |
| 21 | 136.9 | 171.6 | 210.3 | 253.0 | 299.4 | 349.8 | 404.2 |
| 22 | 137.4 | 172.2 | 211.0 | 253.6 | 300.2 | 350.7 | 405.1 |
| 23 | 138.0 | 172.9 | 211.7 | 254.4 | 301.0 | 351.6 | 406.0 |
| 24 | 138.5 | 173.5 | 212.3 | 255.1 | 301.8 | 352.5 | 407.0 |
| 25 | 139.1 | 174.1 | 213.0 | 255.9 | 302.6 | 353.3 | 408.0 |
| 26 | 139.6 | 174.7 | 213.7 | 256.6 | 303.5 | 354.2 | 408.9 |
| 27 | 140.2 | 175.3 | 214.4 | 257.4 | 304.3 | 355.1 | 409.9 |
| 28 | 140.7 | 175.9 | 215.1 | 258.1 | 305.1 | 356.0 | 410.8 |
| 29 | 141.3 | 176.6 | 215.8 | 258.9 | 305.9 | 356.9 | 411.7 |

Reduction to the Meridian ; Values of $k = \dfrac{2 \sin^2 \frac{1}{2} p}{\sin 1''}$

| Sec. | $8^m$ | $9^m$ | $10^m$ | $11^m$ | $12^m$ | $13^m$ | $14^m$ |
|---|---|---|---|---|---|---|---|
|    | $''$ | $''$ | $''$ | $''$ | $''$ | $''$ | $''$ |
| 30 | 141.8 | 177.2 | 216.4 | 259.6 | 306 7 | 357.7 | 412.7 |
| 31 | 142.4 | 177.8 | 217.1 | 260.4 | 307.5 | 358.6 | 413.6 |
| 32 | 143.0 | 178.4 | 217.8 | 261.1 | 308.4 | 359.5 | 414.6 |
| 33 | 143.5 | 179.0 | 218.5 | 261.9 | 309.2 | 360.4 | 415.5 |
| 34 | 144.1 | 179.7 | 219.2 | 262.6 | 310.0 | 361.3 | 416.5 |
| 35 | 144.6 | 180.3 | 219.9 | 263.4 | 310.8 | 362.2 | 417.5 |
| 36 | 145.2 | 180.9 | 220.6 | 264.1 | 311.6 | 363.1 | 418.4 |
| 37 | 145.8 | 181.6 | 221.3 | 264.9 | 312.5 | 364.0 | 419.4 |
| 38 | 146.3 | 182.2 | 222.0 | 265.7 | 313.3 | 364.8 | 420.3 |
| 39 | 146.9 | 182.8 | 222.7 | 266.4 | 314.1 | 365.7 | 421.3 |
| 40 | 147.5 | 183.5 | 223.4 | 267.2 | 315.0 | 366.6 | 422.2 |
| 41 | 148.0 | 184.1 | 224.1 | 267.9 | 315.8 | 367.5 | 423.2 |
| 42 | 148.6 | 184.7 | 224.8 | 268.7 | 316.6 | 368.4 | 424.2 |
| 43 | 149.2 | 185.4 | 225.5 | 269.5 | 317.4 | 369.3 | 425.1 |
| 44 | 149.7 | 186.0 | 226.2 | 270.3 | 318.3 | 370.2 | 426.1 |
| 45 | 150.3 | 186.6 | 226.9 | 271.0 | 319.1 | 371.1 | 427.0 |
| 46 | 150.9 | 187.3 | 227.6 | 271.8 | 319.9 | 372.0 | 428.0 |
| 47 | 151.5 | 187.9 | 228.3 | 272.6 | 320.8 | 372.9 | 429.0 |
| 48 | 152.0 | 188.5 | 229.0 | 273.3 | 321.6 | 373.8 | 429.9 |
| 49 | 152.6 | 189.2 | 229.7 | 274.1 | 322.4 | 374.7 | 430.9 |
| 50 | 153.2 | 189.8 | 230.4 | 274.9 | 323.3 | 375.6 | 431.9 |
| 51 | 153.8 | 190.5 | 231.1 | 275.6 | 324.1 | 376.5 | 432.8 |
| 52 | 154.4 | 191.1 | 231.8 | 276.4 | 325.0 | 377.4 | 433.8 |
| 53 | 154.9 | 191.8 | 232.5 | 277.2 | 325.8 | 378.3 | 434.8 |
| 54 | 155.5 | 192.4 | 233.2 | 278.0 | 326.7 | 379.3 | 435.8 |
| 55 | 156.1 | 193.1 | 234.0 | 278.8 | 327.5 | 380.2 | 436.7 |
| 56 | 156.7 | 193.7 | 234.7 | 279.5 | 328.4 | 381.1 | 437.7 |
| 57 | 157.3 | 194.4 | 235.4 | 280.3 | 329.2 | 382.0 | 438.7 |
| 58 | 157.8 | 195.0 | 236.1 | 281.1 | 330.0 | 382.9 | 439.7 |
| 59 | 158.4 | 195.7 | 236.8 | 281.9 | 330.9 | 383.8 | 440.6 |

*Reduction to the Meridian; Values of* $k = \dfrac{2 \sin^2 \frac{1}{2} p}{\sin 1''}$

| Sec. | 15$^m$ | 16$^m$ | 17$^m$ | 18$^m$ | 19$^m$ | 20$^m$ | 21$^m$ |
|---|---|---|---|---|---|---|---|
|  | '' | '' | '' | '' | '' | '' | '' |
| 0 | 441. 6 | 502. 5 | 567. 2 | 635. 9 | 708. 4 | 784. 9 | 865. 3 |
| 1 | 442. 6 | 503. 5 | 568. 3 | 637. 0 | 709. 7 | 786. 2 | 866. 6 |
| 2 | 443. 6 | 504. 6 | 569. 4 | 638. 2 | 710. 9 | 787. 5 | 868. 0 |
| 3 | 444. 6 | 505. 6 | 570. 5 | 639. 4 | 712. 1 | 788. 8 | 869. 4 |
| 4 | 445. 6 | 506. 7 | 571. 6 | 640. 6 | 713. 4 | 790. 1 | 870. 8 |
| 5 | 446. 5 | 507. 7 | 572. 8 | 641. 7 | 714. 6 | 791. 4 | 872. 1 |
| 6 | 447. 5 | 508. 8 | 573. 9 | 642. 9 | 715. 9 | 792. 7 | 873. 5 |
| 7 | 448. 5 | 509. 8 | 575. 0 | 644. 1 | 717. 1 | 794. 0 | 874. 9 |
| 8 | 449. 5 | 510. 9 | 576. 1 | 645. 3 | 718. 4 | 795. 4 | 876. 3 |
| 9 | 450. 5 | 511. 9 | 577. 2 | 646. 5 | 719. 6 | 796. 7 | 877. 6 |
| 10 | 451. 5 | 513. 0 | 578. 4 | 647. 7 | 720. 9 | 798. 0 | 879. 0 |
| 11 | 452. 5 | 514. 0 | 579. 5 | 648. 9 | 722. 1 | 799. 3 | 880. 4 |
| 12 | 453. 5 | 515. 1 | 580. 6 | 650. 0 | 723. 4 | 800. 7 | 881. 8 |
| 13 | 454. 5 | 516. 1 | 581. 7 | 651. 2 | 724. 6 | 802. 0 | 883. 2 |
| 14 | 455. 5 | 517. 2 | 582. 9 | 652. 4 | 725. 9 | 803. 3 | 884. 6 |
| 15 | 456. 5 | 518. 3 | 584. 0 | 653. 6 | 727. 2 | 804. 6 | 886. 0 |
| 16 | 457. 5 | 519. 3 | 585. 1 | 654. 8 | 728. 4 | 806. 0 | 887. 4 |
| 17 | 458. 5 | 520. 4 | 586. 2 | 656. 0 | 729. 7 | 807. 3 | 888. 8 |
| 18 | 459. 5 | 521. 5 | 587. 4 | 657. 2 | 730. 9 | 808. 6 | 890. 2 |
| 19 | 460. 5 | 522. 5 | 588. 5 | 658. 4 | 732. 2 | 809. 9 | 891. 6 |
| 20 | 461. 5 | 523. 6 | 589. 6 | 659. 6 | 733. 5 | 811. 3 | 893. 0 |
| 21 | 462. 5 | 524. 6 | 590. 8 | 660. 8 | 734. 7 | 812. 6 | 894. 4 |
| 22 | 463. 5 | 525. 7 | 591. 9 | 662. 0 | 736. 0 | 813. 9 | 895. 8 |
| 23 | 464. 5 | 526. 8 | 593. 0 | 663. 2 | 737. 3 | 815. 2 | 897. 2 |
| 24 | 465. 5 | 527. 9 | 594. 2 | 664. 4 | 738. 5 | 816. 6 | 898. 6 |
| 25 | 466. 5 | 528. 9 | 595. 3 | 665. 6 | 739. 8 | 817. 9 | 900. 0 |
| 26 | 467. 5 | 530. 0 | 596. 5 | 666. 8 | 741. 1 | 819. 2 | 901. 4 |
| 27 | 468. 5 | 531. 1 | 597. 6 | 668. 0 | 742. 3 | 820. 5 | 902. 8 |
| 28 | 469. 5 | 532. 2 | 598. 7 | 669. 2 | 743. 6 | 821. 9 | 904. 2 |
| 29 | 470. 5 | 533. 2 | 599. 9 | 670. 4 | 744. 9 | 823. 2 | 905. 6 |

*Reduction to the Meridian; Values of* $k = \dfrac{2 \sin^2 \frac{1}{2}\, p}{\sin 1''}$

| Sec. | 15$^m$ | 16$^m$ | 17$^m$ | 18$^m$ | 19$^m$ | 20$^m$ | 21$^m$ |
|---|---|---|---|---|---|---|---|
|  | '' | '' | '' | '' | '' | '' | '' |
| 30 | 471.5 | 534.3 | 601.0 | 671.6 | 746.2 | 824.6 | 907.0 |
| 31 | 472.6 | 535.4 | 602.2 | 672.8 | 747.4 | 825.9 | 908.4 |
| 32 | 473.6 | 536.5 | 603.3 | 674.1 | 748.7 | 827.3 | 909.8 |
| 33 | 474.6 | 537.6 | 604.5 | 675.3 | 750.0 | 828.6 | 911.2 |
| 34 | 475.6 | 538.7 | 605.6 | 676.5 | 751.3 | 829.9 | 912.6 |
| 35 | 476.6 | 539.7 | 606.8 | 677.7 | 752.6 | 831.2 | 914.0 |
| 36 | 477.6 | 540.8 | 607.9 | 678.9 | 753.8 | 832.6 | 915.5 |
| 37 | 478.7 | 541.9 | 609.1 | 680.1 | 755.1 | 833.9 | 916.9 |
| 38 | 479.7 | 543.0 | 610.2 | 681.3 | 756.4 | 835.3 | 918.3 |
| 39 | 480.7 | 544.1 | 611.4 | 682.6 | 757.7 | 836.6 | 919.7 |
| 40 | 481.7 | 545.2 | 612.5 | 683.8 | 759.0 | 838.0 | 921.1 |
| 41 | 482.8 | 546.3 | 613.7 | 685.0 | 760.2 | 839.3 | 922.5 |
| 42 | 483.8 | 547.4 | 614.8 | 686.2 | 761.5 | 840.7 | 923.9 |
| 43 | 484.8 | 548.4 | 616.0 | 687.4 | 762.8 | 842.0 | 925.3 |
| 44 | 485.8 | 549.5 | 617.2 | 688.7 | 764.1 | 843.4 | 926.8 |
| 45 | 486.9 | 550.6 | 618.3 | 689.9 | 765.4 | 844.7 | 928.2 |
| 46 | 487.9 | 551.7 | 619.5 | 691.1 | 766.7 | 846.1 | 929.6 |
| 47 | 488.9 | 552.8 | 620.6 | 692.4 | 768.0 | 847.5 | 931.0 |
| 48 | 490.0 | 553.9 | 621.8 | 693.6 | 769.3 | 848.9 | 932.4 |
| 49 | 491.0 | 555.0 | 623.0 | 694.8 | 770.6 | 850.2 | 933.8 |
| 50 | 492.0 | 556.1 | 624.1 | 696.0 | 771.9 | 851.6 | 935.2 |
| 51 | 493.1 | 557.2 | 625.3 | 697.3 | 773.1 | 852.9 | 936.6 |
| 52 | 494.1 | 558.3 | 626.5 | 698.5 | 774.5 | 854.3 | 938.1 |
| 53 | 495.2 | 559.4 | 627.6 | 699.7 | 775.8 | 855.7 | 939.5 |
| 54 | 496.2 | 560.5 | 628.8 | 701.0 | 777.1 | 857.1 | 940.9 |
| 55 | 497.2 | 561.6 | 630.0 | 702.2 | 778.4 | 858.4 | 942.3 |
| 56 | 498.3 | 562.7 | 631.2 | 703.5 | 779.7 | 859.8 | 943.8 |
| 57 | 499.3 | 563.9 | 632.3 | 704.7 | 781.0 | 861.1 | 945.2 |
| 58 | 500.3 | 565.0 | 633.5 | 705.9 | 782.3 | 862.5 | 946.6 |
| 59 | 501.4 | 566.1 | 634.7 | 707.1 | 783.6 | 863.9 | 948.1 |

$$\text{Reduction to the Meridian; Values of } k = \frac{2 \sin^2 \tfrac{1}{2} p}{\sin 1''}$$

| Seconds. | $22^m$ | $23^m$ | $24^m$ | Seconds. | $22^m$ | $23^m$ | $24^m$ |
|---|---|---|---|---|---|---|---|
| | '' | '' | '' | | '' | '' | '' |
| 0 | 949.6 | 1037.8 | 1129.9 | 30 | 993.2 | 1083.3 | 1177.5 |
| 1 | 951.0 | 1039.3 | 1131.4 | 31 | 994.7 | 1084.8 | 1179.1 |
| 2 | 952.4 | 1040.8 | 1133.0 | 32 | 996.2 | 1086.4 | 1180.7 |
| 3 | 953.8 | 1042.3 | 1134.6 | 33 | 997.6 | 1087.9 | 1182.3 |
| 4 | 955.3 | 1043.8 | 1136.2 | 34 | 999.1 | 1089.5 | 1183.9 |
| 5 | 956.7 | 1045.3 | 1137.8 | 35 | 1000.6 | 1091.0 | 1185.5 |
| 6 | 958.2 | 1046.8 | 1139.3 | 36 | 1002.1 | 1092.6 | 1187.1 |
| 7 | 959.6 | 1048.3 | 1140.9 | 37 | 1003.5 | 1094.1 | 1188.7 |
| 8 | 961.1 | 1049.8 | 1142.5 | 38 | 1005.0 | 1095.7 | 1190.3 |
| 9 | 962.5 | 1051.3 | 1144.0 | 39 | 1006.5 | 1097.2 | 1191.9 |
| 10 | 963.9 | 1052.8 | 1145.6 | 40 | 1008.0 | 1098.8 | 1193.5 |
| 11 | 965.4 | 1054.3 | 1147.2 | 41 | 1009.4 | 1100.3 | 1195.1 |
| 12 | 966.9 | 1055.9 | 1148.8 | 42 | 1010.9 | 1101.9 | 1196.7 |
| 13 | 968.3 | 1057.4 | 1150.4 | 43 | 1012.4 | 1103.4 | 1198.3 |
| 14 | 969.8 | 1058.9 | 1152.0 | 44 | 1013.9 | 1105.0 | 1199.9 |
| 15 | 971.2 | 1060.4 | 1153.6 | 45 | 1015.4 | 1106.5 | 1201.5 |
| 16 | 972.7 | 1062.0 | 1155.2 | 46 | 1016.9 | 1108.1 | 1203.1 |
| 17 | 974.1 | 1063.5 | 1156.8 | 47 | 1018.4 | 1109.6 | 1204.7 |
| 18 | 975.5 | 1065.0 | 1158.3 | 48 | 1019.9 | 1111.2 | 1206.4 |
| 19 | 977.0 | 1066.5 | 1159.9 | 49 | 1021.4 | 1112.7 | 1208.0 |
| 20 | 978.5 | 1068.1 | 1161.5 | 50 | 1022.8 | 1114.3 | 1209.6 |
| 21 | 979.9 | 1069.6 | 1163.1 | 51 | 1024.3 | 1115.8 | 1211.2 |
| 22 | 981.4 | 1071.1 | 1164.7 | 52 | 1025.8 | 1117.4 | 1212.9 |
| 23 | 982.9 | 1072.6 | 1166.3 | 53 | 1027.3 | 1118.9 | 1214.5 |
| 24 | 984.4 | 1074.2 | 1167.9 | 54 | 1028.8 | 1120.5 | 1216.1 |
| 25 | 985.8 | 1075.7 | 1169.5 | 55 | 1030.3 | 1122.0 | 1217.7 |
| 26 | 987.3 | 1077.2 | 1171.1 | 56 | 1031.8 | 1123.6 | 1219.4 |
| 27 | 988.8 | 1078.7 | 1172.7 | 57 | 1033.3 | 1125.1 | 1221.0 |
| 28 | 990.3 | 1080.3 | 1174.3 | 58 | 1034.8 | 1126.7 | 1222.6 |
| 29 | 991.8 | 1081.8 | 1175.9 | 59 | 1036.3 | 1128.3 | 1224.2 |

### Second Part of the Reduction to the Meridian.

$$\text{Values of } m = \frac{2 \sin^4 \frac{1}{2} p}{\sin 1''}$$

| Minutes. | 0ˢ | 10ˢ | 20ˢ | 30ˢ | 40ˢ | 50ˢ |
|---|---|---|---|---|---|---|
| | '' | '' | '' | '' | '' | '' |
| 5 | 0,01 | 0,01 | 0,01 | 0,01 | 0,01 | 0,01 |
| 6 | 0,01 | 0,01 | 0,01 | 0,02 | 0,02 | 0,02 |
| 7 | 0,02 | 0,02 | 0,03 | 0,03 | 0,03 | 0,04 |
| 8 | 0,04 | 0,04 | 0,05 | 0,05 | 0,05 | 0,06 |
| 9 | 0,06 | 0,07 | 0,08 | 0,08 | 0,08 | 0,09 |
| 10 | 0,09 | 0,10 | 0,11 | 0,11 | 0,12 | 0,13 |
| 11 | 0,14 | 0,15 | 0,15 | 0,16 | 0,17 | 0,18 |
| 12 | 0,19 | 0,20 | 0,22 | 0,23 | 0,24 | 0,25 |
| 13 | 0,27 | 0,28 | 0,30 | 0,31 | 0,33 | 0,34 |
| 14 | 0,36 | 0,38 | 0 39 | 0,41 | 0,43 | 0,45 |
| 15 | 0,47 | 0,49 | 0,52 | 0,54 | 0,56 | 0,59 |
| 16 | 0,61 | 0,64 | 0,67 | 0,69 | 0,72 | 0,75 |
| 17 | 0,78 | 0,81 | 0,84 | 0,88 | 0,91 | 0,95 |
| 18 | 0,98 | 1,02 | 1,06 | 1,09 | 1,13 | 1,18 |
| 19 | 1,22 | 1,26 | 1,30 | 1,35 | 1,40 | 1,44 |
| 20 | 1,49 | 1,54 | 1,60 | 1,65 | 1,70 | 1,76 |
| 21 | 1,82 | 1,87 | 1,93 | 1,99 | 2,06 | 2,12 |
| 22 | 2,19 | 2,25 | 2,32 | 2,39 | 2,46 | 2,54 |
| 23 | 2,61 | 2,69 | 2,77 | 2,85 | 2,93 | 3,01 |
| 24 | 3,10 | 3,18 | 3,27 | 3,36 | 3,45 | 3,55 |
| 25 | 3 64 | 3,74 | 3,84 | 3,94 | 4,05 | 4,15 |
| 26 | 4,26 | 4,37 | 4,48 | 4,60 | 4,72 | 4,83 |
| 27 | 4,96 | 5,08 | 5,20 | 5,33 | 5,46 | 5,60 |
| 28 | 5,73 | 5,87 | 6,01 | 6,15 | 6,30 | 6,44 |
| 29 | 6,59 | 6,75 | 6,90 | 7,06 | 7,22 | 7,38 |
| 30 | 7,55 | 7,72 | 7,89 | 8,06 | 8,24 | 8,42 |
| 31 | 8,61 | 8,79 | 8,98 | 9,17 | 9,37 | 9,57 |
| 32 | 9,77 | 9,97 | 10,18 | 10,39 | 10,61 | 10,82 |
| 33 | 11,04 | 11,27 | 11,50 | 11,73 | 11,96 | 12,20 |
| 34 | 12,44 | 12,69 | 12,94 | 13,19 | 13,45 | 13,71 |
| 35 | 13,97 | 14,24 | 14,51 | 14,78 | 15,06 | 15,35 |

FORM FOR

SURVEY OF ——————— DETERMINATION OF THE LATITUDE,

and South of

DATE AND STATION.—1843, *October* 13.—*Mouth of the Big Black River,*

NAME OF STAR, γ *Pegasi, South of the Zenith.*

INSTRUMENTS ... { Sextant No. 2197, by *Troughton & Simms,* and

{ *Mean Solar* Chronometer No. 76, by *Charles*

| No. for reference. | Times of observation by chronometer. | MERIDIAN DISTANCES, $= p$. | | $\dfrac{2 \sin^2 \frac{1}{2} p}{\sin 1''} = k$ | $\dfrac{\cos l \cdot \cos D}{\text{Co sine } a}$ | Reduction to the meridian, (in arc,) $= x$. |
|---|---|---|---|---|---|---|
| | | In mean solar time. | In sidereal time. | | | |
| | *h. m. s.* | *m. s.* | *m. s.* | *''* | | *' ''* |
| 1 | 10 18 40.4 | 9 44.2 | 9 45.8 | 187.3 | | 3 49.8 |
| 2 | 19 44.4 | 8 40.2 | 8 41.6 | 148.3 | | 2 51.9 |
| 3 | 20 48 | 7 36.5 | 7 37.7 | 114.2 | | 2 27.3 |
| 4 | 21 46.4 | 6 38.2 | 6 39.3 | 86.9 | | 1 47.6 |
| 5 | 22 44.4 | 5 40.2 | 5 41.1 | 63.4 | | 1 17.8 |
| 6 | 23 54 | 4 30.5 | 4 31.2 | 40.1 | | 0 49.2 |
| 7 | 25 12 | 3 12.6 | 3 13.1 | 20.3 | Constant multiple, 1.227. | 0 24.9 |
| 8 | 26 46 | 1 38.6 | 1 38.8 | 5.2 | | 0 06.3 |
| 9 | 28 16.4 | 0 08.2 | 0 08.2 | 0.0 | | 0 00.0 |
| 10 | 29 42 | 1 17.4 | 1 17.6 | 3.2 | | 0 03.9 |
| 11 | 31 42 | 3 17.4 | 3 17.9 | 21.4 | | 0 26.2 |
| 12 | 32 54.4 | 4 29.8 | 4 30.5 | 40.0 | | 0 49 |
| 13 | 34 18 | 5 53.4 | 5 54.3 | 68.5 | | 1 24 |
| 14 | 36 14.2 | 7 49.6 | 7 50.9 | 123.5 | | 2 31.5 |
| 15 | 38 32.2 | 10 07.6 | 10 09.2 | 202.3 | | 4 08.2 |
| 16 | 40 06 | 11 41.4 | 11 43.3 | 269.9 | | . 5 31.1 |

Observer, *Major J. D. Graham.*
Computer, *do.* *do.*

RECORD AND COMPUTATION.

*from Observed Double Circum-Meridian Altitudes of Stars, North the Zenith.*

*a tributary to the river Saint John, Maine.*

Artificial Horizon of Mercury.

*Young.*

| Observed double circum-meridian altitudes of star. | True circum-meridian altitude of star, as corrected for refraction and errors of instrument, $= a$. | True meridian altitudes deduced, $=(a + x) = A$. | Latitude deduced from each observation, $= L = (90° + D - A)$. |
|---|---|---|---|
| ° ′ ″ | ° ′ ″ | ° ′ ″ | ° ′ ″ |
| 114 34 15 | 57 18 38. 5 | 57 22 28. 3 | 46 56 42. 55 |
| 36 15 | 57 19 38. 5 | 57 22 30. 4 | 56 40. 45 |
| 37 10 | 57 20 06 | 57 22 33. 3 | 56 37. 55 |
| 38 10 | 57 20 36 | 57 22 23. 6 | 56 47. 25 |
| 39 30 | 57 21 16 | 57 22 33. 8 | 56 37. 05 |
| 40 30 | 57 21 46 | 57 22 35. 2 | 56 35. 65 |
| 41 05 | 57 22 03. 5 | 57 22 28. 4 | 56 42. 45 |
| 41 50 | 57 22 26 | 57 22 32. 3 | 56 34. 55 |
| 41 50 | 57 22 26 | 57 22 26 | 56 40. 85 |
| 41 50 | 57 22 26 | 57 22 29. 9 | 56 36. 95 |
| 41 00 | 57 22 01 | 57 22 27. 2 | 56 39. 65 |
| 39 45 | 57 21 23. 5 | 57 22 12. 5 | 56 58. 35 |
| 38 40 | 57 20 51 | 57 22 15 | 56 55. 85 |
| 36 30 | 57 19 46 | 57 22 17. 5 | 56 53. 55 |
| 33 20 | 57 18 11 | 57 22 19. 2 | 56 51. 85 |
| 30 50 | 57 16 56 | 57 22 27. 1 | 46 56 39. 75 |

LATITUDE—Deduced from a mean of 16 altitudes of star γ
Pegasi ....................................... 46° 56′ 43″.4
Deduced from a mean of 10 altitudes of star γ
Cephei, observed this night with same sextant .. 46 57 10 . 7

Mean, or latitude adopted ............................. 46° 56′ 57″

*Form for Record and Computation*—Continued.

D=apparent declinat'n of star=14°19′10″.85 N. log cos   9.98629
*l*=approximate latitude of place = 46°57′ ..log cos   9.83418

     Sum ............. ...................... .... 19.82048
a=approximate merid. alt. of star=57°22′10″ .log cos   9.73176

$$\frac{\cos l \cos D}{\cos a} = \text{constant multiple} = 1.227 \dots \log \qquad 0.08872$$

Refraction (ther. 28°, bar. 29.14 in.) for mean obs'd alts.  — 0′ 39″
Index-error of sextant ......................... + 2 40
*Error of eccentricity, &c., of sextant............ + 1 40

Apparent AR. of the star γ Pegasi ............ 0ʰ05ᵐ14ˢ.09
Sidereal time at mean noon at this station ...... 13 26 20.83
Sidereal interval from mean noon of star's culmi-
    nation ............................... 10 38 53.16
Retardation of mean on sidereal time.......... — 1 44.96

Mean time of culmination of star γ Pegasi...... 10 37 08.2
Chronometer (C. Y. 76) *slow* of mean time at time
    of observation ......................... — 08 43.6

Time by chronometer of culmination of star γ Pegasi 10ʰ28ᵐ24ˢ.6


On this night, October 13, 1843, Major Graham
    obtained for the latitude of this station, from 75
    observations on 5 stars south of the zenith, com-
    bined with 21 observations on γ Cephei and
    Polaris, to the north....................... 46° 56′ 56″.3
On the night of October 24, by 43 observations on
    4 southern stars, combined with 2 observations
    on γ Cephei, the latitude deduced was........ 46 56 57.2
On September 17, 1844, 66 observations on north
    and south stars gave for the latitude of this station 46 56 60.4

---

* The error of eccentricity is approximately ascertained by comparing
latitudes, well determined by observations on north and south stars, with
that which will result from north or south stars individually of various meridional altitudes. It varies with the altitudes observed; that is to say, it
is different for different parts of the limb of the instrument.

LVI.—*To Determine the Latitude by an Altitude of a Star Near the Pole, at any hour.*

$$L = A - (\Delta \cos p) + a (\Delta \sin p)^2 \tan A - \beta (\Delta \sin p)^2 (\Delta \cos p)$$

where—

A = the observed altitude, corrected for refraction, &c.;

$\Delta$ = the polar distance of the star in seconds of arc;

$a = \frac{1}{2} \sin 1''$; $\log a = 4.3845449$;

$\beta = \frac{1}{3} \sin^2 1''$; $\log \beta = 8.89403$; and

$p$ = the hour-angle of the star.

$\pm p$ = sidereal time — AR. $\ast$ = solar time + AR. $\odot$ — AR. $\ast$

$p$ is *plus* when the star is west, and *minus* when it is east of the meridian.

The sign of cos $p$ should also be attended to, for when $p$ is greater than $6^h$, or $90°$, the cosine is negative, and the second and fourth terms change the sign *minus* to *plus*.

. The fourth term may be generally omitted; its greatest value being only $0''.55$.

This formula is only applicable to stars within a very few degrees of the pole.

For other circumpolar stars—

$$\tan x = \tan \Delta \cos p$$

$$\sin y = \frac{\cos x \sin A}{\cos \Delta}$$

$$L = y \mp x$$

in which the upper sign is used when the star is above the pole; the under when below the pole.

SURVEY OF ........................................ DETERMINATION OF THE

DATE AND STATION.—1843, *September* 6—*Woodstock,*

NAME OF STAR.—*Polaris,* (*a Ursæ Minoris,*) *observed on*

INSTRUMENTS ... $\begin{cases} \text{Sextant No. 2197, by } \textit{Troughton \&} \\ \textit{Mean Solar} \text{ Chronometer, No. 2440,} \end{cases}$

| No. for reference. | Times of observation by *mean solar chronometer No.* 2440. | True sidereal times of observation. | MERIDIAN DISTANCES. | | $-\triangle \cos p$ |
|---|---|---|---|---|---|
| | | | In sid'l time, $= p$. | In arc, $= p$. | |
| | *h. m. s.* | *h. m. s.* | *h. m. s.* | ° ′ ″ | ′ ″ |
| 1 | 1 33 02. 5 | 20 05 34. 1 | 4 58 23. 2 | 74 35 48 | — 24 18. 1 |
| 2 | 1 34 28 | 20 06 59. 8 | 4 56 57. 5 | 74 14 22. 5 | — 24 54. 5 |
| 3 | 1 35 42. 7 | 20 08 14. 7 | 4 55 42. 6 | 73 55 39 | — 25 19. 8 |
| 4 | 1 36 38. 2 | 20 09 10. 4 | 4 54 46. 9 | 73 41 43. 5 | — 25 41. 4 |
| 5 | 1 39 07. 5 | 20 11 40. 1 | 4 52 17. 2 | 73 04 18 | — 26 34. 7 |
| 6 | 1 41 11. 2 | 20 13 44. 1 | 4 50 13. 2 | 72 33 22. 5 | — 27 27. 1 |
| 7 | 1 44 28. 2 | 20 17 01. 7 | 4 46 55. 6 | 71 43 54 | — 28 40. 8 |

Observer, *Major J. D. Graham.*
Computer,       *Do.*

AND COMPUTATION.

## LATITUDE, *from Observed Double Altitudes of Polaris.*

*New Brunswick,* (*Grover's Inn.*)

*between four and five hours before its upper meridian passage.*

*Simms,* and Artificial Horizon of Mercury.
by *Parkinson & Frodsham.*

| $+ a(\triangle \sin p)^2 .$ tan $A$ | $- \beta(\triangle \sin p)^2$ . $(\triangle \cos p)$ | Observed double altitudes of *Polaris* out of the meridian. | True altitudes of star, as corrected for refraction and errors of instrument, $= A.$ | Latitude deduced from each observation, $= L.$ |
|---|---|---|---|---|
| $\prime \quad \prime\prime$ | $\prime$ | $\circ \quad \prime \quad \prime\prime$ | $\circ \quad \prime \quad \prime\prime$ | $\circ \quad \prime \quad \prime\prime$ |
| $+$ 1 11.63 | $-$ 0.32 | 93 01.30 | 46 31 58.6 | 46 08 51.8 |
| $+$ 1 11.41 | $-$ 0.33 | 93 02.45 | 46 32 36 | 46 08 52.6 |
| $+$ 1 11.20 | $-$ 0.33 | 93 03.50 | 46 33 08.6 | 46 08 59.7 |
| $+$ 1 11.04 | $-$ 0.33 | 93 04.40 | 46 33 33.6 | 46 09 02.9 |
| $+$ 1 10.63 | $-$ 0.34 | 93 06.15 | 46 34 21 | 46 08 56.6 |
| $+$ 1 10.28 | $-$ 0.35 | 93 08.20 | 46 35 23.5 | 46 09 06.3 |
| $+$ 1 09.68 | $-$ 0.37 | 93 10.50 | 46 36 38.5 | 46 09 07 |

LATITUDE, deduced from a mean of 7 altitudes of star *Polaris,* 46° 08′ 59″.4.

## *Form for Record and Computation*—Continued.

Apparent declination of star $= 88°\ 28'\ 30''.5$.
Apparent N. P. D. of star $= 1°\ 31'\ 29''.5 = 5489''.5 = \triangle$
Refraction (ther. 57°; bar. 30.013 inches) ......................... $- \quad 55''.4$
Index-error of sextant....................................................... $+ 2'\ 50''$
Errors of eccentricity, &c., of sextant .......................... $+ 1'\ 28''$
Apparent AR. of the star *Polaris (α Ursæ Minoris)* .......... $1^h\ 03^m\ 57^s.3$
Sidereal time at mean noon at this station..................... 11  00  27 .1

Sidereal interval from mean noon of star's culmination ........ 10  03  30.2
Retardation of mean on sidereal time........................ $-$   2  18.2

Mean time of culmination of star *Polaris*..................... 14  01  12
Chronometer No. 2440 fast of mean time at time of observation. 4  29  24 .8

Time by chronometer of culmination of star *Polaris* .......... $6^h\ 30^m\ 36^s.8$

    The reduction of the mean time of observation to sidereal time, in the preceding example, might have been omitted by using table of *AR. in Arc into Mean Time*, pages 198, &c. Thus:

Mean time of observation....................................... $1^h\ 33^m\ 02^s.5$
Mean time of culmination of *Polaris* ......................... 6  30  36 .8

Hour-angle, *p*, in intervals of mean time..................... $4^h\ 57^m\ 34^s.3$
Sidereal equivalents, in arc .......................... $4^h\ = 60°\ 09'\ 51''.39$
$\qquad\qquad\qquad\qquad\qquad\qquad\qquad\qquad 57^m = 14\ \ 17\ 20\ .45$
$\qquad\qquad\qquad\qquad\qquad\qquad\qquad\qquad 34^s = \quad\ \ 8\ 31\ .40$
$\qquad\qquad\qquad\qquad\qquad\qquad\qquad\qquad 0^s.3 = \qquad\ \ 4\ .51$

    *p*, in arc........................................ $= 74°\ 35'\ 47''.75$

## *Computation, First Observation.*

| 1st term. | 2d term. | 3d term. |
|---|---|---|
| $\log \cos p\,(+) = \quad 9.4242480$ | $\sin p \quad = 9.98411$ | |
| $\log \quad \triangle \quad = \quad 3.7395327$ | $\triangle \quad\quad = 3.73953$ | |
| $\qquad\qquad\qquad\quad 3.1637807$ | $\triangle \sin p \ = 3.72364$ | $\ldots = 3.16378$ |
| $\triangle \cos p \quad = \qquad 1458''.1$ | | |
| 1st term $\quad = - \quad 24'\ 18''.1$ | $(\triangle \sin p)^2 = 7.44728$ | $\ldots = 7.44728$ |
| | $\log a \quad\ = 4.38454$ | $\log \beta \ = 8.89403$ |
| A $\qquad = 46°\ 31'\ 58''.6$ | $\tan A \quad = 0.02325$ | |
| | | $9.50509$ |
| $\qquad\qquad\quad 46\ 07\ 40\ .5$ | $\quad\quad\quad\quad 1.85507$ | 3d term $= - 0''.32$ |
| 2d term $\quad = + \quad 1\ 11\ .63$ | $\quad\quad\quad = 71''.63$ | |
| $\qquad\qquad\quad 46\ 08\ 52\ .13$ | 2d term $= + 1'\ 11''.63$ | |
| 3d term $\quad = - \qquad 0\ .32$ | | |
| Latitude $\ = 46°\ 08'\ 51''.81$ | | |

LVI.—*Determination of the Latitude by Transits over the Prime Vertical.*

Suppose a transit-instrument so placed that the transit-axis is on the meridian, or very nearly so, and that the axis is horizontal, and the collimation nothing:

1. Call the time T at which a star whose declination is D passes the middle wire of the instrument on the eastern side of the meridian, the clock-correction to reduce the observed time to the true E, and the right ascension of the star AR.; and let T' and E' denote the corresponding quantities for the western transit. Then the two hour angles, in sidereal time, will be, the eastern negative,

$$t = T + E - AR.; \qquad t' = T' + E' - AR.$$

Let the unknown latitude of the place be L, and the azimuth of the line of collimation $a$. The spherical triangle, formed by great circles connecting the zenith, the pole, and the place of the star, gives the following relations:

$$\cot a = \frac{\cos t \cos D \sin L - \sin D \cos L}{\cos D \sin t}$$

$$= \frac{\cos t' \cos D \sin L - \sin D \cos L}{\cos D \sin t'}$$

Whence—

$$\tan L = \tan D \frac{\cos \frac{1}{2}(t' + t)}{\cos \frac{1}{2}(t' - t)}$$

If the instrument is very nearly on the prime vertical,

$$\cos \frac{1}{2}(t' + t) = \cos 0° = 1$$

and—

$$\tan L = \tan D \sec \frac{1}{2}(t' - t)$$

for the passage over the middle wire of the instrument.

2. Call the time of passage of the star, from a side wire to the middle wire, $\tau$.

Let the distance, in arc, of one of the lateral wires from the middle wire, measured on a great circle, be 15 $f$; $f$ being the equatorial interval of the wire, in time.

## LVI.—*Prime Vertical Transits*—Continued.

Then, to reduce the transit over a side wire to the center wire,

$$\tau = \frac{f}{[\sin (L + D) \cdot \sin (L - D) \pm \frac{1}{2}f]^{\frac{1}{2}}}$$

The upper sign of the term $\pm \frac{1}{2} f$ is to be used for wires crossed by the star earlier than the middle wire in the eastern transit, and later in the western transit, and the lower sign in the opposite cases. An approximate latitude may be used for L.

3. Should the optical axis not coincide with the middle wire, substitute $f \pm c$ for $f$ in the above, according as the error of collimation, $c$, lies on the same or opposite sides of $f$.

4. The preceding formula gives the latitude on the supposition that the axis of the instrument is parallel to the horizon. If the instrument is on the prime vertical, but the north end of the axis is, for instance, $n$ seconds too high, the axis is parallel to the horizon of a place whose latitude is $n$ seconds less than where the instrument is placed, and the true latitude is, therefore,

$$L + n$$

5. But should the instrument not be on the prime vertical, the true latitude becomes—

$$L + n \sin a$$

$a$ being the azimuth of the center wire of the telescope, supposed in collimation.

This may be found from the time elapsed between the east and west transits of the same star: thus—

$$\cot u = \tan \tfrac{1}{2} (t' - t) \sin D$$

$$\sin a = \cos D \frac{\sin u}{\cos L}$$

$a$ is taken between 0° and 90° when the north end of the transit-axis is between the north and west, and between 90° and 180° when the same end is between the north and east.

If $n$ is called *plus* when the north end of the axis is too high, and *vice versa*, the signs of the corrections are indicated by those of the quantities resulting from the formula.

When $a$ is nearly 90°, the correction is exceedingly small; so that, when the instrument is placed nearly east and west, we may proceed in all the computations as if it were exactly so.

LVI.—*Prime Vertical Transits*—Continued.

6. The instrument should be set up in the firmest manner. A change of azimuth between the east and west transits of a star will affect the result much less than an equal change of level.

It is better, in order to obtain a close result in the shortest time, to observe several stars on the same evening, and between the first and last observations to determine with the level the inclination of the axis several times, and then to interpolate for transits between the times of observation of the level. It is of course understood that the changes of inclination must be small, which will be the case if the instrument is properly placed.

7. In order to point the telescope rightly, the hour-angles and zenith-distances of the stars to be observed must be computed for the time of transit.

When the telescope is on the prime vertical, calling $p$ the hour-angle, and $z$ the zenith-distance of the star, then—

$$\cos p = \tan D \cot L$$

$$\cos z = \frac{\sin D}{\sin L}$$

An allowance must be made for the time of crossing the first wire, and for change of zenith-distance from the first to the middle wire.

8. To correct, for errors of collimation, irregularity in the pivots, &c., the instrument may be reversed between the transits over each vertical; *i. e.*, the wires on one side of the center wire are observed, the instrument reversed in its Y's, and the transit over the same wires continued, but in an inverse order; so that, in each vertical the same wire is at one time as far north as it is at another south of the optical axis.

Then let—

$L$ = the latitude sought;

$D$ = the apparent declination of the star;

$t$ = the hour-angle, illuminated axis *north;*

= $\frac{1}{2}$ diff. of sidereal time of transit over the same wire, for same position of axis; and

$t'$ = hour-angle, illuminated axis *south.*

$$\tan L = \frac{\tan D}{\cos \frac{1}{2}\,(t' + t)\,.\,\cos \frac{1}{2}\,(t' - t)}$$

LVII.—*To determine the Latitude of a Place by observing the Difference of the Meridional Zenith-Distances of Two Stars on Opposite Sides of the Zenith, with the* Zenith and Equal-Altitude Telescope.

Compute an approximate latitude by the formula—

$$L = \tfrac{1}{2} \left[ 180° - ( \triangle + \wedge') \right] + \tfrac{1}{2} (z - z')$$

where $\triangle$ and $\triangle'$ are the polar distances of the south and north stars, respectively, and $(z - z')$ the quantity measured by the micrometer.   Then—

1. The *correction for level* is applied by  adding the angle which the vertical axis of the instrument makes with the zenith when the inclination is *southward*, or subtracting it when to the northward.   This correction is found by multiplying the value of one division of the level-scale, in arc, by one-half the mean change, in level-divisions, which any one end of the bubble undergoes by reversing the instrument on the meridian; or, if $o$ and $e$, $o'$ and $e'$, denote the readings of the object and eye-ends of the bubble, for south and north stars, respectively; corrections for level = $\tfrac{1}{4} (o' - e') - \tfrac{1}{4} (o - e) \times$ the value of one division of the level-scale in arc.

2. The correction for *error of meridional position* of the central vertical wire is found by computing the usual "reduction to the meridian" for each star; then the difference between the reductions for the northern and southern stars is taken, and one-half that difference added or subtracted, according as the reduction for the *northern* star is *greater* or less than that for the southern; or,

$$\text{correction for position,} = \frac{m' - m}{2}$$

$m$ being the reduction for stars south, and $m'$ for stars north of the zenith.

## LVII.—*Zenith Telescope*—Continued.

3. When the star is observed off the line of collimation, the instrument remaining in the plane of the meridian,

$$m = \frac{2 \sin^2 \frac{1}{2} p}{\sin 1''} \times \tfrac{1}{2} \sin 2 \, D$$

The correction to the latitude is one-half of this quantity, whether the star be north or south; and if the two stars forming a pair are observed off the line of collimation, two such corrections, separately computed, must be added to the latitude. D *is minus when south.*

Values of *m* are given in the following table:

| D. | 10ª | 15ª | 20ª | 25ª | 30ª | 35ª | 40ª | 45ª | 50ª | 55ª | 60ª | D. |
|---|---|---|---|---|---|---|---|---|---|---|---|---|
| o | ″ | ″ | ″ | ″ | ″ | ″ | ″ | ″ | ″ | ″ | ″ | o |
| 5 | .00 | .01 | .02 | .03 | .04 | .06 | .08 | .10 | .12 | .14 | .17 | 85 |
| 10 | .01 | .02 | .04 | .06 | .08 | .11 | .15 | .19 | .23 | .28 | .34 | 80 |
| 15 | .01 | .03 | .05 | .09 | .12 | .17 | .22 | .28 | .34 | .41 | .49 | 75 |
| 20 | .02 | .04 | .07 | .11 | .16 | .22 | .28 | .36 | .44 | .53 | .63 | 70 |
| 25 | .02 | .05 | .08 | .13 | .19 | .26 | .34 | .42 | .52 | .63 | .75 | 65 |
| 30 | .02 | .05 | .09 | .15 | .21 | .29 | .38 | .48 | .59 | .71 | .85 | 60 |
| 35 | .03 | .06 | .10 | .16 | .23 | .31 | .41 | .52 | .64 | .77 | .92 | 55 |
| 40 | .03 | .06 | .11 | .17 | .24 | .33 | .43 | .54 | .67 | .81 | .97 | 50 |
| 45 | .03 | .06 | .11 | .17 | .25 | .33 | .44 | .55 | .68 | .82 | .98 | 45 |

4. The *correction for refraction* is applied similarly to reduction to meridian (2) but with a contrary sign; or,

$$\text{Correction for refraction} = \frac{r - r'}{2}$$

$r - r'$ being small, no note need be taken of the state of the barometer and thermometer at the time of observation.

## LVII.—*Zenith Telescope*—Continued.

The following table gives the correction to the latitude for differential refraction; arguments, one-half difference of zenith-distances on one side and zenith-distance on the top:

| $\dfrac{z - z'}{2}$ | Zenith-distance. | | | | | |
|---|---|---|---|---|---|---|
| | 0° | 10° | 20° | 25° | 30° | 35° |
| ′   ″ | ″ | ″ | ″ | ″ | ″ | ″ |
| 0   0 | .00 | .00 | .00 | .00 | .00 | .00 |
| 0  30 | .01 | .01 | .01 | .01 | .01 | .01 |
| 1 | .02 | .02 | .02 | .02 | .02 | .02 |
| 1  30 | .02 | .03 | .03 | .03 | .03 | .03 |
| 2 | .03 | .03 | .04 | .04 | .04 | .05 |
| 2  30 | .04 | .04 | .05 | .05 | .05 | .06 |
| 3 | .05 | .05 | .06 | .06 | .07 | .08 |
| 3  30 | .06 | .06 | .07 | .07 | .08 | .09 |
| 4 | .07 | .07 | .08 | .08 | .09 | .10 |
| 4  30 | .08 | .08 | .09 | .09 | .10 | .11 |
| 5 | .08 | .09 | .10 | .10 | .11 | .13 |
| 5  30 | .09 | .10 | .10 | .11 | .12 | .14 |
| 6 | .10 | .10 | .11 | .12 | .13 | .15 |
| 6  30 | .11 | .11 | .12 | .13 | .14 | .16 |
| 7 | .12 | .12 | .13 | .14 | .15 | .18 |
| 7  30 | .13 | .13 | .14 | .15 | .16 | .19 |
| 8 | .13 | .14 | .15 | .16 | .18 | .21 |
| 8  30 | .14 | .15 | .16 | .17 | .19 | .22 |
| 9 | .15 | .16 | .17 | .18 | .20 | .23 |
| 9  30 | .16 | .17 | .18 | .20 | .21 | .24 |
| 10 | .17 | .18 | .19 | .21 | .23 | .26 |
| 10  30 | .18 | .19 | .20 | .22 | .24 | .27 |
| 11 | .18 | .19 | .21 | .23 | .25 | .28 |
| 11  30 | .19 | .20 | .22 | .24 | .26 | .30 |
| 12 | .20 | .21 | .23 | .25 | .27 | .31 |

The sign of the correction is the same as that of the micrometer-difference.

## LVII.—*Zenith Telescope*—Continued.

5. To find the value $a$ of one division of the micrometer, note the time by chronometer of the transit of Polaris or other close circumpolar star over the movable wire placed vertically and set successively before the star for each turn or half-turn of the screw. Then let $x$ be the angular distance *from the meridian* at any reading of the screw; $p$, the hour-angle of the star at the same instant; and D, its declination :

$$\sin x = \cos D \sin p$$

The value of $x$ is computed for each reading, and the differences of these values divided by the differences of the corresponding micrometer-readings give values for the screw.

A better method, as it avoids displacing the micrometer, is to observe a close circumpolar star near its elongation, when rapidly rising or falling, with but slight motion in azimuth. The level should be carefully noted in order to allow for possible changes, and a correction applied for differential refraction.

The sidereal time of elongation and the azimuth of the star can be determined from LXI.

About 40 or more minutes before elongation, transits are noted, the micrometer being set in advance consecutively by whole or half turns of the screw throughout its length. A correction for rate of chronometer should be applied if sensible.

Let—

$t$ = the difference between the time of observation and the time of elongation of the star; and

$s''$ = the number of seconds of arc from elongation in the direction of the vertical.

$$s'' = 15 \cos \cdot D \left[ t - \tfrac{1}{6} \left( 15 \sin 1'' \right)^2 t^3 \right]$$

where $t$ is expressed in seconds of time.

LVII.—*Zenith Telescope*—Continued.

Values of $\frac{1}{6}$ $(15 \sin 1'')^2 t^3$ for minutes of time from elongation are given in the following table :

| *t.* | Term. | *t.* | Term. | *t.* | Term. | *t.* | Term. |
|---|---|---|---|---|---|---|---|
| *m.* | *s.* | *m.* | *s.* | *m.* | *s.* | *m.* | *s.* |
| 5 | 0.0 | 15 | 0.6 | 25 | 3.0 | 35 | 8.2 |
| 6 | 0.0 | 16 | 0.8 | 26 | 3.3 | 36 | 8.9 |
| 7 | 0.0 | 17 | 0.9 | 27 | 3.7 | 37 | 9.6 |
| 8 | 0.1 | 18 | 1.1 | 28 | 4.2 | 38 | 10.4 |
| 9 | 0.1 | 19 | 1.3 | 29 | 4.6 | 39 | 11.3 |
| 10 | 0.2 | 20 | 1.5 | 30 | 5.1 | 40 | 12.2 |
| 11 | 0.2 | 21 | 1.8 | 31 | 5.7 | 41 | 13.1 |
| 12 | 0.3 | 22 | 2.0 | 32 | 6.2 | 42 | 14.1 |
| 13 | 0.4 | 23 | 2.3 | 33 | 6.8 | 43 | 15.1 |
| 14 | 0.5 | 24 | 2.6 | 34 | 7.5 | 44 | 16.2 |

It is convenient to apply these values to the observed time of noting, additive to the observed time before, and subtractive after, either elongation.

The correction to be applied to the observed times of noting for change of level is given by the formula—

$$\pm \left\{ \tfrac{1}{2} (N - S) - \tfrac{1}{2} (N_0 - S_0) \right\} \frac{b}{15 \cos D}$$

where $N_0$, $S_0$, the north and south readings for a selected state of level, N, S, the readings for any other state, and $b$ the value of one division of level-scale in seconds of arc; the upper sign to be used for western, the lower sign for eastern elongation.

After these two corrections have been applied we have in one column the readings of the micrometer, and in another the corresponding times, such as would have been obtained if the star had moved uniformly in a vertical line, leaving out of consideration for the present the change in refraction and the rate of the chronometer.

Various methods of combination might be adopted for the determination of the turn of the screw. That of subtracting the values resulting from the first operation from those of the middle one; next, those of the second from those of the middle one, *plus* one, and so on, is recommended for its simplicity.

## LVII.—*Zenith Telescope*—Continued.

A number of values are thus obtained for the time of a given number of turns or half-turns, from which is deduced the value of one turn: thus—

| | | |
|---|---|---|
| Mean time for one turn, | 116ˢ.774 | log = 2.06735 |
| | cos D | log = 8.40750 |
| | 15 | log = 1.17609 |
| One turn ................. 44″.765 | | = 1.65094 |
| Correction for refraction.... — .025 | | |
| Correction for rate......... — .003 | | |
| Resulting value ....... 44 .737 | | |

The correction for refraction, in seconds of arc, is negative for either eastern or western elongation, and equals the change of refraction for the space equal to one turn; equal to the value of one turn times the difference of refraction for 1′ at star's altitude divided by 60.

6. The value $b$ of 1 division of the level-scale will be best found by using, in conjunction with the micrometer, a distant point as a mark, or the central wire of another instrument used as a collimator; for the space above or below the mark, passed over by the horizontal wire of the micrometer, during the bubble's run over the scale, as the telescope's elevation is gradually altered, may afterward be measured by the micrometer-screw. The temperature should be noted, since the result may change with a change of temperature.

Including all corrections, the general expression for latitude is—

$$L = \frac{180-(\triangle+\triangle')}{2} + \frac{(z-z')}{2} \times a + \frac{(d'+c)-(o+c')}{4} \times b + \frac{(m'-m)}{2} + \frac{(r-r')}{2}$$

$a$ and $b$ being the arc values of one division of the micrometer and level-scale respectively.

To correct as much as possible an erroneous determination of the value of the micrometer-screw, select stars for observation, such, if practicable, that the greatest zenith-distance of a pair will belong as often to the north star as to the south star; because if the zenith-distance of the north star is the greatest, the observed quantity is subtractive; if least, additive. For, as a general rule, the error of latitude arising from an erroneous value to the micrometer-screw will be the least when in a set of stars—

$$\Sigma z - \Sigma z' = 0$$

FORM FOR RECORD AND COMPUTATION.

*Northwestern Boundary Survey.*

Latitude, Station Kootenay West,—Observations with Zenith-Telescope by ..............................

| Date, 1860. | Star No. B.A.C. | N./S. | Micrometer Reading (T. D.) | Diff. Z.-D. (T. D.) | Level N. | Level S. | Diff. N.−S. | Meridian distance. | Declination. | Sum and half-sum. | Corrections Microm. | Level. | Ref. | Merid. | Latitude. | Remarks. |
|---|---|---|---|---|---|---|---|---|---|---|---|---|---|---|---|---|
| July 9 | 6289 | N. | 10 90.2 | | 19.7 | 11.3 | + 4.0 | 23 | 58 43 15.69 | 98 11 25.37 | − 5 47.76 | + 1.13 | − 0.10 | + 0.09 | 48 59 56.04 | " |
|  | 6391 | S. | 21 69.2 | − 10 79.0 | 13.6 | 18.0 |  | 14 | 39 28 09.68 | 49 05 42.68 |  |  |  |  | " | " |
| 14 |  | N. | 10 89.3 | | 27.1 | 26.1 | + 2.6 | 10 | 17.29 | 28.44 | 49.02 | + 0.73 | − 0.10 | + 0.01 | 55.84 |  |
|  |  | S. | 21 72.2 | 83.9 | 27.8 | 26.2 |  |  | 11.15 | 44.22 |  |  |  |  |  |  |
| 15 |  | N. | 11 94.4 | | 27.8 | 29.9 | − 0.2 | 9 | 17.60 | 29.04 | 48.54 | + 0.73 | − 0.10 | + 0.01 | 55.83 |  |
|  |  | S. | 22 75.8 | 81.4 | 30.3 | 28.4 |  | 8 | 11.44 | 44.52 |  |  |  |  |  |  |
| 16 |  | N. | 11 03.3 | | 39.7 | 13.6 | + 10.5 | 9 | 17.92 | 29.66 | 51.92 | − 0.06 | − 0.10 | + 0.01 | 55.80 |  |
|  |  | S. | 21 95.2 | 91.9 | 19.3 | 34.9 |  | 9 | 11.74 | 44.83 |  | + 2.97 | − 0.10 | + 0.02 |  |  |
|  |  |  |  |  |  |  |  |  |  |  | Mean.... |  |  |  | 48 59 55.88 |  |

LVIII.—*Knowing the Time and the Latitude of the Place, to find the Azimuth of the Sun or a Star.*

$$\tan \tfrac{1}{2}(A + S) = \cot \tfrac{1}{2} p \, \frac{\cos \tfrac{1}{2}(\Delta - \lambda)}{\cos \tfrac{1}{2}(\Delta + \lambda)}$$

$$\tan \tfrac{1}{2}(A - S) = \cot \tfrac{1}{2} p \, \frac{\sin \tfrac{1}{2}(\Delta - \lambda)}{\sin \tfrac{1}{2}(\Delta + \lambda)}$$

$$A = \tfrac{1}{2}(A + S) \mp \tfrac{1}{2}(A - S)$$

the *upper* or *negative* sign is used when $\lambda$ is greater than $\Delta$;

where

$A =$ the azimuth counted from the *north*, which must be subtracted from 180° if counted from the south;

$S =$ the angle at the star, called the angle of variation;

$\lambda =$ the co-latitude of the place;

$\Delta =$ the north-polar distance of the sun or star; and

$p =$ the hour-angle at the pole.

*Without the Use of a Chronometer, by observing the Altitude of the Sun or Star at the Same Instant with the Observation of the Azimuth.*

Let $Z =$ the zenith-distance, corrected for refraction, parallax, and semi-diameter; then—

$$\cos^2 \tfrac{1}{2} A = \frac{\sin k \, . \, \sin (k - \Delta)}{\sin Z \sin \lambda}$$

$$2\,k = Z + \Delta + \lambda$$

LIX.—*To find the* AMPLITUDE *of a Celestial Object at its Rising or Setting;* by " Amplitude " is meant the complement of the azimuth, or distance from the east or west points of the horizon.

This is a particular case of the preceding problem. When the object appears to be in the horizon, its zenith-distance, instead of being 90°, is, on account of refraction and parallax, 90° + $k$, where—

$$k = \text{horizontal refraction} - \text{horizontal parallax}$$

$$= 36' \ 29'' - \text{horizontal parallax.}$$

For stars, the horizontal parallax = 0 and $k = 90° 36' 29''$; for the sun, $k = 36' 29'' - 8''.8 = 90° \ 36' \ 20''.2$. The mean refraction and mean horizontal parallax are here used, as these observations are not susceptible of much accuracy.

LX.—*To find the True Meridian by the Method of Equal Altitudes of the Sun.*

The instrument remaining stationary, observe the readings of the horizontal limb when the altitude of the sun's center is the same in the forenoon and afternoon.

Then, the correction to the mean of these two readings for the change in the sun's declination in the interval is—

$$c = \frac{\frac{1}{2} (D - D')}{\cos L \cdot \sin \frac{1}{2} (t - t')}$$

where—

     $D - D' =$ the change in the sun's declination in the interval of the observations;

     $(t - t') =$ this interval of time expressed in arc; and

     $L =$ the latitude of the place.

LXI.—*To find the Azimuth of* POLARIS, *or other Close Circumpolar Star, at its Greatest Eastern or Western Elongation.*

$$\cos p = \tan \triangle \cot \lambda = \cot D \tan L = \tan L \tan \triangle$$

$$\cos L \sin A = \sin \triangle = \cos D$$

where—

$p$ = the hour-angle of the star ;

$D$ = its declination ;

$\triangle$ = its polar distance ;

$A$ = the required azimuth ;

$L$ = the latitude of the place ; and

$\lambda$ = the co-latitude.

The first equations give the hour-angle of the star at its greatest elongation ; hence the sidereal time of elongation ;

The second, the azimuth of the star at its greatest elongation.

The azimuth at *any* hour-angle is found by the methods in LVIII, or by the formula—

$$A \text{ (in seconds)} = \frac{\sin p}{\cos L} \left\{ \triangle + \triangle^2 \sin 1'' \cos p \tan L \right.$$

$$\left. + \tfrac{1}{3} \triangle^3 \sin^2 1'' \left[ (1 + 4 \tan^2 L) \cos^2 p - \tan^2 L \right] \right\}$$

If the hour-angle is counted from the lower culmination, change the sign of the second term.

The most approved method is to observe a series of azimuths of the star *about* the elongation, say for not more than 30 minutes before and after, and to reduce them to the elongation. To do this, compute, from the known latitude, the azimuth of the star at its greatest elongation = A, and call the sidereal time from elongation $t$; the correction to the azimuth will be—

$$c = (112.5) \, t^2 \sin 1'' \tan A$$

$$\log (112.5) \sin 1'' = 6.7367274$$

The quantities found in the table for "Reduction to the meridian"—

$$\left( 2 \, \frac{\sin^2 \tfrac{1}{2} p}{\sin 1''} \right)$$

correspond very nearly to—

$$(112.5) \, t^2 \sin 1''$$

so, by entering the table with the time from elongation, and

LXI.—*To find the Azimuth of Polaris, &c.*—Continued.

multiplying the tabular quantities by tan A, we obtain the required correction in seconds of arc.   This will be found a convenient substitute for the more rigorous method.

The formula may be separately applied to each observation, or
the work may be shortened by computing only the azimuth
corresponding to the mean hour angle and applying to it a
*correction to mean azimuth.*

Let $n$ be the number of observations on the star; $A_1$, the azimuth corresponding to the mean hour-angle; and let, also, $\tau =$
the difference between the time of any observation and the mean
of the times; then, for a circumpolar star,

$$\text{Correction to mean azimuth} = A_1 - \tan A_1 \frac{1}{n} \, \Sigma \frac{2 \sin^2 \tfrac{1}{2} \tau}{\sin 1''}$$

*Example.*

Means of the times of observations by chronometer $= 3^h \, 48^m \, 12^s .3$.

| Chronometer time of observation. | $\tau$ | Tab. quantities, (page 240.) | |
|---|---|---|---|
| *h.  m.  s.* | *m.  s.* | '' | |
| 3  42  30. 5 | 5  41. 8 | 63. 7 | log 27''.4 .... $= 1.4380$ |
| 44  08. 0 | 4  04. 3 | 32. 5 | log tan A..... $= 8.5144$ |
| 45  52. 0 | 2  30. 3 | 10. 7 | |
| 47  15. 0 | 57. 3 | 1. 8 | |
| 48  59. 0 | 46. 7 | 1. 2 | 9. 9524 |
| 50  34. 0 | 2  21. 7 | 11. 0 | Correction. $= - 0''.90$ |
| 52  28. 0 | 4  15. 7 | 35. 7 | |
| 3  53  51. 5 | 5  39. 2 | 62. 8 | |
| | | $\frac{1}{n} \Sigma = 27.4$ | |

| | |
|---|---|
| Azimuth of star at elongation..$1° \, 52' \, 42''. 8$ | log $t^2$........... $6.0734174$ |
| Chronometer time of elongation.$4^h \, 06^m \, 20^s. 5$ | log $(112. 5) \sin 1''$..$6.7367274$ |
| Mean time of obs'n by chron..$3^h \, 48^m \, 12^s. 3$ | log tan A ........ $8.5162425$ |
| $t = 1088''.2 = \quad 18^m \, 8^s. 2$ | Cor. $= 21''.2 = .. \; 1.3263873$ |

Azimuth corresponding to mean hour-angle $= 1° \, 52' \, 21''.6$
Correction to mean azimuth ............. $= \qquad 0 \,.9$
Mean azimuth ................. $= 1° \, 52' \, 20''.7$

LXI.—*To find the Azimuth of Polaris, &c.*—Continued.

Azimuths are usually reckoned from the south and in the direction of south to west. When circumpolar stars are observed, it is more convenient to reckon from the north meridian. The determination of primary azimuths supposes the local time to be known; for secondary azimuths observations for time and azimuth may be made together. The sun is only employed in connection with the inferior class of azimuths.

For the purpose of referring azimuths observed at night to the direction of any geodetic signal, a mark is set up, consisting of a perforated box, (about ¾ of a foot cube,) through the front face of which the light of a bull's-eye lantern is shown, appearing of about the size and brilliancy of the star observed upon. The distance of this mark from the place of observation is generally determined by local circumstances, but should not, if possible, be nearer than about a mile, in order that the sidereal focus of the telescope may not require changing. For day observations a vertical black stripe, of the same width as the aperture, is painted upon a white wand placed vertically above it. If the diameter of the aperture is a quarter of an inch, it will subtend at the distance of a mile an angle of a little more than $0''.8$. The horizontal angle between the mark and any trigonometrical station is measured in connection with the triangulation by combining it with all other directions radiating from the station observed from.

Observations for azimuth are usually made in sets, commencing with a number on the mark, followed by about an equal number of readings on the star preceded and followed by level-readings. The instrument is then reversed, and the preceding operations are repeated in the reverse order, the number of readings on the star and mark being as before.

In these observations the optical axis of the telescope of the theodolite must be made to describe a truly vertical plane.

If the axis of the telescope is not horizontal, the correction to the azimuth will be—

$$\pm \frac{d}{4}\left\{(w + w') - (e - e')\right\}\tan L$$

where $d =$ the value of one division of the level-scale in seconds

LXI.—*To find the Azimuth of Polaris, &c.*—Continued.

of arc, $w$, $e$, and $w'$, $e'$, the west and east readings of the level before and after reversal.

The circumpolar stars $a$, $\delta$, $\lambda$ Ursæ Minoris and 51 Cephei are those almost exclusively used. When $\delta$ Ursæ Minoris and 51 Cephei culminate on either side of the pole, Polaris is not far from its elongation; and, on the contrary, when the pole-star culminates, the other two are not far from their elongations on either side of the meridian. $\lambda$ Ursæ Minoris, from its greater proximity to the pole, and its small size, presents to the larger instruments a finer and steadier object than Polaris.

LXII.—*Correction for* RUN *in reading Microscopes.*

As it is difficult to adjust the microscopes so that five revolutions of the micrometer-screw shall carry the wire exactly over one of the five-minute spaces on the limb of the instrument, (if it be so graduated,) it is preferred to observe the number of revolutions and the part of a revolution made by the screw while the wire passes over the space; then, let—

$m$ = the mean of *first readings;* that is, the readings obtained by turning the screw in the direction of increasing numbers from the zero of the comb; and

$m'$ = the mean of second, or reverse, readings;

then—

$$\text{(mean) run} = r = m - m' + 300$$

and—

$$\text{true (mean) reading} = \frac{300 \cdot m}{r} = \frac{300\,(r + m' - 300)}{r}$$

$$= \text{the number of minutes and seconds to be added to the degrees and minutes of the limb.}$$

### LXIII.—*Longitude by Lunar Culminations.*

1. Make—

$l$ = true longitude sought;

$l'$ = approximate longitude;

$m$ = observed change, in right ascension, of the moon's bright limb between the first meridian and that sought;

$m'$ = computed change in same, by interpolation;

$V$ = rate of motion, in right ascension, of the moon's bright limb, when on the meridian $l'$; and

$I$ = the constant difference between the values of the independent variable, or arguments, corresponding to the consecutive tabulated values of the right ascension of the moon;

then—

$$l = l' + I \cdot \frac{m - m'}{V} + \ \cdot \ \cdot \ \cdot \ \cdot \ \cdot \ (1)$$

2. *Interpolation.*—Take the following scheme, viz:

| I | F | $\Delta_1$ | $\Delta_2$ | $\Delta_3$ | $\Delta_4$ | $\Delta_5$ |
|---|---|---|---|---|---|---|
| $t'''$ | $a'''$ | | | | | |
| | | $b''$ | | | | |
| $t''$ | $a''$ | | $c''$ | | | |
| | | $b'$ | | $d'$ | | |
| $t'$ | $a'$ | | $c'$ | | $e'$ | |
| | | $b$ | | $d$ | | $f$ |
| $t_,$ | $a_,$ | | $c_,$ | | $e_,$ | |
| | | $b_,$ | | $d_,$ | | |
| $t_{,,}$ | $a_{,,}$ | | $c_{,,}$ | | | |
| | | $b_{,,}$ | | | | |
| $t_{,,,}$ | $a_{,,,}$ | | | | | |

In which the column I contains the independent variable, or argument, as time, terrestial longitude, degrees, and the like; F, the value of the function of this variable, as found in any set of tables; $\Delta_1$, $\Delta_2$, &c., the first, second, &c., order of differences of these functions.

LXIII.—*Longitude by Lunar Culminations*—Continued.

Make—

$s =$ the value of the function corresponding to any value $t_{\bullet}$ between $t'$ and $t_{,}$;

$$
\left.\begin{aligned}
t &= \frac{t_{\bullet} - t''}{t_{,} - t'} \\[2mm]
a &= \frac{a' + a_{,}}{2} \\[2mm]
c &= \frac{c' + c_{,}}{2} \\[2mm]
e &= \frac{e' + e_{,}}{2}
\end{aligned}\right\} \quad \ldots \ldots \ldots \ldots \quad (2) \quad \cdot
$$

Then, according to Bessel, Ast. Nach. No. 30—

$$
s = a + \frac{t - \frac{1}{2}}{1} \cdot b + \frac{t\,(t - 1)}{1 \cdot 2} \cdot c + \frac{t\,(t - 1)\,(t - \frac{1}{2})}{1 \cdot 2 \cdot 3} \cdot d
$$

$$
+ \frac{(t + 1) \cdot t \cdot (t - 1)\,(t - 2)}{1 \cdot 2 \cdot 3 \cdot 4} \cdot e + \&c.
$$

Or, making—

$$
\left.\begin{aligned}
\Delta_1 &= b \\[2mm]
\Delta_2 &= \frac{c' + c_{,}}{2} \\[2mm]
\Delta_3 &= d \\[2mm]
\Delta_4 &= \frac{e' + e_{,}}{2}
\end{aligned}\right\} \quad \ldots \ldots \ldots \ldots \quad (3)
$$

$$
s = a' + t\,\Delta_1 + \frac{t\,(t - 1)}{1 \cdot 2} \cdot \Delta_2 + \frac{t\,(t - 1)\,(t - \frac{1}{2})}{1 \cdot 2 \cdot 3} \cdot \Delta_3
$$

$$
+ \frac{(t + 1) \cdot t \cdot (t - 1)\,(t - 2)}{1 \cdot 2 \cdot 3 \cdot 4} \cdot \Delta_4 + \&c.
$$

or, by the ascending powers of $t$,

$$
s = a' + At + Bt^2 + Ct^3 + Dt^4 + \&c. \quad \ldots \quad (4)
$$

LXIII.—*Longitude by Lunar Culminations*—Continued.

in which, stopping at the fourth differences,

$$
\begin{aligned}
A &= \Delta_1 - \tfrac{1}{2}\Delta_2 + \tfrac{1}{12}\Delta_3 + \tfrac{1}{12}\Delta_4 \\
B &= \tfrac{1}{2}\Delta_2 - \tfrac{1}{4}\Delta_3 - \tfrac{1}{24}\Delta_4 \\
C &= \tfrac{1}{6}\Delta_3 - \tfrac{1}{12}\Delta_4 \\
D &= \tfrac{1}{24}\Delta_4
\end{aligned}
\quad\Bigg\} \quad \dots \quad (5)
$$

Also,

$$
V = \frac{ds}{dt} = A + 2\,B\,t + 3\,C\,t^2 + 4\,D\,t^3 + \ . \quad . \quad (6)
$$

$$
m' = s - a' = A\,t + B\,t^2 + C\,t^3 + D\,t^4 + \ . \quad .\,(7)
$$

The value of $m$ may be obtained either from observations on the two meridians, or by observations on one and the tabulated results under the head of Moon Culminations in the Nautical Almanac, which may be used as actual observations on the meridian of the ephemeris.

Note.—If the lunar tables were perfectly accurate, the true longitude given by the observation would be found at once by comparing the observed right ascension with that of the ephemeris. There are two methods of avoiding or eliminating the errors of the ephemeris. In the first, the observation is compared with a corresponding one on the same day at the first meridian, or at some meridian the longitude of which is well established. In this method the increase of the right ascension in passing from one meridian to the other is directly observed, and the error of the ephemeris on the day of observation is consequently avoided; but observations at the unknown meridian are frequently rendered useless by a failure to obtain the corresponding observations at the first meridian.

In the second method, the ephemeris is first corrected by means of all the observations taken at the fixed observatories during the semi-lunation within which the observation for longitude falls. The corrected ephemeris then takes the place of the corresponding observation, and is even better than the single corresponding observation, since it has been corrected by means of *all* the observations at the fixed observatories during the semi-lunation.—*Chauvenet's Practical Astronomy.*

LXIII.—*Longitude by Lunar Culminations*—Continued.

3. *Example.*—Let—

$$l' = 4^h\ 55^m\ 50^s \text{ west from Greenwich,}$$

and suppose the following transits with a chronometer marking sidereal time. The *error* of the time-keeper is not material, and the transit is very nearly in the meridian, viz:

```
Feb. 18. ζ  Geminorum 6ʰ 54ᵐ 41ˢ.75
         δ  Geminorum 7  10  38.97
         ☽'s first limb .. — —   — —   7ʰ 38ᵐ 06ˢ.76
         ζ  Cancri..... 8  03  06.11

              3)22  08  26.83    7  22  48.943

                                 0  15  17.817
Chronometer rate + 3ˢ daily.........  —0.0318   0ʰ 15ᵐ 17ˢ.785
```

The corresponding observations at Greenwich, as given by the Nautical Almanac, are:

```
Feb. 18. ζ  Geminorum 6ʰ 54ᵐ 57ˢ.41
         δ  Geminorum 7  10  54.36
         ☽'s first limb .. — —   — —   7ʰ 27ᵐ 47ˢ.66
         ζ  Cancri..... 8  03  21.44

              3)22  09  13.21    7  23  04.403   0ʰ 04ᵐ 43ˢ.257

                          m = 634ˢ.528=0  10  34.528
```

Next compute this increase from Nautical Almanac. The right ascensions of the moon are given in that work for the upper and lower passages over the meridian of Greenwich. The independent variable is, therefore, terrestrial longitude, of which the unit is one hour, and the intervals between the consecutive tabulated values of its function, 12. The increase to be computed is for the interval of passage from the upper meridian of Greenwich to that $4^h\ 55^m\ 50^s$ west. Now, according to the scheme and equations (2), (3), and (5),

LXIII.—*Longitude by Lunar Culminations*—Continued.

| Day. | Culm. | AR. ☽'s first limb, Nautical Almanac. | $\Delta_1$ | $\Delta_2$ | $\Delta_3$ | $\Delta_4$ | |
|---|---|---|---|---|---|---|---|
| 17 | U | $6^h\ 35^m\ 55^s.73$ | $+26^m\ 00^s.54$ | | | | $l = \dfrac{h_1 - l'}{l_1 - l'} = \dfrac{4^h\ 55^m\ 50^s}{12^h}$ |
|  | L | 7 01 56.57 | $+25\ 51.39$ | $9^s.15$ | $-1^s.06$ | | |
| 18 | U | 7 27 47.66 | $b = +25\ 41.18$ | $l = -10.21$ | $d = -0.25$ | $l' = +0^s.81$ | $\log t = 9.6137147$ |
|  | L | 7 53 28.84 | $+25\ 30.72$ | $l_1 = -10.46$ | $+0.59$ | $l_1' = +0.84$ | |
| 19 | U | 8 18 59.56 | $+25\ 20.85$ | $-\ 9.87$ | | | $\Delta_1 = 25^m\ 41^s.180$ |
|  | L | 8 44 20.41 | | | | | $\Delta_2 =\ --\ 10.335$ |
| | | | | | | | $\Delta_3 =\ --\ 0.250$ |
| Sum of differences | | | $2^h\ 08^m\ 24^s.68$ | $-\ 0^m\ 39^s.69$ | $-\ 0^s.72$ | $+\ 1^s.65$ | $\Delta_4 =\ +\ 0.825$ |
| Top of left column | | | 6 35 55.73 | 26 00.54 | $-\ 9.15$ | $-\ 1.06$ | |
| Check-sum = bottom of left column | | | 8 44 20.41 | 25 20.85 | $-\ 9.87$ | $+\ 0.59$ | |

$A = 25^m\ 41^s.180 + 5^s.167 - 0^s.021 + 0^s.068 = +\ 1546^s.394$

$B =\quad\quad\ -\ 5.167 + 0.062 - 0.034 =\ --\quad 5.139$

$C =\quad\quad\quad\quad\quad -\ 0.041 - 0.068 =\ --\quad 0.109$

$D =\quad\quad\quad\quad\quad\quad\quad\quad + 0.034 =\ +\quad 0.034$

$\log A = 3.1893180$

$\log B = \bar{0}.7108786$

$\log C = \bar{9}.0374265$

$\log D = 8.5910646$

LXIII.—*Longitude by Lunar Culminations*—Continued.

Then equation 7—

| | | |
|---|---|---|
| A log | 3.1893180 | |
| $t$ log | 9.6137147 | |
| | 2.8030327 | Nos.. + 635ˢ.337 |
| B log | $\bar{0}$.7108786 | |
| $t^2$ log | 9.2274294 | |
| | 9.9383080 | Nos.. —    0.867 |
| C log | $\bar{9}$.0374265 | |
| $t^3$ log | 8.8411441 | |
| | $\bar{7}$.8785706 | Nos.. —    0.007 |
| D log | 8.5910646 | |
| $t^4$ log | 8.4548588 | |
| | 7.0459234 | Nos.. +    0.001 |
| $m' =$ | .................. | 634 .464 |

And equation 6—

| | | |
|---|---|---|
| A | ..................... | 1546ˢ.394 |
| B log | $\bar{0}$.7108786 | |
| $t$ log | 9.6137147 | |
| 2 log | 0.3010300 | |
| | $\bar{0}$.6256233 | Nos.. —  4.222 |
| C log | $\bar{9}$.0374265 | |
| $t^2$ log | 9.2274294 | |
| 3 log | 0.4771213 | |
| | $\bar{8}$.7419772 | Nos.. —  0.055 |
| D log | 8.5910646 | |
| $t^3$ log | 8.8411441 | |
| 4 log | 0.6020600 | |
| | 8.0342687 | Nos.. +  0.011 |
| V = | .................. | 1542 .128 |

LXIII.—*Longitude by Lunar Culminations*—Continued.

And equation (1)—

$$I = 12^h \ldots \ldots \ldots \ldots \ldots \ldots \ldots \ldots \ldots \ldots \ldots \quad \log \quad 4.6354837$$

$$m - m' = 0^s.064 \ldots \ldots \ldots \ldots \ldots \ldots \quad \log \quad 8.8061800$$

$$V = 1542^s.128 \ldots \ldots \ldots \ldots \ldots \ldots \quad \log \text{ ac } 6.8118797$$

$$1^s.792 \ldots \ldots \ldots \ldots \ldots \ldots \ldots \ldots \quad 0.2535434$$

Whence—

$$l = 4^h \ 55^m \ 50^s + 1^s.792 = 4^h \ 55^m \ 51^s.792$$

If $m$ be the *observed* increase of right ascension between any meridian not the first, (but of which the longitude is well known,) and the meridian sought, interpolate the increase $m_{,}$ for the known meridian as well as $m'$ for that sought. Then, for $m - m'$, in equation (1), substitute $m - (m' - m_{,})$, and the result will be the corrected longitude from the first meridian, as before.

It often happens that two observers do not use the same number of wires, or do not observe the same number of stars at the two places. In such cases the observed increase of the right ascension of the moon's limb requires a correction, which Mr. Walker deduces as follows, from Gauss's method:

For the eastern observatory and western station respectively, let—

A' and A = the observed AR. of a star;

E = A' — A for the same star;

E' = a similar value for another star;

$l$ and $l'$ = the number of wires on which each limb was observed;

$a$ and $a'$ = similar values for a star;

$$\lambda = \frac{l\,l'}{l + l'} \text{ for the moon's limb;}$$

$$u = \frac{a\,a'}{a + a'} \text{ for one star;}$$

$u'$ = a similar value for another star;

$\Sigma$ = symbol to denote the aggregate of similar quantities; and

$\epsilon$ = the correction required;

LXIII.—*Longitude by Lunar Culminations*—Continued.

then—

$$\varepsilon = \frac{\Sigma\left(E \cdot \frac{\lambda u}{\lambda + u}\right)}{\Sigma \frac{\lambda u}{\lambda + u}}$$

$$L = l + I \cdot \frac{m - m' + \varepsilon}{V}$$

Also, calling W the *weight* of each day's comparison,

$$W = \frac{\sigma \lambda}{(\sigma + \lambda)\, z^2}$$

in which $z$ is the same as as $\frac{I}{V}$ and $\sigma = u + u' + u'' + \&c.$

For the weight of the result of all the comparisons, we have—

$$\frac{\sigma \lambda}{(\sigma + \lambda)\, z^2}$$

Let $e$ denote the probable error of observation, and E the probable error of the final result; then,

$$E = \frac{e}{\sqrt{\Sigma \frac{\sigma \lambda}{(\sigma + \lambda)\, z^2}}}$$

It also frequently happens that the moon cannot be observed on the middle wire, in which case she is far enough from the meridian to have a sensible parallax in right ascension; and, as it may be very desirable not to lose the observation, this parallax must be computed and applied to the hour-angle from the middle wire, which is supposed to be nearly coincident with the meridian.

Denoting this parallax in right ascension by $p$, the horizontal parallax by $w$, the latitude of the place of observation by $\varphi$, and the true declination of the moon by $\delta$, we have from the ordinary series for the parallax in right ascension, neglecting the terms after the first, which would in this case be insignificant,

$$p = \theta \sin w \cos \varphi \sec \delta$$

in which $\theta$ is the hour-angle, or equatorial interval in sidereal

LXIII.—*Longitude by Lunar Culminations*—Continued.

time from the lateral wire on which the moon is observed to the central wire; so that, at the instant of observation, the actual distance of the moon's limb from the central wire is—

$$0 - 0 \sin w \cos \varphi \sec \delta$$

and the reduction to meridian or middle wire will be—

$$\pm \frac{0}{\cos \delta} \cdot \frac{1 - \sin w \cos \varphi \sec \delta}{1 - 0.00277m}$$

in which $m$ is the motion of the moon in right ascension in one day, expressed in degrees. The upper sign is to be used when the observation is on a wire *before* and the lower *after* the middle wire.

In what precedes the approximate longitude $l''$ is supposed to be known. When this is not the case, it may be found from—

$$l' = 12^h \frac{m}{\Delta'}$$

and the interpolation is then to be made for this value of $l'$ to obtain the value of $m'$.

## LXIV.—*Longitude by the Electric Telegraph.*

[From Chauvenet's Practical Astronomy.]

It is evident that the clocks at two stations, A and B, may be compared by means of signals communicated through an electro-telegraphic wire which connects the stations. Suppose at a time T by the clock at A, a signal is made which is perceived at B at the time T' by the clock at that station. Let $\Delta$T and $\Delta$T' be the clock-corrections on the times at these stations respectively, (both being solar or both sidereal.) Let $x$ be the time required by the electric current to pass over the wire, then, A being the more easterly station, we have the difference of longitude $\lambda$ by the formula—

$$\lambda = (T + \Delta T) - (T' + \Delta T') + x = \lambda_1 + x$$

Since $x$ is unknown we must endeavor to eliminate it. For this purpose let a signal be made at B at the clock-time T'', which is perceived at A at the clock-time T''', then we have—

$$\lambda = (T''' + \Delta T''') - (T'' + \Delta T'') - x = \lambda_2 - x$$

In these formulæ $\lambda_1$ and $\lambda_2$ denote the approximate values of the difference of longitude, found by signals east-west and west-east, respectively, when the transmission-time $x$ is disregarded, and the true value is—

$$\lambda = \tfrac{1}{2} (\lambda_1 + \lambda_2)$$

Such is the simple and obvious application of the telegraph to the determination of longitudes; but the degree of accuracy of the result depends greatly—more than at first appears—upon the manner in which the signals are communicated and received.

Suppose the observer at A taps upon a signal-key at an exact second by his clock, thereby producing an audible click of the armature of the electro-magnet at B. The observer at B may not only determine the nearest second by his clock when he hears this click, but may also estimate the fraction of a second; and it would seem that we ought in this way to be able to determine a longitude within one-tenth of a second. But before even this

LXIV.—*Longitude by the Electric Telegraph*—Continued.

degree of accuracy can be secured, we have yet to eliminate, or reduce to a minimum, the following sources of error :

1. The personal error of the observer who gives the signal;
2. The personal error of the observer who receives the signal and estimates the fraction of a second by the ear;
3. The small fraction of time required to complete the galvanic circuit after the finger touches the signal-key;
4. The *armature time*, or the time required by the armature at the station where the signal is received, to move through the space in which it plays and to give the audible click;
5. The errors of the supposed clock-corrections, which involve errors of observation and errors in the right ascensions of the stars employed.

For the means of contending successfully with these sources of error we are indebted to our Coast Survey, which, under the superintendence of Professor Bache, not only called into existence the chronographic instruments, but has given us the most efficient method of using them. The "Method of Star-Signals," as it is called, was originally suggested by the distinguished astronomer, Mr. S. C. Walker, but its full development in the form now employed by the Coast Survey is due to Dr. B. A. Gould.

*Method of Star-Signals.*

The difference of longitude between the two stations is merely the time required by a star to pass from one meridian to the other, and this interval may be measured by means of a single clock placed at either station, but in the main galvanic circuit extending from one station to the other. Two chronographs, one at each station, are also in the circuit, and, when the wires are suitably connected, the clock-seconds are recorded upon both. A good transit-instrument is carefully mounted at each station.

When the star enters the field of the transit-instrument at A, (the eastern station,) the observer, by a preconcerted signal with his signal-key, gives notice to the assistants at both A and B, who at once set the chronographs in motion, and the clock then records its seconds upon both. The instants of the star's transits

## LXIV.—*Longitude by the Electric Telegraph*—Continued.

over the several threads of the reticule are also recorded upon both chronographs by the taps of the observer upon his signal-key. When the star has passed all the threads the observer indicates it by another preconcerted signal, the chronographs are stopped, and the record is suitably marked with date, name of the star, and place of observation, to be subsequently identified and read off accurately by a scale. When the star arrives at the meridian of B, the transit is recorded in the same manner upon both chronographs.

Suitable observations having been made by each observer to determine the errors of his transit-instrument and the rate of the clock, let us put—

$T_1$ = the mean of the clock-times of the eastern transit of the star over all the threads, as read from the chronograph at A;

$T_2$ = the same, as read from the chronograph at B;

$T_1'$ = the mean of the clock-times of the western transit of the star over all the threads, as read from the chronograph at A;

$T_2'$ = the same, as read from the chronograph at B;

$e, e'$ = the personal equations of the observers at A and B, respectively;

$\tau, \tau'$ = the corrections of $T_1$ and $T_1'$ (or of $T_2$ and $T_2'$) for the state of the transit-instruments at A and B, or the respective "reductions to the meridian;"

$\delta T$ = the correction for clock-rate in the interval $T_1' - T_1$;

$x$ = the transmission-time of the electric current between A and B; and

$\lambda$ = the difference of longitude;

then it is easily seen that we have, from the chronographic records at A,

$$\lambda = T_1' + \delta T + \tau' + e' - x - (T_1 + \tau + e)$$

and from the chronographic records at B,

$$\lambda = T_2' + \delta T + \tau' + e' + x - (T_2 + \tau + e)$$

LXIV.—*Longitude by the Electric Telegraph*—Continued.

and the mean of these values is

$$\lambda = [\tfrac{1}{2}(T_1' + T_2') + \tau'] - [\tfrac{1}{2}(T_1 + T_2) + \tau] + \delta T + e' - e$$

which we may briefly express thus:

$$\lambda = \lambda_1 + e' - e$$

in which—

$\lambda_1 =$ the approximate difference of longitude found by the exchange of star-signals, when the personal equations of the observers are neglected.

This equation would be final if $e' - e$, or the relative personal equation of the observers, were known; however, if the observers now exchange stations and repeat the above process, we shall have, provided the relative personal equation is constant,

$$\lambda = \lambda_2 + e - e'$$

in which $\lambda_2$ is the approximate difference of longitude found as before; and hence the final value is—

$$\lambda = \tfrac{1}{2}(\lambda_1 + \lambda_2)$$

I have not here introduced any consideration of the armature-time, because it affects clock-signals and star-signals in the same manner; and therefore the time read from the chronographic fillet or sheet is the same as if the armature acted instantaneously. It is necessary, however, that this time should be constant from the first observation at the first station to the last observation at the second, and therefore it is important that no changes should be made in the adjustment of the apparatus during the interval.

As the observer has only to tap the transit of the star over the threads, the latter may be placed very close together. The reticules prepared by Mr. W. Würdemann for the Coast Survey have generally contained twenty-five threads, in groups or "tallies" of five, the equatorial intervals between the threads of a group being $2^s.5$, and those between the groups, $5^s$; with an additional thread on each side at the distance of $10^s$, for use in observations by "eye and ear." Except when clouds intervene and render it necessary to take whatever threads may be available, only the

LXIV.—*Longitude by the Electric Telegraph*—Continued.

three middle tallies, or fifteen threads, are used. The use of more has been found to add less to the accuracy of a determination than is lost in consequence of the greater fatigue from concentrating the attention for nearly twice as long.

A large number of stars may thus be observed on the same night; and it will be well to record half of them by the clock at one station and the other half by the clock at the other station, upon the general principle of varying the circumstances under which several determinations are made, whenever practicable, without a sacrifice of the integrity of the method. For this reason, also, the transit-instrument should be reversed during a night's work at least once, an equal number of stars being observed in each position, whereby the results will be freed from any undetermined errors of collimation and inequality of pivots. Before and after the exchange of the star-signals, each observer should take at least two circumpolar stars to determine the instrumental constants, upon which $\tau$ and $\tau'$ depend. This part of the work must be carried out with the greatest precision, employing only standard stars, as the errors of $\tau$ and $\tau'$ come directly into the difference of longitude. The right ascensions of the "signal-stars" do not enter into the computation, and the result is, therefore, wholly free from any error in their tabular places; hence, any of the stars of the larger catalogues may be used as signal-stars, and it will always be possible to select a sufficient number which culminate at moderate zenith-distances at both stations, (unless the difference of latitude is unusually great,) so that instrumental errors will have the minimum effect.

A single night's work, however, is not to be regarded as conclusive, although a large number of stars may have been observed and the results appear very accordant; for experience shows that there are always errors which are constant, or nearly so, for the same night, and which do not appear to be represented in the corrections computed and applied. Their existence is proved when the mean results of different nights are compared. Moreover, it is necessary to interchange the observers in order to eliminate their personal equations. The rule of the Coast Survey has been that when fifty stars have been exchanged on not less than three nights, the observers exchange stations, and fifty stars

### LXIV.—*Longitude by the Electric Telegraph*—Continued.

are again exchanged on not less than three nights. The observers should also meet and determine their relative personal equation, if possible, before and after each series, as it may prove that this equation is not absolutely constant.

Before entering upon a series of star-signals, each observer will be provided with a list of the stars to be employed. The preparation of this list requires a knowledge of the approximate difference of longitude, in order that the stars may be so selected that transits at the two stations may not occur simultaneously.

*Example.*—For the purpose of finding the difference of longitude between the Seaton station of the United States Coast Survey and Raleigh, a list of stars was prepared, from which I extract the following for illustration. The latitudes are:

Seaton station, (Washington) .............. $\varphi = + 38^\circ 53'.4$
Raleigh station, (North Carolina) .......... $\varphi = + 35\ 47\ .0$
and Raleigh is assumed to be west from Washington $6^m 30^s$.

| Star. | Mag. | $\alpha$ | | | $\delta$ | | Seaton sidereal time of Raleigh transit. | | |
|---|---|---|---|---|---|---|---|---|---|
| | | *h.* | *m.* | *s.* | $\circ$ | $'$ | *h.* | *m.* | *s.* |
| No. 5036, B. A. C. | 3 | 15 | 09 | 36 | + 33 | 52 | 15 | 16 | 06 |
| 5084 | 4.3 | | 18 | 58 | 37 | 54 | | 25 | 28 |
| 5131 | 4½ | | 27 | 02 | 31 | 51 | | 33 | 32 |
| 5192 | 5 | | 36 | 35 | 26 | 46 | | 43 | 05 |
| 5259 | 5 | | 45 | 43 | 36 | 07 | | 52 | 13 |
| 5322 | 4½ | | 55 | 59 | 23 | 12 | 16 | 02 | 29 |
| 5388 | 5 | 16 | 04 | 09 | 45 | 19 | | 10 | 39 |
| 5463 | 3.4 | | 15 | 21 | 46 | 40 | | 21 | 51 |

The following table contains the observations made on one of these stars at the above-named stations by the United States Coast-Survey telegraphic party in 1853—April 28—under the direction of Dr. B. A. Gould.

In this table "Lamp W" expresses the position of the rotation-axes of the transit-instruments. The first column contains the symbols by which the fifteen threads of the three middle tallies

## LXIV.—*Longitude by the Electric Telegraph*—Continued.

were denoted; the second column, the times of transit of the star over each thread at Seaton, as read from the chronographs at Seaton; the third column, the times of these transits as read from the chronographs at Raleigh; the fourth column, the mean of the second and third columns; the fifth column, the reduction of each thread to the mean of all, computed from the known equatorial intervals of the threads; the sixth column, the time of the star's transit over the mean of the threads, being the algebraic sum of the numbers in the fourth and fifth columns; and the remaining columns, the Raleigh observations similarly recorded and reduced.

Seaton–Raleigh, 1853, April 28.—Star No. 5259, B. A. C.

| Thread | Seaton observations, lamp W. | | | | | Raleigh observations, lamp W. | | | | |
|---|---|---|---|---|---|---|---|---|---|---|
| | $T_1$ | $T_2$ | Mean | Red. | $\frac{T_1+T_2}{2}$ | $T_1'$ | $T_2'$ | Mean | Red. | $\frac{T_1'+T_2'}{2}$ |
| | h. m. s. | s. | s. | s. | s. | h. m. s. | s. | s. | s. | s. |
| $C_1$ | 37.97 | 38.00 | 37.98 | +25.49 | 3.47 | | 11.00 | 11.00 | +25.45 | 36.45 |
| $C_2$ | 41.37 | 41.34 | 41.36 | 22.21 | 3.57 | 14.58 | 14.50 | 14.54 | 22.25 | 36.79 |
| $C_3$ | 44.03 | 44.21 | 44.12 | 19.06 | 3.18 | 17.60 | 17.55 | 17.58 | 19.05 | 36.63 |
| $C_4$ | 47.81 | 47.74 | 47.78 | 15.71 | 3.49 | 20.88 | 20.79 | 20.84 | 15.85 | 36.69 |
| $C_5$ | 50.76 | 50.70 | 50.73 | 12.71 | 3.44 | 23.90 | 23.87 | 23.89 | 12.70 | 36.59 |
| $D_1$ | 56.96 | 57.10 | 57.03 | 6.21 | 3.24 | 30.19 | 30.05 | 30.12 | 6.32 | 36.44 |
| $D_2$ | 0.06 | 0.04 | 0.05 | 3.25 | 3.30 | 33.34 | 33.25 | 33.30 | 3.18 | 36.48 |
| $D_3$ | 15 46 3.40 | 3.38 | 3.39 | + 0.05 | 3.44 | 15 52 36.40 | 36.30 | 36.35 | + 0.07 | 36.42 |
| $D_4$ | 6.70 | 6.70 | 6.70 | − 3.03 | [3.67] | 39.61 | 39.53 | 39.57 | − 3.16 | 36.41 |
| $D_5$ | 9.58 | 9.58 | 9.58 | 6.28 | 3.30 | 43.00 | 43.00 | 43.00 | 6.36 | 36.64 |
| $E_1$ | 16.03 | 15.93 | 15.98 | 12.54 | 3.44 | 49.04 | 48.81 | 48.92 | 12.75 | [36.17] |
| $E_2$ | 19.26 | 19.30 | 19.28 | 15.83 | 3.45 | 52.30 | 52.33 | 52.32 | 15.90 | 36.42 |
| $E_3$ | 22.47 | 22.45 | 22.46 | 18.99 | 3.47 | 55.50 | 55.41 | 55.46 | 19.10 | 36.36 |
| $E_4$ | 25.60 | 25.60 | 25.60 | 22.23 | 3.38 | 58.73 | 58.60 | 58.67 | 22.20 | 36.47 |
| $E_5$ | 28.60 | 28.70 | 28.65 | 25.33 | 3.32 | 2.08 | 2.08 | 2.08 | 25.38 | 36.70 |
| | | | | Mean. | 3.392 | | | | Mean. | 36.535 |

The numbers in the last column for each station would be equal if the observations and chronographic apparatus were perfect; and by carrying them out thus individually we can estimate their accuracy. The numbers 3.67 at Seaton and 36.17 at Raleigh are rejected by the application of Peirce's Criterion, (Method of Least Squares,) and the given means are found from the remaining numbers.

LXIV.—*Longitude by the Electric Telegraph*—Continued.

The corrections of the transit-instruments for this star ($\delta =$ + 36° 6'.9) were—

for the Seaton instrument.. $\tau = -$ 0.028
for the Raleigh instrument $\tau' = -$ 0.193

The rate of the clock was insensible in the brief interval $T_1' - T$. Hence, neglecting the personal equations of the observers, the difference of longitude is found as follows:

$$\tfrac{1}{2}\,(T_1' + T_2') + \tau' = 15^h\ 52^m\ 36^s.342$$
$$\tfrac{1}{2}\,(T_1 + T_2) + \tau = 15\ \ 46\ \ \ 3\,.364$$
$$\lambda_1 = \ \ \ \ \ \ \ 6\ \ 32\,.978$$

In this manner seven other stars were observed on the same night, and the results were as follows:

| Star. | $\lambda_1$ | Diff. from mean, $= v$. |
|---|---|---|
| | *m. s.* | *s.* |
| 5036, B. A. C. | 6 33.03 | + 0.04 |
| 5084, B. A. C. | 33.09 | + 0.10 |
| 5131, B. A. C. | 32.91 | — 0.08 |
| 5192, B. A. C. | 33.00 | + 0.01 |
| 5259, B. A. C. | 32.98 | — 0.01 |
| 5322, B. A. C. | 33.00 | + 0.01 |
| 5388, B. A. C. | 33.02 | + 0.03 |
| 5463, B. A. C. | 32.91 | — 0.08 |
| Mean, $\lambda_1 = 6$ 32.99 | | |

From the residuals, $v$, we deduce the mean error of a single determination by one star,

$$\epsilon = \sqrt{\left(\frac{vv}{m-1}\right)} = \sqrt{\left(\frac{.0256}{7}\right)} = \pm\ 0^s.06$$

and hence the mean error of the value $6^m\ 32^s.99$ is—

$$\epsilon_0 = \pm\ \frac{0^s.06}{\sqrt{8}} = \pm\ 0^s.02$$

But this error will be somewhat increased by those errors of the instruments which are constant for the night, and not represented in $\tau$ and $\tau'$, and by the errors of the personal equations yet to be applied. Moreover, a greater number of determinations should be compared in order to arrive at a just evaluation of the mean error.

## LXV.—*Formulæ for Probable Error and Precision.*

[Contributed by Lieutenant Mercur, Corps of Engineers.]

1. Let—

$m$ = the number of observations;

$n, n'$, &c. = results found by observation;

$x$ = their arithmetical mean;

$v, v'$, &c. = $(x-n)$, $(x-n')$, &c. = the residual errors of observation;

$\epsilon$ = the mean error of $n, n'$, &c.;

$\eta$ = the mean of errors, (arithmetical;)

$E_0$ = the mean error of final result;

$E_1$ = the mean error of the observation assumed as the standard of excellence;

$r$ = the probable error of a single observation, $n, n'$, &c.;

$R$ = the probable error of the final result, $x$;

$h$ = the measure of exactness of a single observation, $n, n'$, &c.;

$H$ = the measure of exactness of the final result, $x$;

$p, p'$, &c. = the weights of different observations or sets of observations;

$\epsilon', \epsilon''$, &c. = the mean errors of observations corresponding to $p, p'$, &c.;

$\Sigma$ = symbol representing the sum;

$$\epsilon = \sqrt{\frac{\Sigma v^2}{(m-1)}}; \quad \eta = \frac{\Sigma v}{m}; \quad E_0 = \sqrt{\frac{\Sigma v^2}{m(m-1)}}$$

$$r = 0.6745 \, \epsilon = 0.8453 \, \eta = 0.8453 \, \frac{\Sigma v}{\sqrt{m(m-1)}}$$

$$R = 0.6745 \, E_0 = 0.6745 \sqrt{\frac{\Sigma v^2}{m(m-1)}} = \sqrt{\frac{0.45495 \, \Sigma v^2}{m(m-1)}} = \frac{r}{\sqrt{m}}$$

$$h = \frac{0.46936}{r}; \quad H = h \sqrt{m}$$

2. By "the weight of any determination" is meant its relative approximation to the true value.

It may be measured by the number of observations (each of which may be considered as good as the other, and of which one is assumed to represent the unit of excellence) necessary to give a result equally near the true value.

LXV.—*Formulæ for Probable Error, &c.*—Continued.

Then, since the weights of observations are inversely as the squares of their mean or probable errors,

$$ p' = \frac{\varepsilon^2}{\varepsilon'^2}; \quad \frac{p'}{p''} = \frac{\varepsilon''^2}{\varepsilon'^2} $$

and when we arbitrarily assign weights to each observation or set of observations,

$$ \varepsilon' = \sqrt{\frac{\Sigma p \, v^2}{(m-1)}} $$

in which $m$ = the number of observations or sets of observations whose weights are $p$, $p'$, &c.

When observations are combined by weights, the probable error is given by the formula—

$$ R = 0.6745 \sqrt{\frac{\Sigma p \, v^2}{(m-1)\,\Sigma p}} $$

3. PEIRCE's Criterion for the rejection of doubtful observations.

To apply this to any set of observations involving but one unknown quantity: *

Let—

$m$ = the number of observations taken;

$n$ = the number of doubtful observations to be rejected, (to be found by trial;)

$\varepsilon$ = the mean error of one observation in the set of $m$ observations;

$v'$, $v''$, &c. = the residual errors of the observations; and

$x$ = the ratio of the required limit of error for the rejection of $n$ observations to the mean error $\varepsilon$, so that $x\varepsilon$ is the limiting error.

Find from the table the value of $x^2$ for $n = 1$, $n = 2$, $n = 3$, &c., in succession, and reject all observations in which $x\varepsilon > v'$; stopping, however, when the value of $x\varepsilon$ found for any particular value of $n$ does not reject any observations (not already rejected) for a value of $n$ numerically one less.

---

* For rules for determining mean and probable errors, and for applying the Criterion to cases involving more than one unknown quantity, see Chauvenet's Manual of Spherical and Practical Astronomy, vol. II, pages 469 to 566.

## LXV.—*Formulæ for Probable Error, &c.*—Continued.

*Example.*—To determine the value of one turn of the micrometer of a zenith-telescope, the following "reduced intervals" were obtained, each corresponding to the ten turns of the micrometer:

| Between observa-tions. | Reduced intervals. | $v'$ | $v'^2$ | |
|---|---|---|---|---|
| | *m. s.* | | | |
| 1 and 10 | 20 35.1 | 11.0 | 121.00 | $m = 11$ |
| 2 and 12 | 43.9 | 2.2 | 4.84 | when $n = 1$   $\kappa^2 = 3.707$ |
| 3 and 13 | 41.2 | 4.9 | 24.01 | $\kappa^2 \epsilon^2 = 204.148$ |
| 4 and 14 | 51.2 | 5.1 | 26.01 | $\kappa \epsilon = 14.28$ |
| 5 and 15 | 46.6 | 0.5 | 00.25 | which rejects (11 and 21) |
| 6 and 16 | 44.0 | 2.1 | 4.41 | |
| 7 and 17 | 38.2 | 7.9 | 62.41 | $m = 11$ |
| 8 and 18 | 54.2 | 8.1 | 65.61 | when $n = 2$   $\kappa^2 = 2.621$ |
| 9 and 19 | 47.9 | 1.8 | 3.24 | $\kappa^2 \epsilon^2 = 144.341$ |
| 10 and 20 | 43.9 | 2.2 | 4.84 | $\kappa \epsilon = 12.01$ |
| 11 and 21 | 21 01.4 | 15.3 | 234.09 | which rejects none other. |
| | | | | |
| Sum = 228 27.6 | | $\Sigma v^2 = 550.71$ | | |
| | | | | |
| x = 20 46.1 | | $\epsilon^2 = 55.071$ | | |

After rejecting the interval between (11 and 21) the probable error is found as follows:

| Between observa-tions. | Reduced intervals. | $v'$ | $v'^2$ |
|---|---|---|---|
| | *m. s.* | | |
| 1 and 11 | 20 35.1 | 9.5 | 90.25 |
| 2 and 12 | 43.9 | 0.7 | .49 |
| 3 and 13 | 41.2 | 3.4 | 11.56 |
| 4 and 14 | 51.2 | 6.6 | 43.56 |
| 5 and 15 | 46.6 | 2.0 | 4.00 |
| 6 and 16 | 44.0 | 0.6 | .36 |
| 7 and 17 | 38.2 | 6.4 | 40.96 |
| 8 and 18 | 54.2 | 9.6 | 92.16 |
| 9 and 19 | 47.9 | 3.3 | 10.89 |
| 10 and 20 | 43.9 | 0.7 | .49 |
| | | | |
| Sum = 207 26.2 | | $\Sigma v^2 = 294.72$ | |
| | | | |
| x = 20 44.62 | | $\epsilon^2 = 32.749$ | |

$$\epsilon = \sqrt{\frac{294.72}{10 - 1}} = 5.72$$

Probable error of single set $= 0.6745 \times 5.72$ or

$$r = \pm 3^s.86$$

Probable error of final result $= \frac{r}{\sqrt{m}}$ or

$$R = \pm \frac{3^s.86}{\sqrt{10}} = 1^s.220, \text{ or}$$

$$R = \sqrt{\frac{294.72}{10 \times 9}} \times 0.6745 = 1^s.220$$

This is for ten revolutions; for one revolution the probable error is $\pm 0^s.122$.

## Peirce's Criterion.

Values of $\kappa^2$ for $\mu = 1$.

| $m$ | 1 | 2 | 3 | 4 | 5 | 6 | 7 | 8 | 9 |
|---|---|---|---|---|---|---|---|---|---|
| 3 | 1.480 | | | | | | | | |
| 4 | 1.912 | 1.163 | | | | | | | |
| 5 | 2.278 | 1.439 | | | | | | | |
| 6 | 2.592 | 1.687 | 1.208 | | | | | | |
| 7 | 2.866 | 1.910 | 1.400 | 1.045 | | | | | |
| 8 | 3.109 | 2.112 | 1.589 | 1.229 | | | | | |
| 9 | 3.327 | 2.295 | 1.753 | 1.388 | 1.091 | | | | |
| 10 | 3.526 | 2.464 | 1.904 | 1.531 | 1.242 | | | | |
| 11 | 3.707 | 2.621 | 2.045 | 1.662 | 1.373 | 1.122 | | | |
| 12 | 3.875 | 2.766 | 2.176 | 1.785 | 1.492 | 1.249 | 1.018 | | |
| 13 | 4.029 | 2.902 | 2.299 | 1.901 | 1.604 | 1.362 | 1.145 | | |
| 14 | 4.173 | 3.030 | 2.416 | 2.009 | 1.709 | 1.465 | 1.255 | 1.053 | |
| 15 | 4.309 | 3.151 | 2.526 | 2.111 | 1.807 | 1.561 | 1.354 | 1.163 | |
| 16 | 4.436 | 3.264 | 2.630 | 2.207 | 1.898 | 1.651 | 1.445 | 1.259 | 1.080 |
| 17 | 4.555 | 3.371 | 2.729 | 2.300 | 1.985 | 1.736 | 1.529 | 1.347 | 1.176 |
| 18 | 4.668 | 3.475 | 2.824 | 2.389 | 2.069 | 1.817 | 1.609 | 1.428 | 1.261 |
| 19 | 4.776 | 3.571 | 2.914 | 2.474 | 2.150 | 1.895 | 1.685 | 1.504 | 1.341 |
| 20 | 4.878 | 3.664 | 3.001 | 2.556 | 2.227 | 1.970 | 1.757 | 1.576 | 1.415 |
| 21 | 4.975 | 3.755 | 3.084 | 2.634 | 2.301 | 2.041 | 1.827 | 1.644 | 1.483 |
| 22 | 5.068 | 3.840 | 3.164 | 2.709 | 2.373 | 2.109 | 1.893 | 1.710 | 1.549 |
| 23 | 5.157 | 3.923 | 3.240 | 2.782 | 2.442 | 2.176 | 1.957 | 1.773 | 1.612 |
| 24 | 5.242 | 4.002 | 3.315 | 2.852 | 2.509 | 2.240 | 2.019 | 1.833 | 1.671 |
| 25 | 5.324 | 4.078 | 3.387 | 2.920 | 2.573 | 2.302 | 2.079 | 1.892 | 1.729 |
| 26 | 5.403 | 4.151 | 3.456 | 2.986 | 2.636 | 2.362 | 2.137 | 1.948 | 1.784 |
| 27 | 5.479 | 4.222 | 3.523 | 3.049 | 2.697 | 2.420 | 2.194 | 2.003 | 1.838 |
| 28 | 5.552 | 4.291 | 3.588 | 3.111 | 2.756 | 2.477 | 2.249 | 2.056 | 1.891 |
| 29 | 5.622 | 4.358 | 3.651 | 3.171 | 2.813 | 2.532 | 2.302 | 2.108 | 1.941 |
| 30 | 5.690 | 4.422 | 3.712 | 3.229 | 2.869 | 2.586 | 2.354 | 2.158 | 1.990 |
| 31 | 5.756 | 4.484 | 3.772 | 3.285 | 2.923 | 2.638 | 2.404 | 2.207 | 2.038 |
| 32 | 5.820 | 4.545 | 3.829 | 3.340 | 2.976 | 2.689 | 2.454 | 2.255 | 2.085 |
| 33 | 5.882 | 4.604 | 3.884 | 3.394 | 3.028 | 2.738 | 2.502 | 2.302 | 2.130 |
| 34 | 5.942 | 4.661 | 3.939 | 3.446 | 3.078 | 2.787 | 2.549 | 2.347 | 2.175 |
| 35 | 6.001 | 4.717 | 3.992 | 3.497 | 3.127 | 2.834 | 2.594 | 2.392 | 2.218 |
| 36 | 6.058 | 4.771 | 4.044 | 3.547 | 3.174 | 2.880 | 2.639 | 2.436 | 2.261 |
| 37 | 6.113 | 4.823 | 4.095 | 3.595 | 3.221 | 2.926 | 2.683 | 2.478 | 2.302 |
| 38 | 6.167 | 4.874 | 4.144 | 3.643 | 3.267 | 2.970 | 2.726 | 2.520 | 2.343 |
| 39 | 6.219 | 4.925 | 4.192 | 3.689 | 3.312 | 3.013 | 2.768 | 2.561 | 2.383 |
| 40 | 6.270 | 4.974 | 4.239 | 3.734 | 3.356 | 3.055 | 2.809 | 2.601 | 2.422 |
| 41 | 6.320 | 5.022 | 4.285 | 3.779 | 3.398 | 3.097 | 2.849 | 2.640 | 2.460 |
| 42 | 6.369 | 5.069 | 4.331 | 3.822 | 3.440 | 3.138 | 2.888 | 2.678 | 2.497 |
| 43 | 6.416 | 5.114 | 4.375 | 3.865 | 3.481 | 3.178 | 2.927 | 2.716 | 2.534 |
| 44 | 6.463 | 5.159 | 4.418 | 3.906 | 3.521 | 3.217 | 2.965 | 2.753 | 2.570 |
| 45 | 6.508 | 5.202 | 4.460 | 3.947 | 3.561 | 3.255 | 3.002 | 2.789 | 2.606 |
| 46 | 6.552 | 5.245 | 4.501 | 3.987 | 3.600 | 3.293 | 3.039 | 2.825 | 2.641 |
| 47 | 6.596 | 5.287 | 4.542 | 4.026 | 3.638 | 3.330 | 3.075 | 2.860 | 2.675 |
| 48 | 6.639 | 5.328 | 4.581 | 4.065 | 3.675 | 3.366 | 3.110 | 2.894 | 2.708 |
| 49 | 6.681 | 5.368 | 4.620 | 4.103 | 3.712 | 3.401 | 3.145 | 2.928 | 2.741 |
| 50 | 6.720 | 5.408 | 4.657 | 4.140 | 3.748 | 3.436 | 3.179 | 2.962 | 2.774 |
| 51 | 6.761 | 5.447 | 4.695 | 4.176 | 3.784 | 3.471 | 3.213 | 2.994 | 2.806 |
| 52 | 6.800 | 5.484 | 4.732 | 4.212 | 3.819 | 3.505 | 3.246 | 3.027 | 2.838 |
| 53 | 6.838 | 5.522 | 4.768 | 4.247 | 3.853 | 3.538 | 3.279 | 3.059 | 2.869 |
| 54 | 6.876 | 5.559 | 4.804 | 4.282 | 3.887 | 3.571 | 3.311 | 3.090 | 2.899 |
| 55 | 6.913 | 5.595 | 4.839 | 4.316 | 3.920 | 3.603 | 3.342 | 3.121 | 2.929 |
| 56 | 6.950 | 5.630 | 4.873 | 4.349 | 3.952 | 3.635 | 3.373 | 3.151 | 2.959 |
| 57 | 6.986 | 5.665 | 4.907 | 4.382 | 3.984 | 3.666 | 3.404 | 3.181 | 2.988 |
| 58 | 7.021 | 5.699 | 4.941 | 4.415 | 4.016 | 3.697 | 3.434 | 3.210 | 3.017 |
| 59 | 7.056 | 5.733 | 4.974 | 4.447 | 4.047 | 3.728 | 3.463 | 3.239 | 3.046 |
| 60 | 7.090 | 5.766 | 5.006 | 4.478 | 4.078 | 3.758 | 3.492 | 3.268 | 3.074 |

## Geographical Positions.

| | Latitude. | Longitude west from Greenwich. |
|---|---|---|
| | ° ′ ″ | h. m. s. |
| Cambridge, Observatory..................... | 42 22 48. 1 | 4 44 31.04 |
| Quebec, Citadel ...... .................. | 46 48 17. 3 | 4 44 49. 42 |
| New York, Observatory.................... | 40 43 48. 5 | 4 55 56. 65 |
| Oswego, Court-House ...... ............. | 43 27 49. 1 | 5 06 01. 05 |
| WASHINGTON, Observatory................... | 38 53 38. 8 | 5 08 12. 12 |
| Buffalo, Michigan and Exchange streets....... | 42 52 41. 8 | 5 15 27. 58 |
| Detroit, New L. S. Observatory.............. | 42 19 58. 6 | 5 32 12. 24 |
| Chicago, City Hall......................... | 41 53 06. 2 | 5 50 32. 08 |
| Saint Louis, Washington University .......... | 38 37 | 6 00 49. 02 |
| Saint Paul, Custom-House.................. | 44 53 | 6 12 21. 84 |
| Fort Leavenworth, Engineer Observatory ..... | 39 21 | 6 19 39. 35 |
| Omaha, Coast-Survey Observatory............ | 41 16 | 6 23 46. 33 |
| Denver, Mint ............................ | 39 45 01. 8 | 6 59 58. 72 |
| Salt Lake, Coast-Survey Observatory ......... | 40 46 | 7 27 35. 45 |
| San Francisco, Washington Square........... | 37 47 55. 3 | 8 09 38. 23 |

# TABLES AND FORMULÆ.

PART IV.

# APPENDIX.

## LXVI.—*Field Magnetic Observations.*

[By Captain CHAS. W. RAYMOND, Corps of Engineers.]

These observations have for their object the determination of the magnetic declination, dip, and intensity, at any given time and place.

The following instructions have reference to the determination of the magnetic elements on land. It will be supposed that the theodolite magnetometer and dip-circle are the instruments employed; the first to determine the declination and the horizontal component of the intensity, and the second to determine the inclination or dip.

### *Magnetic Declination.*

The magnetometer having been mounted upon its tripod, or upon a sound post firmly imbedded in the ground, the horizontal limb and the rotation-axis of the telescope must be leveled, the vertical wire of the telescope made truly vertical, and its collimation-error reduced. The magnet must then be suspended by as few filaments as possible; four or five are usually required. The magnet is then made horizontal by adjusting the balancing-ring, the position of which should be carefully preserved throughout the experiments. In order to adjust the magnet-scale to the stellar focus of its lens, the telescope must be turned upon the sun or a star, and adjusted to perfectly distinct vision. The telescope must then be turned upon the suspended magnet, and the scale-ring screwed in or out until the scale is seen with perfect distinctness.

The lines of detorsion and collimation must now be brought into the plane of the magnetic meridian by the following method: The magnet being suspended, turn the instrument in azimuth until the scale is seen through the telescope. Remove the magnet and suspend the brass detorsion-cylinder. Bring the axis of the cylinder, by estimation, into the plane of the magnetic meridian by turning the torsion-circle. Remove the cylinder and suspend the declination-magnet. Turn the instrument in azimuth until the vertical wire of the telescope bisects the scale-zero. Remove the declination and suspend the short magnet.

LXVI.—*Field Magnetic Observations, &c.*—Continued.

Turn the torsion-circle until the vertical wire of the telescope coincides with the scale-zero. Exchange the magnets and adjust as before. The time, temperature, and readings of the verniers and torsion-circle should be recorded. This method requires that the magnets and detorsion-cylinder should be of equal weight. With the cylinder alone, the plane of detorsion may be determined to within about one degree, an error which does not seriously affect the accuracy of declinations observed in ordinary field-work.

The instrument is now in adjustment for observations of declination.

The *angular value of one scale-interval* and the *scale-zero* of each magnet employed must be determined at some convenient time. The former is unchangeable. The latter must be occasionally redetermined, as it is liable to change through accident. To determine the angular scale-value, fix the magnet in the position which it occupies when suspended, in such a way that the instrument may be moved in azimuth without disturbing it. Turn the instrument in azimuth until the vertical wire of the telescope coincides with a scale-division. Record the vernier and scale readings. Turn the instrument until the vertical wire coincides with another division, and record as before. Repeat over different parts of the scale until a sufficient number of observations have been obtained. Each pair of observations furnishes a single determination of the required value. The probable error of the mean value may be determined by the method of least squares.

To determine the *scale-zero*, or reading of the magnetic axis, suspend the magnet, turn the instrument in azimuth until the vertical wire of the telescope coincides with a division near the middle of the scale, and record the scale-reading. Invert the magnet, and record the reading corresponding to this position. Move the instrument slightly in azimuth, and repeat this operation until three or four readings with the scale erect, and as many with the scale inverted, have been obtained. The scale-zero may then be determined by the method of alternate means, (see Form A.) These observations should, if practicable, be made about the epoch of the day when the magnet is stationary.

LXVI.—*Field Magnetic Observations, &c.*—Continued.

The *co-efficient of torsion* may be determined as follows: The declination-magnet being suspended and the instrument in adjustment for observations of declination, record the readings of the scale and torsion-circle. Turn the torsion-circle through an angle of 90°, and record this difference of arc and the corresponding scale-reading. Turn the torsion-circle 180° in the reverse direction, and record as before. Finally, turn the torsion-circle back to its original position, and repeat the readings and record.

*Computation.*

$$\frac{H}{F} = \frac{u}{90° - u}$$

$1 + \dfrac{H}{F}$ = co-efficient of torsion; and

$u$ = difference in scale-readings (reduced to arc) corresponding to a change of direction of the magnet caused by twisting the suspension-thread through an angle of 90°.

The mean value of $u$ deduced from the observations is employed. For convenience the co-efficient of torsion is usually applied to the angular value [$a$] of the scale-interval, (see Form B.)

At some convenient time, while the instrument is in position, the vernier-readings corresponding to the astronomical meridian must be determined, either by turning the telescope upon a point of which the azimuth is known, or directly, by any suitable method.

The preliminary adjustments and determinations having been made, the instrument is left in position for observations of the diurnal variations in declination. At some time early in the morning, the north end of the magnet attains its most easterly position, which is called the *morning eastern elongation*. The record of scale-readings must be commenced early enough to include this elongation. When this point is fairly passed, or the north end of the magnet has fully commenced its westerly motion, the readings may be discontinued until about noon, when they must be resumed and continued until the western elongation has been observed, and the easterly motion has fairly set in.

LXVI.—*Field Magnetic Observations, &c.*—Continued.

The telescope should be reversed at each observation, and a mean of the readings in the two positions should be taken. The temperature should be noted. The readings should be made half or quarter hourly during the periods of observation. The observations of the first day will determine the approximate times of elongation, **or** *turning-hours*, in accordance with which the periods of observation are to be subsequently regulated.

*Computation.*

$$\delta = \delta' \pm a \left( 1 + \frac{H}{F} \right) (e - s + fr)$$

$\delta$ = value of declination for the day;

$\delta'$ = declination at instant of final instrumental adjustment, which is $\left\{ \begin{matrix} + \\ - \end{matrix} \right\}$ when the magnetic meridian is $\left\{ \begin{matrix} west \\ east \end{matrix} \right\}$ of the true north meridian;

$a$ = angular scale-value;

$1 + \dfrac{H}{F}$ = co-efficient of torsion;

$e$ = mean of scale-readings at the elongations, which is $\left\{ \begin{matrix} + \\ - \end{matrix} \right\}$ when $\left\{ \begin{matrix} greater \\ less \end{matrix} \right\}$ than $s$;

$s$ = scale-reading of magnetic axis, (zero of magnet-scale,) which is $\left\{ \begin{matrix} + \\ - \end{matrix} \right\}$ when $\left\{ \begin{matrix} less \\ greater \end{matrix} \right\}$ than $e$;

$r$ = difference between scale-readings at elongations, or *daily range.*

$f$ = factor for reduction to the mean of *hourly* observations. It may be taken, with its sign, from the following table:

$$f$$

| | | |
|---|---|---|
| January..... — 0.089 | May ....... — 0.013 | September ... — 0.044 |
| February ... — 0.040 | June ...... + 0.010 | October...... — 0.096 |
| March ...... — 0.019 | July ....... + 0.005 | November ... — 0.096 |
| April ....... — 0.068 | August .... — 0.023 | December.... — 0.154 |

LXVI.—*Field Magnetic Observations, &c.*—Continued.

The correction to $\delta'$ is positive or negative according as it indicates a motion from or toward the true meridian. (See Form C.)

## Magnetic Intensity.

For the determination of the *horizontal intensity* two distinct series of experiments are required—*experiments of deflection* and *experiments of oscillation*.

The *experiments of deflection* are made as follows: The deflection-bar is made fast in its position, and the copper damper placed within the box. The instrument is then adjusted as for observations of declination, the short magnet being suspended.

The experiments should be made, when practicable, at three distances. The first position of the deflector should be at about three times its length from the suspended magnet; the third, at a distance about one-third greater; and the second midway between the other two. These distances are measured from center to center. At the beginning, the verniers are read and time noted, in order to follow changes of declination. The temperature is recorded at each observation. The long magnet is placed upon its carriage on the deflection-bar, on either side of the suspended magnet, and at the nearest distance. The instrument is then turned in azimuth until the vertical wire of the telescope coincides with the scale-zero. The time is then noted and the verniers read. The carriage is then moved to the next distance, and finally to the greatest distance, and the observation is repeated at each position.

The deflector is then reversed on its carriage, and the observations are repeated at the three distances, beginning with the greatest and ending with the least. At the nearest distance the magnet is again reversed, and the complete set of observations is repeated, in order to obtain a double set of results. The deflector is now placed on the opposite side of the suspended magnet at the nearest distance. A double set of experiments similar to that already described is then made.

At suitable intervals special observations should be made to measure changes of declination. For this purpose the deflector is removed, and the instrument is turned in azimuth until the vertical wire of the telescope coincides with the scale-zero. The

LXVI.—*Field Magnetic Observations, &c.*—Continued.

verniers are then read and the temperature and time noted. From these observations we may, by simple interpolation, determine with sufficient accuracy corrections for the reduction of the observed angles of deflection to the same declination.

*Computation.*

$$\frac{m}{X} = \tfrac{1}{2}r^3 \sin u \left(1 - \frac{P}{r^2}\right)$$

$m$ = magnetic moment of the deflector;
$X$ = horizontal intensity;
$r$ = distance between the centers of the magnets; and
$P$ = constant depending on the distribution of magnetism in the magnets.

The value of $\frac{m}{X}$ must be computed for each distance separately and a mean adopted. (See Form D.)

The correction depending upon the constant P may be neglected when there is a considerable difference between the lengths of the magnets. Its value is greatest when the magnets are equal in length, and zero when the lengths are in the proportion of 1 to 1.224.

To determine the value of P, deflections are made alternately at two different distances, which should be in the proportion of 1 to 1.32. About twenty-five corresponding sets should be obtained.

*Computation.*

$$P = \frac{A - A_1}{\dfrac{A}{r^2} - \dfrac{A_1}{r_1^2}}$$

$\left.\begin{array}{c} A \\ A_1 \end{array}\right\}$ = value of $\frac{m}{X}$ for $\left\{\begin{array}{c} \text{shorter} \\ \text{longer} \end{array}\right\}$ distance $\left\{\begin{array}{c} r \\ r_1 \end{array}\right.$

To reduce the angles of deflection determined from different sets to the same temperature:

$$\sin u = \frac{\sin u_0}{1 - (t_0 - t)q}$$

$u_0$ = observed angle at temperature $t_0$;
$u$ = angle reduced to standard temperature $t$; and
$q$ = temperature-constant, determined as explained hereafter.

### LXVI.—*Field Magnetic Observations, &c.*—Continued.

The *experiments of oscillation* are made as follows: The instrument having been adjusted as for observations of declination, with the long magnet suspended, and the co-efficient of torsion having been determined, the magnet is made to oscillate horizontally by attracting or repelling one of its poles. The impulse should be sufficient to make it oscillate beyond the limits of the scale for at least ten minutes, as steadiness of motion is thus acquired. All vertical oscillations must be carefully checked. When the amplitude is sufficiently reduced, the scale is read at the limits of an oscillation, and a division near the mean of these readings is selected as the zero or point at the passage of which the times are to be noted. The intervals of oscillation may now be noted in the following way: The approximate interval corresponding to six oscillations is noted for convenience. The instant of passage is then noted at every sixth oscillation up to the sixtieth; then at the hundredth, two hundreth, and three hundreth; then at every sixth oscillation up to the three hundred and sixtieth; and then at the four hundredth, if it be desirable to prolong the observations to this extent. The number of oscillations timed should depend on the length of the interval. The entire time of observation should not exceed a quarter of an hour. The approximate time of ten oscillations is computed for convenience at the sixtieth. For the semi-oscillations timed the magnet will always move in the same direction. The amplitude of oscillation at the beginning and end of the experiments should be noted. At suitable intervals the mean reading of the scale should be observed, since the zero is liable to changes due to variations of declination.

The temperature should be observed at intervals, the bulb of the thermometer being placed within the box.

### Computation.

$$mX = \frac{\pi^2 K}{T^2}$$

$m =$ magnetic moment of the magnet;

$X =$ horizontal intensity;

$\pi = 3.14159$;

$K =$ moment of inertia of magnet and stirrup;

LXVI. — *Field Magnetic Observations, &c.*—Continued.

T = corrected time of oscillation;

$$T^2 = T'^2 \left( 1 + \frac{H}{F} \right) \left( 1 - (t' - t)\, q \right)$$

T' = observed time of oscillation;

$\left. \begin{array}{c} t' \\ t \end{array} \right\}$ = temperature of magnet when $\left\{ \begin{array}{l} \text{oscillating;} \\ \text{deflecting;} \end{array} \right.$

$q$ = temperature-constant, or change in magnetic moment for a change in temperature of 1° Fahrenheit. (See Form E.)

To determine the *moment of inertia of the magnet and stirrup*, [K], the moment of inertia of the inertia-ring must first be computed. For this purpose accurate measurements of its outer and inner radii (in decimals of a foot) and the weight of the ring (in grains) are required. These data are usually furnished by the maker.

The instrument being in adjustment with the long magnet suspended, the inertia-ring is balanced upon the magnet by means of the balancing-blocks. The time of a single oscillation is then determined as before described. The load is then removed and the time of oscillation is again determined. At least twelve sets of these experiments should be made. A separate determination of the co-efficient of torsion must be made for the loaded magnet. The temperature must be recorded throughout the experiments.

*Computation.*

$$K = K'' \frac{T^2}{T'^2 - T^2}$$

$$K'' = \tfrac{1}{2} (r^2 + r'^2)\, w$$

K = moment of inertia of the suspended mass;

K'' = moment of inertia of the ring;

$\left. \begin{array}{c} T' \\ T \end{array} \right\}$ = corrected time of single oscillation of $\left\{ \begin{array}{l} \text{loaded} \\ \text{unloaded} \end{array} \right\}$ magnet;

$\left. \begin{array}{c} r \\ r' \end{array} \right\}$ = $\left\{ \begin{array}{l} \text{outer} \\ \text{inner} \end{array} \right\}$ radius of ring, in feet; and

$w$ = weight of ring, in grains.

LXVI.—*Field Magnetic Observations, &c.*—Continued.

The values of $\pi^2$ K for different temperatures should be tabulated. The co-efficient of expansion for brass (0.00001) may be employed for their computation.

The *change in magnetic moment for a change in temperature of* 1° *Fahrenheit*, [$q$], is best determined by the method of deflections. The magnet for which $q$ is to be determined is the deflector. At least three consecutive sets of deflections should be made, the first and third being at about the same temperature, and the intermediate set at a very different one. A mean of the results from the first and third sets must be compared with the result from the intermediate set. The required difference of temperature may be produced by a jacket of ice and hot water, or advantage may be taken of extreme natural temperatures.

*Computation.*

$$q = \frac{a\, n \cot u}{t - t_0}$$

$q$ = temperature-constant :

$\left.\begin{array}{c} t \\ t_0 \end{array}\right\} = \left\{\begin{array}{c} \text{higher} \\ \text{lower} \end{array}\right\}$ temperature :

$n$ = difference of scale-readings corresponding to $t - t_0$ ;

$a$ = angular value of one scale-interval ; and

$u$ = angle of deflection corresponding to $t_0$.

*Computation of the Horizontal Intensity and Magnetic Moment.*

$$X = \sqrt{\frac{a}{\beta}}$$

$$m = \sqrt{a\, \beta}$$

$X$ = absolute horizontal intensity ;

$m$ = magnetic moment of deflecting and oscillating magnet ;

$a = mX$, determined by experiments of oscillation ; and

$\beta = \dfrac{m}{X}$, determined by experiments of deflection.

Strictly, the values of $a$ and $\beta$ should be corrected for the effect of induction. These corrections are very small, and require

### LXVI.—*Field Magnetic Observations, &c.*—Continued.

special apparatus for their determination. They are therefore neglected in field-work. (See Form E.)

*To reduce m to a Standard Temperature.*

$$m_0 = m \left(1 + (t - t_0)\, q\right)$$

$m_0$ = value of $m$ at standard temperature $t_0$;

$t$ = temperature at time of experiments; and

$q$ = temperature-constant.

*Computation of the Total Intensity.*

$$\varphi = X \sec \theta$$

$\varphi$ = total intensity;

$X$ = horizontal intensity; and

$\theta$ = inclination or *dip*.

To convert measures of intensity expressed in English units into their equivalents expressed in the metric system, multiply by 0.46108, (log = 9.66378.)

To convert measures of intensity expressed in metric units into their equivalents expressed in English units, multiply by 2.1688, (log = 0.33622.)

### *Magnetic Inclination.*

The dip-circle having been mounted on its tripod or post, the horizontal limb must be leveled. The needle is charged by means of the magnetizing-bars, and then suspended as follows: Raise the Y's, and placing the needle in them, lower it gently upon the agate supports. Turn the vertical circle slowly in azimuth around the entire circle, and see whether the needle plays freely, and whether its face lies in the plane of the face of the vertical circle in all azimuths. If necessary, the agate supports must be re-adjusted. The face of the vertical circle is then brought into the plane of the magnetic meridian by the following method: Turn the vertical circle in azimuth until the needle is vertical. Record the reading of the azimuth-circle. Reverse the needle on its supports; make it again vertical by a slight movement in azimuth, and record as before. Turn the vertical circle 180° in azimuth, and repeat the double observation. A mean of the four readings is the reading of the magnetic prime-vertical, from which the settings of the magnetic meridian are obtained by

LXVI.—*Field Magnetic Observations, &c.*—Continued.

adding and subtracting 90°. Set the vertical circle at one of these readings. (See Form F.)

The observations for the determination of the inclination are made as follows: Record the reading of the azimuth-circle. Record the polarity of the needle, (marked end *north* or *south*,) the position of the vertical-circle, (face *east* or *west*,) a mean of the readings of the vertical circle at the ends of the needle, and the temperature Fahrenheit. Raise the needle, reverse it on its supports, and repeat the observations. Bring the needle back to its original position, and observe as before. Reverse it again, and repeat the observations. Turn the vertical circle 180° in azimuth, and repeat the observations. Remove the needle, and reverse its polarity. Suspend it again, and repeat the observations with the circle and needle in both positions, as before.

*Computation.*

$$\theta = \frac{a + \beta}{2} + c$$

$\theta$ = magnetic inclination;

$\left.\begin{array}{c} a \\ \beta \end{array}\right\}$ = mean of observed values of $\theta$ $\left\{\begin{array}{c} \text{before} \\ \text{after} \end{array}\right\}$ reversal of polarity;

$c$ = constant correction for errors of axle and limb.

The *constant correction* ($c$) is determined as follows: A complete set of experiments must be made in the plane of the magnetic meridian. The vertical circle is then turned in either direction about 45° in azimuth, and a similar series of experiments is made. The vertical circle is then turned 90° in azimuth in the opposite direction, and the experiments are again repeated.

*Computation.*

$$c = \theta - \theta_1; \quad \cot^2 \theta = \cot^2 \theta' + \cot^2 \theta''$$

$\left.\begin{array}{c} \theta' \\ \theta'' \end{array}\right\}$ = observed inclinations in planes at right angles to each other; and

$\theta_1$ = observed inclination in the plane of the magnetic meridian.

## FORM A.

### *Determination of the Zero of the Magnet-Scale.*

Station, Fort Yukon, Alaska.—Date, August 12, (p. m.,) 1869.—Observer, C. W. R.—Recorder, C. W. R.

Scale-zero of declination-magnet.

| Position of scale. | Reading of scale. | Alternate means. | Zeros. | Mean zero. |
|---|---|---|---|---|
| Erect ........... | 45. 00 | | | |
| Inverted ........ | 57. 00 | 45. 50 | 51. 25 | |
| Erect ........... | 46. 00 | 55. 50 | 50. 75 | |
| Inverted ........ | 54. 00 | 44. 75 | 49. 37 | 50. 18 |
| Erect ........... | 43. 50 | 55. 25 | 49. 37 | |
| Inverted ........ | 56. 50 | | | |

## FORM B.

### *Determination of the Co-efficient of Torsion.*

Station, Fort Yukon, Alaska.—Date, August 13, 1869.—Observer, C. W. R.—Recorder, C. W. R.

Declination-magnet suspended.

| Circle-readings. | Scale-readings. | Diff. of arc. | Diff. of scale. | Mean for 90°. |
|---|---|---|---|---|
| 3. 95 | 51. 25 | | | |
| 30. 95 | 30. 50 | 90 | 20. 75 | |
| 12. 95 | 79. 50 | 180 | 49. 00 | 23. 56 |
| 3. 95 | 55. 00 | 90 | 24. 50 | |

### *Computation.*

$$90° = 324000''.0$$
$$u = 23^d.56 = \quad 699 .7 \qquad \log \qquad 2.84491$$
$$90° - u = 323300 .3 \qquad \log \qquad 5.50961$$
$$\frac{H}{F} = 0.00216 \qquad \log \qquad 7.33530$$
$$1 + \frac{H}{F} = 1.00216$$
$$a^* \left(1 + \frac{H}{F}\right) = 29''.76 \qquad \frac{H}{F} = \frac{u}{90° - u}$$

$$^*a = 29''.70$$

## FORM C.

### *Declination—Record of Observations.*

Station, Fort Porter, Buffalo, N. Y., 222 feet north of flag-staff.—Date, June 14, 1872.—Observer, A. N. L.

| | | |
|---|---|---|
| Azimuth-reading at adjustment, ver. B ..................... | 158° 09′ 00″ | |
| ver. A ..................... | 338 09 30 | |
| Reading of mark on flag-staff..ver. B .....:............. | 166 21 00 | |
| ver. A ..................... | 346 21 00 | |
| Azimuth of flag-staff, west of north ..................... | 175 35 53 | |

Magnet $C_2$ suspended; scale-zero, 21.7.

| Time. | Scale. | Remarks. |
|---|---|---|
| *h. m.* | | |
| 6 00 a. m. | 21. 7 | |
| 6 20 | 22. 3 | |
| 6 45 | 22. 6 | |
| 7 00 | 22. 7 | Maximum. |
| 7 30 | 22. 4 | |
| 8 00 | 22. 0 | |
| | | |
| 11 30 | 15. 6 | |
| 12 00 m. | 14. 75 | |
| 12 30 p. m. | 14. 35 | Minimum. |
| 12 45 | 14. 65 | |
| 1 00 | 15. 05 | |
| 1 15 | 15. 7 | |

### *Computation.*

| Correction to $\delta'$. | | | | |
|---|---|---|---|---|
| | | | Astr. merid .. 161° 56′ 53″ | |
| $e = -18.5$ | | | Az. at adjust- | |
| $s = -21.7$ | $e - s = 3.2$ | | ment...... 158 09 15 | |
| | | $e - s + fr = 3.28$ | $\delta' = +$ 3 47 38 | |
| $r = 8.35$ | | | Corr. to $\delta' = +$ 9 04.5 | |
| $f = +$ .01 | $fr = +0.08$ | $a\left(1 + \frac{H}{F}\right) = 166''$ | $\delta = +$ 3 56 42.5 | |

## FORM D.

### *Horizontal Intensity—Experiments of Deflection.*

Station, Fort Yukon, Alaska.—Date, August 14, 1869.—Observer, C. W. R.—Recorder, C. W. R.

Magnets at right angles to each other; long magnet deflecting; short magnet suspended.

Magnet at 1′.09 east and west.

| Position of deflector. | No. of observations. | Marked end. | Temp, F. | Verniers E. and W. | Means of verniers. | Time. | Means corr. for change of declination. | Arc of deflection; semi-arc. |
|---|---|---|---|---|---|---|---|---|
| | | | ° | ° ′ | ° ′ | h. m. | ° ′ | ° ′ |
| E. | 3 | E. | 81.0 | 71 18 / 251 18 | 161 18 | 2 10 | 161 17.5 | |
| | | | | | | | | 15 43.8 |
| | 4 | W. | 82.0 | 55 33 / 235 33 | 145 33 | 2 18 | 145 33.7 | 7 51.9 |
| | | | | | | | | 15 46.0 |
| | 9 | E. | 81.0 | 71 21 / 251 21 | 161 21 | 3 02 | 161 19.7 | 7 53.0 |
| | | | | | | | | 15 43.3 |
| | 10 | W. | 82.0 | 55 35 / 235 35 | 145 35 | 3 08 | 145 36.4 | 7 51.6 |
| | | | | | | | Mean.... | 7 52.2 |
| W. | 15 | E. | 85.5 | 71 34 / 251 34 | 161 34 | 4 13 | 161 34.1 | |
| | | | | | | | | 16 04.3 |
| | 16 | W. | 87.0 | 55 30 / 235 30 | 145 30 | 4 22 | 145 29.8 | 8 02.1 |
| | | | | | | | | 16 06.7 |
| | 21 | E. | 84.0 | 71 36 / 251 36 | 161 36 | 5 22 | 161 36.5 | 8 03.3 |
| | | | | | | | | 15 55.1 |
| | 22 | W. | 85.0 | 55 42 / 235 42 | 145 42 | 5 35 | 145 41.4 | 7 57.5 |
| | Mean for E. and W..... | | | | 7 56.6 | | Mean.... | 8 01.0 |

| Remarks. | Computation. | | |
|---|---|---|---|
| | | | log. |
| 1ʰ 34ᵐ p. m.—Turned instrument on 50.18 declination-scale. Verniers, E., 63° 51′; W., 243° 51′. Temperature, 78°.5 F., (attached.) 3ʰ 45ᵐ p. m.—Deflector away to show changes of declination. Temperature, 87° F. Verniers, E., 63° 49′; W., 243° 49′. 6ʰ 00ᵐ p.m.—End of experiments. Temperature, 82° F.; scale, 58.50. Verniers, E., 63° 54′; W., 243° 54′. | $u = 7°56′.6$ $r = 1′.09$ | $\sin u$ $r^3$ | 9. 14049 0. 11228 |
| | $P = 0$ | $1 - \dfrac{P}{r^2}$ | |
| | | ′ $\frac{1}{2}$ | 9. 69897 |
| | | $\dfrac{m}{X}$ | 8. 95174 |

## FORM E.

### Horizontal Intensity—Experiments of Oscillation.

Station, Fort Yukon, Alaska.—Date, August, 16, 1869.—Chronometer, Bliss & Creighton, 1609, M. T.—Daily rate, unknown.—Observer, C. W. R.—Recorder, J. J. M.

Long magnet suspended without load.

| No. of observations. | Times by chronometer. | Temp., F. | Diff. of times. | Time of 10 oscillations. | Remarks. |
|---|---|---|---|---|---|
| | h. m. s. | ° | s. | s. | |
| 0 | 1 16 05.0 | 82.5 | | | |
| 6 | 16 40.8 | | | | Approximate time of 6 oscilla- |
| 12 | 17 16.5 | | | | tions at the beginning 35ˢ. |
| 18 | 17 52.3 | | | | |
| 24 | 18 28.3 | | | | Scale reading noted just be- |
| 30 | 19 04.0 | | 179.0 | 59.67 | fore 200th oscillation to |
| 36 | 19 39.7 | | 178.9 | 59.63 | detect changes of declina- |
| 42 | 20 15.5 | | 179.0 | 59.67 | tion. Reading, 65°.00; lo- |
| 48 | 20 51.3 | | 179.0 | 59.67 | cal time, 12ʰ 15ᵐ p. m. |
| 54 | 21 27.0 | | 178.7 | 59.57 | |
| 60 | 22 02.8 | | 178.8 | 59.60 | |
| 100 | 26 01.5 | 81.0 | 238.7 | 59.67 | |
| 200 | 35 57.5 | 81.5 | 596.0 | 59.60 | |

Time of 10 oscillations, 59ˢ.64; time of 1 oscillation, 5ˢ.964.

### Computation.

| | | | Logarithms. |
|---|---|---|---|
| $q$ | 0.00015 | $T'$ | |
| $t' - t$ | 0°.06 | $T'^2$ | 0.77554 |
| $(t' - t)q$ | 0.000009 | $1 + \dfrac{H}{F}$ | 1.55108 |
| $1 - (t' - t)q$ | 0.99991 | $1 - (t' - t)q$ | 0.00103 |
| | Logarithms. | $T^2$ | 9.99996 |
| $\overset{*}{\underset{X}{m}}$ | 8.95174 | $\pi^2 K$ | 1.55207 |
| $mX$ | 9.83242 | $mX$ | 1.38449 |
| $m^2$ | 8.78416 | $m$ | 9.83242 |
| | | $X$ | 9.39208 |
| | | | 0.44034 |

* Experiments of deflection.

### Form F.

*Inclination—Determination of the Dip.*

Station, Willet's Point, N. Y.—Date, August 5, 1872.—Observer, C. W. R.—
Recorder, C. W. R.—Dip-circle by Würdemann; Lloyd's needle.

Meridian observations.—Settings for magnetic meridian.

| Face of circle. | Face of needle. | Readings of hor. limb. |
|---|---|---|
| | | ° ′ |
| S. | S. | 157 27 |
| S. | N. | 158 42 |
| N. | S. | 337 01 |
| N. | N. | 338 12 |

Magnetic prime-vertical, 157° 50′; settings, 247° 50′, 67° 50′.

| Marked end. | Face of circle. | Face of needle. | Means of N. and S. ends. | Means. | Means. |
|---|---|---|---|---|---|
| N. | E. | E. | 73° 04′ | 73° 02′ 00″ | |
| | | E. | 73 00 | | 72° 36′ 30″ |
| | | W. | 72 12 | 72 11 00 | |
| | | W. | 72 10 | | |
| | W. | E. | 72 30 | 72 29 00 | |
| | | E. | 72 28 | | 72 39 00 |
| | | W. | 72 49 | 72 49 00 | |
| | | W. | 72 49 | | |
| Mean ... | $\left[\,\alpha\,\right]$ | | | | 72 37 45 |
| S. | E. | E. | 72° 43′ | 72° 44′ 00″ | |
| | | E. | 72 45 | | 72° 55′ 00″ |
| | | W. | 73 00 | 73 06 00 | |
| | | W. | 73 12 | | |
| | | E. | 72 49 | 72 48 00 | |
| | | E. | 72 47 | | 73 01 00 |
| | | W. | 73 14 | 73 14 00 | |
| | | W. | 73 14 | | |
| Mean ... | $\left[\,\beta\,\right]$ | | | | 72 58 00 |
| | | Resulting inclination | $\left[\dfrac{\alpha+\beta}{2}\right]$ | | 72 47 52 |

*Computation:*

$$\theta = \frac{\alpha+\beta}{2} + c; \quad \frac{\alpha+\beta}{2} = 72° \ 47′ \ 52″; \quad c = +18″; \quad \theta = 72° \ 48′.$$

www.ingramcontent.com/pod-product-compliance
Lightning Source LLC
Chambersburg PA
CBHW060537030726
47498CB00004B/1231